Geophysical Monograph Series

Geophysical Monograph Series

- 212 **The Early Earth: Accretion and Differentiation** *James Badro and Michael Walter (Eds.)*
- 213 **Global Vegetation Dynamics: Concepts and Applications in the MC1 Model** *Dominique Bachelet and David Turner (Eds.)*
- 214 **Extreme Events: Observations, Modeling and Economics** *Mario Chavez, Michael Ghil, and Jaime Urrutia-Fucugauchi (Eds.)*
- 215 **Auroral Dynamics and Space Weather** *Yongliang Zhang and Larry Paxton (Eds.)*
- 216 **Low-Frequency Waves in Space Plasmas** *Andreas Keiling, Dong- Hun Lee, and Valery Nakariakov (Eds.)*
- 217 **Deep Earth: Physics and Chemistry of the Lower Mantle and Core** *Hidenori Terasaki and Rebecca A. Fischer (Eds.)*
- 218 **Integrated Imaging of the Earth: Theory and Applications** *Max Moorkamp, Peter G. Lelievre, Niklas Linde, and Amir Khan (Eds.)*
- 219 **Plate Boundaries and Natural Hazards** *Joao Duarte and Wouter Schellart (Eds.)*
- 220 **Ionospheric Space Weather: Longitude and Hemispheric Dependences and Lower Atmosphere Forcing Timothy Fuller-Rowell,** *Endawoke Yizengaw, Patricia H. Doherty, and Sunanda Basu (Eds.)*
- 221 **Terrestrial Water Cycle and Climate Change Natural and Human-Induced Impacts** *Qiuhong Tang and Taikan Oki (Eds.)*
- 222 **Magnetosphere-Ionosphere Coupling in the Solar System** *Charles R. Chappell, Robert W. Schunk, Peter M. Banks, James L. Burch, and Richard M. Thorne (Eds.)*
- 223 **Natural Hazard Uncertainty Assessment: Modeling and Decision Support** *Karin Riley, Peter Webley, and Matthew Thompson (Eds.)*
- 224 **Hydrodynamics of Time-Periodic Groundwater Flow: Diffusion Waves in Porous Media** *Joe S. Depner and Todd C. Rasmussen (Auth.)*
- 225 **Active Global Seismology** *Ibrahim Cemen and Yucel Yilmaz (Eds.)*
- 226 **Climate Extremes** *Simon Wang (Ed.)*
- 227 **Fault Zone Dynamic Processes** *Marion Thomas (Ed.)*
- 228 **Flood Damage Survey and Assessment: New Insights from Research and Practice** *Daniela Molinari, Scira Menoni, and Francesco Ballio (Eds.)*
- 229 **Water-Energy-Food Nexus – Principles and Practices** *P. Abdul Salam, Sangam Shrestha, Vishnu Prasad Pandey, and Anil K Anal (Eds.)*
- 230 **Dawn–Dusk Asymmetries in Planetary Plasma Environments** *Stein Haaland, Andrei Rounov, and Colin Forsyth (Eds.)*
- 231 **Bioenergy and Land Use Change** *Zhangcai Qin, Umakant Mishra, and Astley Hastings (Eds.)*
- 232 **Microstructural Geochronology: Planetary Records Down to Atom Scale** *Desmond Moser, Fernando Corfu, James Darling, Steven Reddy, and Kimberly Tait (Eds.)*
- 233 **Global Flood Hazard: Applications in Modeling, Mapping and Forecasting** *Guy Schumann, Paul D. Bates, Giuseppe T. Aronica, and Heiko Apel (Eds.)*
- 234 **Pre-Earthquake Processes: A Multidisciplinary Approach to Earthquake Prediction Studies** *Dimitar Ouzounov, Sergey Pulinets, Katsumi Hattori, and Patrick Taylor (Eds.)*
- 235 **Electric Currents in Geospace and Beyond** *Andreas Keiling, Octav Marghitu, and Michael Wheatland (Eds.)*
- 236 **Quantifying Uncertainty in Subsurface Systems** *Celine Scheidt, Lewis Li, and Jef Caers (Eds.)*
- 237 **Petroleum Engineering** *Moshood Sanni (Ed.)*
- 238 **Geological Carbon Storage: Subsurface Seals and Caprock Integrity** *Stephanie Vialle, Jonathan Ajo-Franklin, and J. William Carey (Eds.)*
- 239 **Lithospheric Discontinuities** *Huaiyu Yuan and Barbara Romanowicz (Eds.)*
- 240 **Chemostratigraphy Across Major Chronological Eras** *Alcides N.Sial, Claudio Gaucher, Muthuvairavasamy Ramkumar, and Valderez Pinto Ferreira (Eds.)*
- 241 **Mathematical Geoenergy: Discovery, Depletion, and Renewal** *Paul Pukite, Dennis Coyne, and Daniel Challou (Eds.)*
- 242 **Ore Deposits: Origin, Exploration, and Exploitation** *Sophie Decree and Laurence Robb (Eds.)*
- 243 **Kuroshio Current: Physical, Biogeochemical and Ecosystem Dynamics** *Takeyoshi Nagai, Hiroaki Saito, Koji Suzuki, and Motomitsu Takahashi (Eds.)*
- 244 **Geomagnetically Induced Currents from the Sun to the Power Grid** *Jennifer L. Gannon, Andrei Swidinsky, and Zhonghua Xu (Eds.)*
- 245 **Shale: Subsurface Science and Engineering** *Thomas Dewers, Jason Heath, and Marcelo Sánchez (Eds.)*
- 246 **Submarine Landslides: Subaqueous Mass Transport Deposits From Outcrops to Seismic Profiles** *Kei Ogata, Andrea Festa, and Gian Andrea Pini (Eds.)*
- 247 **Iceland: Tectonics, Volcanics, and Glacial Features** *Tamie J. Jovanelly*
- 248 **Dayside Magnetosphere Interactions** *Qiugang Zong, Philippe Escoubet, David Sibeck, Guan Le, and Hui Zhang (Eds.)*
- 249 **Carbon in Earth's Interior** *Craig E. Manning, Jung-Fu Lin, and Wendy L. Mao (Eds.)*
- 250 **Nitrogen Overload: Environmental Degradation, Ramifications, and Economic Costs** *Brian G. Katz*
- 251 **Biogeochemical Cycles: Ecological Drivers and Environmental Impact** *Katerina Dontsova, Zsuzsanna Balogh-Brunstad, and Gaël Le Roux (Eds.)*
- 252 **Seismoelectric Exploration: Theory, Experiments, and Applications** *Niels Grobbe, André Revil, Zhenya Zhu, and Evert Slob (Eds.)*
- 253 **El Niño Southern Oscillation in a Changing Climate** *Michael J. McPhaden, Agus Santoso, Wenju Cai (Eds.)*
- 254 **Dynamic Magma Evolution** *Francesco Vetere (Ed.)*
- 255 **Large Igneous Provinces: A Driver of Global Environmental and Biotic Changes** *Richard. E. Ernst, Alexander J. Dickson, Andrey Bekker (Eds.)*
- 256 **Coastal Ecosystems in Transition: A Comparative Analysis of the Northern Adriatic and Chesapeake Bay** *Thomas C. Malone, Alenka Malej, Jadran Faganeli (Eds.)*
- 257 **Hydrogeology, Chemical Weathering, and Soil Formation** *Allen Hunt, Markus Egli, Boris Faybishenko (Eds.)*
- 258 **Solar Physics and Solar Wind** *Nour E. Raouafi and Angelos Vourlidas (Eds.)*
- 259 **Magnetospheres in the Solar System** *Romain Maggiolo, Nicolas André, Hiroshi Hasegawa, and Daniel T. Welling (Eds.)*
- 260 **Ionosphere Dynamics and Applications** *Chaosong Huang and Gang Lu (Eds.)*
- 261 **Upper Atmosphere Dynamics and Energetics** *Wenbin Wang and Yongliang Zhang (Eds.)*

Space Physics and Aeronomy Collection Volume 5
Geophysical Monograph 262

Space Weather Effects and Applications

Anthea J. Coster
Philip J. Erickson
Louis J. Lanzerotti
Editors

Yongliang Zhang
Larry J. Paxton
Collection Editors in Chief

This Work is a co-publication of the American Geophysical Union and John Wiley and Sons, Inc.

AGU
ADVANCING EARTH AND SPACE SCIENCE

WILEY

This edition first published 2021

Published under the aegis of the AGU Publications Committee

Brooks Hanson, Executive Vice President, Science
Carol Frost, Chair, Publications Committee
For details about the American Geophysical Union visit us at www.agu.org.

The right of Anthea J. Coster, Philip J. Erickson, Louis J. Lanzerotti to be identified as the authors of the editorial material in this work has been asserted in accordance with law.

Registered Office
John Wiley & Sons, Inc., 111 River Street, Hoboken, NJ 07030, USA

Editorial Office
111 River Street, Hoboken, NJ 07030, USA

For details of our global editorial offices, customer services, and more information about Wiley products visit us at www.wiley.com.

Wiley also publishes its books in a variety of electronic formats and by print-on-demand. Some content that appears in standard print versions of this book may not be available in other formats.

Library of Congress Cataloging-in-Publication Data

Names: Coster, A. J. (Anthea J.), editor. | Erickson, Philip, J. 1965– editor. | Lanzerotti, Louis J., editor.
Title: Space weather effects and applications / Anthea J. Coster, Philip J. Erickson, Louis J. Lanzerotti, editors.
Description: Hoboken, NJ : Wiley-American Geophysical Union, 2021. | Series: Geophysical monograph series | Includes bibliographical references and index.
Identifiers: LCCN 2020042802 | ISBN 9781119507574 (cloth) | ISBN 9781119815594 (adobe pdf) | ISBN 9781119815587 (epub)
Subjects: LCSH: Technology–Environmental aspects. | Space environment.
Classification: LCC TD174 .S673 2021 | DDC 600–dc23
LC record available at https://lccn.loc.gov/2020042802

Cover Design: Wiley
Cover Image: © John Oertel, Green Pepper Media LLC, provided courtesy of Dr. Lanzerotti

Set in 10/12pt Times New Roman by SPi Global, Pondicherry, India

CONTENTS

LIST OF CONTRIBUTORS

Daniel N. Baker
Laboratory for Atmospheric and Space Physics
University of Colorado at Boulder
Boulder, Colorado, USA

Timothy S. Bastian
National Radio Astronomy Observatory
Charlottesville, Virginia, USA

Michael Bodeau
Northrup Grumman Aerospace Systems
Redondo Beach, California, USA (ret.)

Gary S. Bust
Johns Hopkins University
Applied Physics Laboratory
Laurel, Maryland, USA

Anthea J. Coster
Haystack Observatory
Massachusetts Institute of Technology
Westford, Massachusetts, USA

Philip J. Erickson
Haystack Observatory
Massachusetts Institute of Technology
Westford, Massachusetts, USA

Dale E. Gary
Center for Solar-Terrestrial Research
New Jersey Institute of Technology
Newark, New Jersey, USA

John Kappenman
Storm Analysis Consultants
Duluth, Minnesota, USA

Louis J. Lanzerotti
New Jersey Institute of Technology and
Alcatel Lucent Bell Laboratories
New Jersey, USA (ret.)

William Liles
HamSCI Community
Virginia, USA

Christopher J. Mertens
Space Radiation Group
NASA Langley Research Center
Hampton, Virginia, USA

Cathryn Mitchell
Department of Electronic and Electrical Engineering
University of Bath
Bath, UK

Marcin D. Pilinski
Laboratory for Atmospheric and Space Physics
University of Colorado at Boulder
Boulder, Colorado, USA

William Radasky
Metatech Corporation
Goleta, California, USA

Eric K. Sutton
Space Weather Technology, Research, and Education
Center
University of Colorado at Boulder
Boulder, Colorado, USA

Jeffrey P. Thayer
Ann and H. J. Smead Aerospace Engineering Sciences
Department
University of Colorado at Boulder
Boulder, Colorado, USA

W. Kent Tobiska
Space Environment Technologies
Pacific Palisades, California, USA

Lawrence W. Townsend
Department of Nuclear Engineering
The University of Tennessee
Knoxville, Tennessee, USA

Endawoke Yizengaw
Space Science Application Laboratory
The Aerospace Corporation
El Segundo, California, USA

PREFACE

Since the advent of the electrical telegraph about 170 years ago, human technologies have greatly expanded in type and in purpose for civilian, commercial, and national security uses. These include electrical grids, pipelines, radar, wireless signaling, navigation, flying spacecraft, and telephony: technologies that cross continents, oceans, and now space. Regardless of specific application, successful operational use of these technologies has determined that compelling needs exist to take into account Sun and Earth space phenomena and processes in both design and implementation. Increasingly sophisticated technical systems require increasingly detailed understanding of solar and terrestrial space phenomena. Achieving this detailed understanding has been aided by the access to space provided by reliable launch vehicles, and by ever more sophisticated instrumentation deployed to measure Earth's space environment. The data acquired can be incorporated into ever better models to describe and even forecast the environment and its changes. This volume contains nine chapters, written by experts, describing current-day technologies and how solar and terrestrial space processes can affect them. Without these technologies, contemporary life in civil, commercial, and national security realms would be very different, and arguably impossible. One chapter in this volume outlines a number of issues related to human survival in the space radiation environment inside and outside Earth's magnetosphere. An epilogue closes by looking to the future in this broad area of applied geophysics. As the historical record demonstrates, despite specific qualities such as form and function, there is a high likelihood that some electrical technologies yet to be implemented or invented will always require design features whose goals are to ensure successful operations under all levels of solar and terrestrial conditions. The study of these environmental conditions in both basic and applied form will thus remain essential for the future.

Anthea J. Coster
Haystack Observatory
Massachusetts Institute of Technology
Westford, Massachusetts, USA

Philip J. Erickson
Haystack Observatory
Massachusetts Institute of Technology
Westford, Massachusetts, USA

Louis J. Lanzerotti
New Jersey Institute of Technology and
Alcatel Lucent Bell Laboratories
New Jersey, USA (ret.)

Introduction: Space Weather Underlies Reliable Technologies

Louis J. Lanzerotti[1], Philip J. Erickson[2], and Anthea J. Coster[2]

Descriptions and understandings of the space environment around Earth have grown exponentially over the centuries since William Gilbert described the Earth as a "great magnet" in his classic book *De Magnete* (1600). Gilbert used a model Earth, called a terrella, in his work, and studied several aspects of what can be called "electricity," including static electricity using amber because of its attractive properties. The invention of the telescope concept by Hans Lippershey and the use of the telescope by Galileo Galilei for astronomical (and thus space environment) purposes (including studies of the Moon and Jupiter's four major moons) occurred in the decade following the publication of *De Magnete*.

Initial use of electrical phenomena for practical purposes by humans can perhaps be attributed to the development of the lightning rod in about 1749 by Benjamin Franklin, and of the telegraph system patented in 1837 by Samuel F. B. Morris. The telegraph system revolutionized long-distance communications for personal, commercial, and military purposes. The long grounded wires of the first telegraph systems in the eastern United States and in western and southern Europe formed the detector arrays that first gave evidence of the coupling of Earth's space environment to human technologies. The engineering superintendent of the Midland Railway Company, William Henry Barlow, first documented "spontaneous" currents in the electrical circuits of railway telegraph systems (Barlow, 1849). His data, purposefully taken over a two-week period to study the subject, showed clear diurnal variations in the electrical currents. Barlow also wrote that "in every case which has come under my observation, the telegraph needles have been deflected whenever aurora has been visible."

The large geomagnetic storm that occurred following the discovery by amateur solar astronomer Richard Carrington of the first white light solar flare on 1 September 1859, caused havoc in the telegraph systems of Europe and the U.S. (Prescott, 1866). One example of the havoc of this singular "Carrington event" was that for lengthy intervals the telegraph between Boston and Portland, Maine, could be operated solely on the basis of the "spontaneous" electrical currents flowing in the wires; batteries were not needed at each end of the telegraph line to send messages. Disruptions of telegraphic communications occurred in systems in the U.S. and Europe throughout the extensive geomagnetic storm interval of 1–2 September 1859.

It seemed clear that there was some type of significant coupling between Sun and Earth (including the generation of aurora) and the telegraph systems. But no authorities had any insight of what such couplings might be. Debate waged in the scientific and engineering literatures for several decades in the late 19th century. Indeed, an authority with the eminence of Lord Kelvin (William Thomson), whose analysis work was key in the implementation of the first trans-Atlantic cable, argued in his presidential address to the Royal Society in 1892 that such Sun–Earth coupling was not physically possible (Kelvin, 1892).

Since the days of the advent of the electrical telegraph about 170 years ago, human technologies have greatly expanded in type and in purpose for civilian, commercial, and national security uses. Many of these important contemporary technology developments are illustrated in the figure. These depicted elements, relied on heavily by current society, must by necessity take into account phenomena and processes in Sun and near-Earth space for their design, implementation, and ultimate successful operations. Since the time of the telegraph, several additional

[1] *New Jersey Institute of Technology and Alcatel Lucent Bell Laboratories, New Jersey, USA (ret.)*

[2] *Haystack Observatory, Massachusetts Institute of Technology, Westford, Massachusetts, USA*

Space Physics and Aeronomy Collection Volume 5: Space Weather Effects and Applications, Geophysical Monograph 262, First Edition.
Edited by Anthea J. Coster, Philip J. Erickson, and Louis J. Lanzerotti.

DOI: 10.1002/9781119815600.introduction

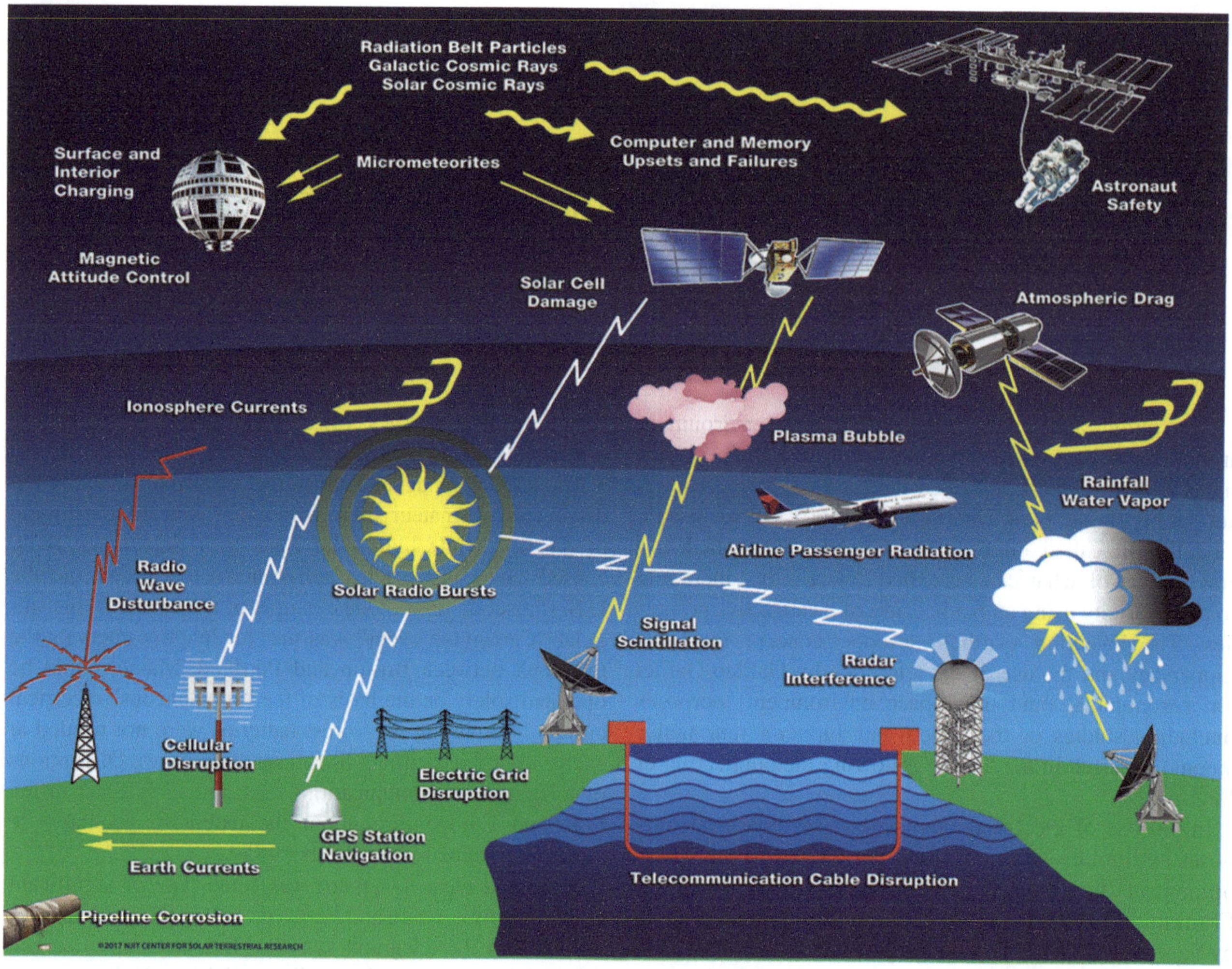

technologies have developed that use long conductors and are therefore susceptible to induced ground currents. These include electrical grids, pipelines, and telephony, both continental and transoceanic.

Successful operations of increasingly sophisticated technical systems require increasingly detailed understanding of solar and terrestrial space phenomena. This has been accomplished by the access to space provided by increasingly reliable launch vehicles since the late 1950s. Ever more sophisticated instrumentation has been deployed to measure Earth's space environment. The data acquired can be incorporated into ever better models to describe and even forecast the environment and its changes. This also has the complementary result of better understanding the effects of the environment on contemporary technologies.

This volume was designed to provide a topical discussion of current day technologies and how they are affected by solar and terrestrial space processes. Without these technologies, contemporary life in civil, commercial, and national security realms would be very different, and indeed impossible. Access to space has also meant that humans can now live, with appropriate support systems, above the sensible atmosphere. Thus, a closely related topic, also covered in this volume, involves the many issues related to human survival in the space radiation environment inside and outside Earth's magnetosphere. These issues deserve serious consideration prior to the planning and execution of projects involving a significant human presence in interplanetary space.

REFERENCES

Barlow, W. H. (1849). On the spontaneous electrical currents observed in the wires of the electric telegraph. *Proc. Royal Soc.*, *139*(61), 1849.

Kelvin, Lord W. T. (1892). *Nature*, *47*(109).

Prescott, G. B. (1866). *History, theory, and practice of the electric telegraph*. Boston: Ticknor & Fields.

1

Effects of Space Radiation on Contemporary Space-Based Systems I: Single Event Upsets, Spacecraft Charging, Degradation of Electronics, and Attenuation on Fiber Cabling

Daniel N. Baker[1] and Michael Bodeau[2]

ABSTRACT

Exposure of space systems to solar energetic particles, galactic cosmic rays, and radiation belt fluxes can cause temporary operational anomalies, damage critical electronics, degrade solar arrays, and blind optical systems such as imagers and star trackers. Moreover, intense solar particle events present a significant radiation hazard for astronauts during the high-latitude segment of the International Space Station orbit as well as for future human exploration of the Moon and Mars. In addition to such direct effects as spacecraft anomalies, a thorough assessment of the impact of space radiation on present-day space operations must include the collateral effects of space-weather-driven technology failures. For example, space radiation can degrade and, during severe events, completely incapacitate various communication and reconnaissance platforms. A complete picture of the impact of space radiation must include both direct as well as collateral effects of incapacitation on susceptible space structures and systems. It is also imperative that we as a technological society develop a truly operational understanding of space radiation in which the benefits of accurate forecasts are clearly established.

1.1. INTRODUCTION

Space systems on which modern society depends mostly operate in the region from altitudes of a few hundred km to ~40,000 km above Earth's surface. This region is filled with various populations of energetic particles. The fact that the Earth is surrounded by belts of very energetic protons and electrons was the first major discovery of the space age in 1958 (see Van Allen et al., 1958, 1959). From the initial realization that the terrestrial magnetic field could "trap" high-energy particles, today there is a much more complete understanding of what are now called the Van Allen radiation belts. There has long been awareness of high-energy solar and galactic cosmic rays as well.

More or less from the beginning of the space age, it was realized that intense populations of penetrating particles could be quite damaging to electronic systems in space (see Gombosi et al., 2017). There also were concerns about spacecraft structural materials and human space travelers (Van Allen, 1966). Thus, from the earliest days, it was realized that the terrestrial space environs were a problem to be reckoned with when it came to flying robotic and human missions in near-Earth regions.

[1] *Laboratory for Atmospheric and Space Physics, University of Colorado Boulder, Boulder, Colorado, USA*

[2] *Northrup Grumman Aerospace Systems, Redondo Beach, California, USA (ret.)*

Space Physics and Aeronomy Collection Volume 5: Space Weather Effects and Applications, Geophysical Monograph 262, First Edition.
Edited by Anthea J. Coster, Philip J. Erickson, and Louis J. Lanzerotti.

DOI: 10.1002/9781119815600.ch1

Today it is recognized that space radiation is one of the most pervasive and concerning threats that constitute what we comprehensively term *space weather* (Baker & Lanzerotti, 2016).

This chapter is intended to provide a brief overview of space radiation sources and their effects. Related impacts are treated in the companion chapter by Bodeau and Baker (chapter 2, this volume). A second goal of this chapter is to describe from an operational perspective the implications of radiation damage to systems in various parts of the geospace domain. In providing such a brief survey, the goal is to characterize in a succinct way the increasing importance of radiation damage on emerging technological systems.

1.2. OVERVIEW OF SPACE RADIATION PROPERTIES

Much of the concern for operational space systems arises from the Earth's radiation belts. A modern view of the radiation belts derived from the Van Allen Probes observations of Baker et al. (Baker, Kanekal, Hoxie, Henderson, et al., 2013) is shown here as Figure 1.1. Closest to the Earth's surface is the inner Van Allen belt. This belt extends from just above the dense atmosphere out to an equatorial altitude of about 10,000 km above the Earth's surface. The inner Van Allen belt is comprised dominantly of very energetic protons (ranging up to multiple GeV energies). Recent results demonstrate that protons with energies from ~10 MeV to ~100 MeV are quite stable in time near the geocentric radial distance of r ~ $1.5R_E$ (Earth radii = 6372 km), at which the inner zone proton fluxes peak. However, an outer "shoulder" of the radial distribution from $1.7 \leq r \leq 2.5\ R_E$ shows tremendous temporal variability for protons with $E \leq 60$ MeV. These variable proton fluxes are probably due primarily to evolution of trapped solar energetic protons (Selesnick et al., 2014).

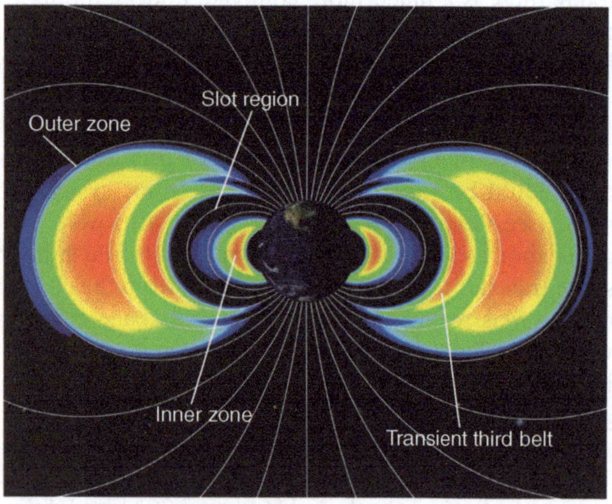

Figure 1.1 A modern-day view of the Earth's radiation belts as observed by the Van Allen Probes mission. (Adapted from Baker, Kanekal, Hoxie, Henderson, et al., 2013.)

The inner Van Allen belt also has copious fluxes of low-and medium-energy electrons (Fennell et al., 2015; Li et al., 2015) as revealed by Van Allen Probes and other spacecraft data sets (Baker, Kanekal, Hoxie, Batiste, et al., 2013). However, the Van Allen Probes era (September 2012–present) has provided many new discoveries about inner magnetospheric ultrarelativistic electrons with energies $E \gtrsim 5$ MeV. In particular, initial results after major storm intervals have shown essentially no detectable prompt ultrarelativistic electron fluxes in the region $r \leq 2.8\ R_E$ (Baker et al., 2014). The paucity of very energetic electrons in the inner magnetosphere immediately following major magnetic storms is quite striking (Foster, 2016; Baker et al., 2016) with a fascinating dependence on plasmasphere conditions before the storm and perhaps even on in-situ radio signals of terrestrial origin (Foster et al., 2016). The consequences of this energetic electron paucity for space radiation effects will be discussed further in this chapter.

The space weather concerns for the inner radiation zone are several. The intense, high-energy trapped protons are extremely damaging to space systems (Vette et al., 1966) and are quite hard to shield against. There are also more variable, trapped solar energetic ions in the inner zone that can cause dose and single-event effects (see Lorentzen et al., 2002; Baker, 2002). The hazardous proton populations of the inner zone and slot regions have two fundamental sources. In the inner zone, the primary source is the neutron albedo decay process, which has been well defined (Selesnick et al., 2007, and references therein). The outer edges of the inner zone and the lower part of the slot region also host energetic protons and ions that consist of entrained solar particles (Mazur et al., 2005; Selesnick et al., 2007). The inner zone also has trapped galactic cosmic rays (see Klecker et al., 1995; Cummings et al., 1993). Finally, the trapped energetic electrons with $E \leq 1$MeV (Claudepierre et al., 2017; Fennell et al., 2015) also represent a further significant source causing total dose effects.

The so-called "slot region" of the radiation belts extends from roughly L ~ 2.0 to L ~ 3.0 depending on particle energy and species. (L is the geocentric distance in Earth radii at which a dipole magnetic field line crosses the magnetic equatorial plane.) The slot is a region often relatively devoid of energetic electrons. However, during strong geomagnetic storm periods, the gap between the inner and outer zone can be filled to a large degree by moderate (and even high) energy electrons (Fennell et al., 2005). For example, in the intense "Halloween" storm period of late October and November 2003, the slot region was filled with multi-MeV electrons for several weeks (Baker et al., 2004). Thus, the slot region can present several space weather concerns including low- and medium-energy electron enhancements, multi-MeV electrons (on rare occasions), and strong solar energetic particle events (again on relatively rare occasions).

Finally, the outer Van Allen radiation belt represents in many ways the most pervasive space weather risks to operating spacecraft. The outer radiation belt is broad in spatial extent (from $r \sim 3R_E$ to $r \gtrsim 6.5\ R_E$). It is comprised of mildly to highly energetic electrons (~100keV to $\gtrsim$10 MeV) and varies widely in particle intensity. Commercial, military, and scientific satellites operating in medium-Earth orbit and geostationary Earth orbit number in the multiple hundreds worldwide. All of these operating spacecrafts are subject to outer Van Allen belt space radiation impacts.

Obviously, from the above brief description it is clear that the space radiation environment can cause a wide variety of impacts on space systems. Having a deeper understanding of radiation properties including dynamics and temporal trends is crucial for our technological society. This chapter describes the current knowledge of these space radiation aspects.

1.3. SPACE RADIATION EFFECTS

Figure 1.2 shows a schematic diagram of a representative spacecraft presumed to be at some location within the Earth's space environment. At high altitudes, such a spacecraft would be subject to a variety of space effects. Both solar energetic particles and trapped energetic protons can cause significant energy deposition for sensitive electronics on board a given spacecraft. Similarly, galactic cosmic ray nuclei may pass through electrical components and induce radiation degradation effects. As illustrated in Figure 1.2, these ion interactions effects are termed "single event effects." There also can be surface and deep-dielectric (bulk) charging due to energetic electrons.

The record of operating spacecraft over the space age is filled with examples of anomalies and complete satellite

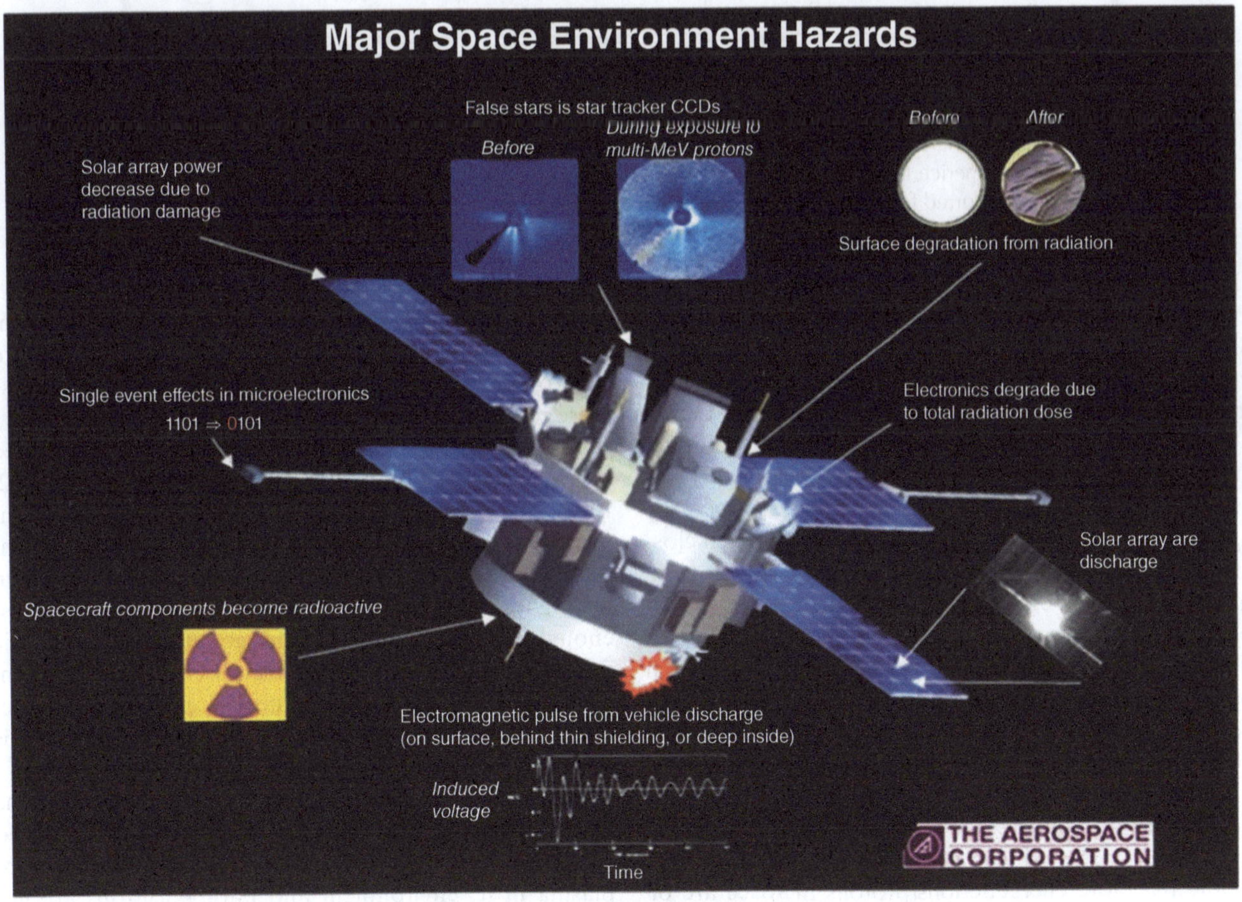

Figure 1.2 A schematic diagram showing spacecraft impacts due to the space environments including single-event ion effects, deep-dielectric charging due to high-energy electrons, and surface charging due to low-to-medium energy electrons. (Source: Aerospace Corporation.)

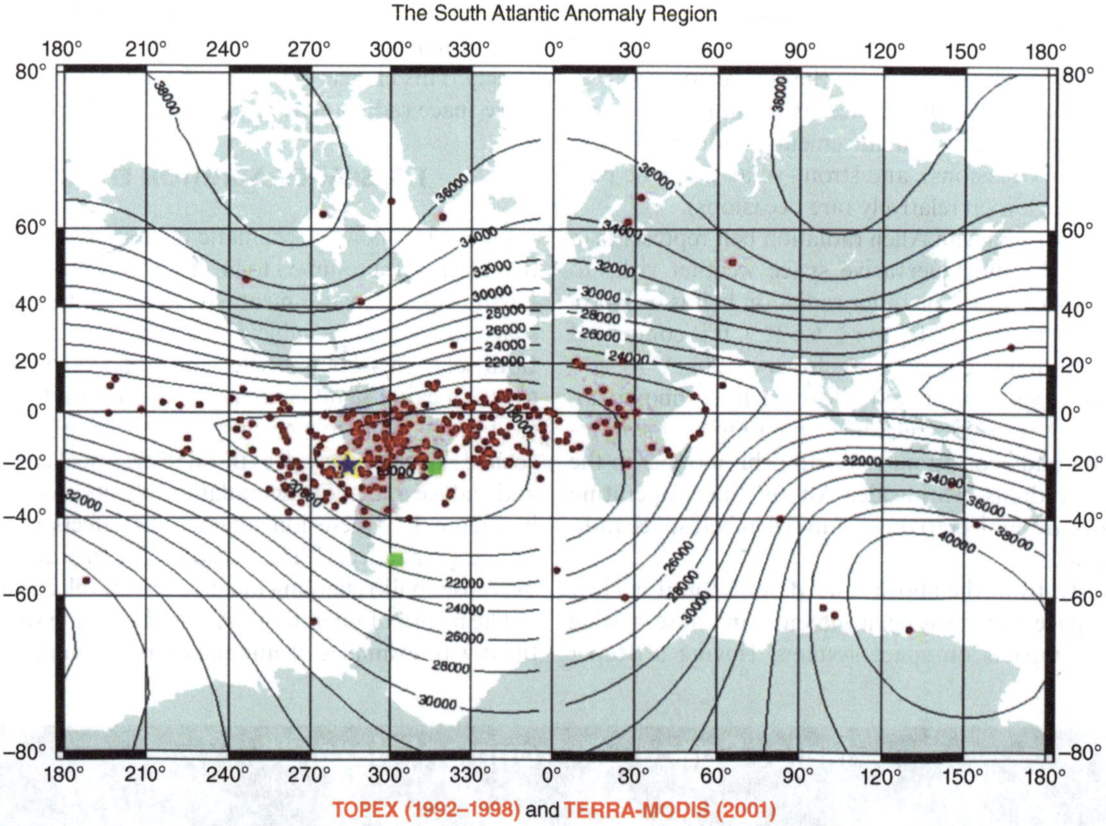

Figure 1.3 A map of the Earth showing contours of constant surface magnetic field strength. The weakest field region over South America is called the South Atlantic Anomaly. The red symbols show where significant operational anomalies occurred for the TOPEX and TERRA spacecraft.

failures due to space weather impacts (Allen, 2002). Many operational issues occur in the South Atlantic Anomaly (SAA) region, as illustrated in Figure 1.3. As shown in the diagram, the Earth's magnetic field is weakest over Brazil and eastward into the South Atlantic Ocean region. This results from the offset, tilted dipole of Earth's intrinsic magnetic field (Cain, 1966). Trapped particles mirroring along terrestrial magnetic field lines, especially high-energy protons in the inner zone, can approach closest to the Earth's surface in the weak SAA field region. Thus, this is the region where low-Earth orbit (LEO) spacecraft encounter the most intense particle radiation. Figure 1.3 demonstrates that the TOPEX (U.S./French Ocean Surface Topography mission) and TERRA (Earth Observing System-1 satellite of NASA at 705 km altitude) spacecraft suffered many electronic anomalies and upsets in the SAA (Allen, 2002), probably due to single-event effects from inner zone protons. The mechanism of single-event upsets is illustrated in Figure 1.4a.

Only the more energetic ions/protons in space are of major concern from a space weather hazard perspective. Very high energy protons and ions can penetrate shielding, depositing their energy in electronic devices. Protons with E > 3 MeV are a primary contributor to solar array degradation, while >15 MeV protons contribute to single-event effects and nonionizing radiation damage to electronic parts within satellites that orbit in or traverse these regions.

Another impact of the space environment is called deep-dielectric charging and is also noted in Figure 1.2. Electrons with energies of hundreds of keV up to multiple MeV energies may penetrate shielding and bury themselves in dielectric materials inside spacecraft. As shown by Figure 1.4b, if the buildup of buried charge is fast enough (and the leakage of charge out of the dielectric material is slow), then a powerful discharge can happen. This can severely damage materials and sensitive electronic components (see Bodeau & Baker, chapter 2, this volume).

Yet another type of space radiation impact within the Earth environment is shown in Figure 1.4c. This is called surface charging. If the space system is subject to hot plasma in its environment and photo-electrons are not able to carry away charge promptly from insulator and dielectric material (as happens in solar eclipse conditions or on the nonilluminated side of a space vehicle), there

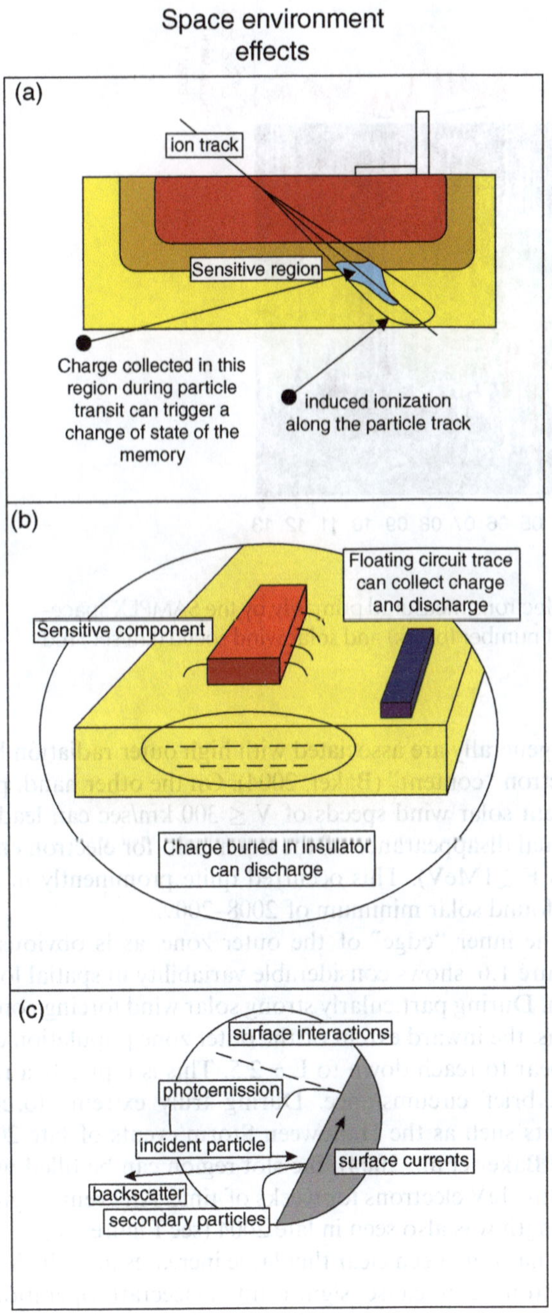

Figure 1.4 Illustration of the (a) single event upset, (b) deep-dielectric discharge, and (c) surface charging mechanisms that can cause severe spacecraft operational anomalies. (Source: Baker, 2005. Reproduced with permission from Springer Nature.)

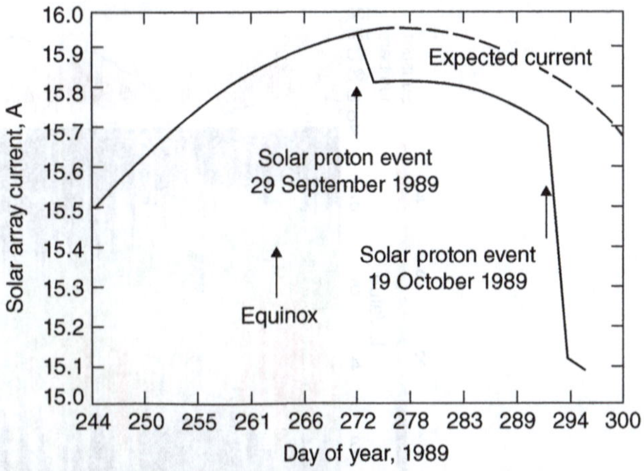

Figure 1.5 Observed solar array current reduction on a high-altitude spacecraft following solar energetic particle events over a three-week period in 1989. (Source: Marvin & Gorney, 1991); reprinted by permission of the American Institute of Aeronautics and Astronautics, Inc.

can be a strong buildup of charge on key areas of the spacecraft surface. When the charge buildup is sufficient, a sudden discharge can result. This may produce material damage and/or a significant electromagnetic discharge around the spacecraft. Such effects cause phantom electromagnetic signals or changes of state in memory systems. There can even be permanent damage to electronics (see Bodeau and Baker, chapter 2, this volume).

In addition to the trapped population, there also are episodic solar energetic particle events, with enhanced energetic solar heavy ions. Those ions that eventually reach the Earth can penetrate deeply into the inner magnetosphere. At such locations, they can cause single-event upsets and enhance the radiation damage to satellite solar arrays and electronic parts, similar to that shown in Figure 1.5 (Marvin & Gorney, 1991). The Solar, Anomalous, and Magnetospheric Particle Explorer (SAMPEX) mission (see Baker et al., 1993) at LEO data showed multiple incidences of >15 MeV solar protons penetrating to low L combined with the presence of the protons from cosmic ray albedo neutron decay (CRAND) at very low L, and inherent variability of the proton fluxes in the $2.5 < L < 4$ region in response to magnetic storms (Mazur et al., 2005). This CRAND mechanism was the first idea put forth to explain the existence of trapped energetic particles in the magnetosphere (see Van Allen, 1983).

Solar energetic particle events are the mechanism by which very high-energy protons appear in the outer zone region of the radiation belts. However, these events are very short lived (hours in duration), as the high-energy protons with such large gyro-radii are not able to be stably trapped at these locations. As noted, they can often penetrate to the inner zone, where stronger magnetic fields in a more rigid magnetic topology can entrain these particles for long times. In general, highly energetic protons (multi-MeV energies) are not considered to be a regular feature in the outer zone (see Selesnick et al., 2014).

The characteristics and dynamics of relativistic and ultrarelativistic electron fluxes in Earth's outer radiation

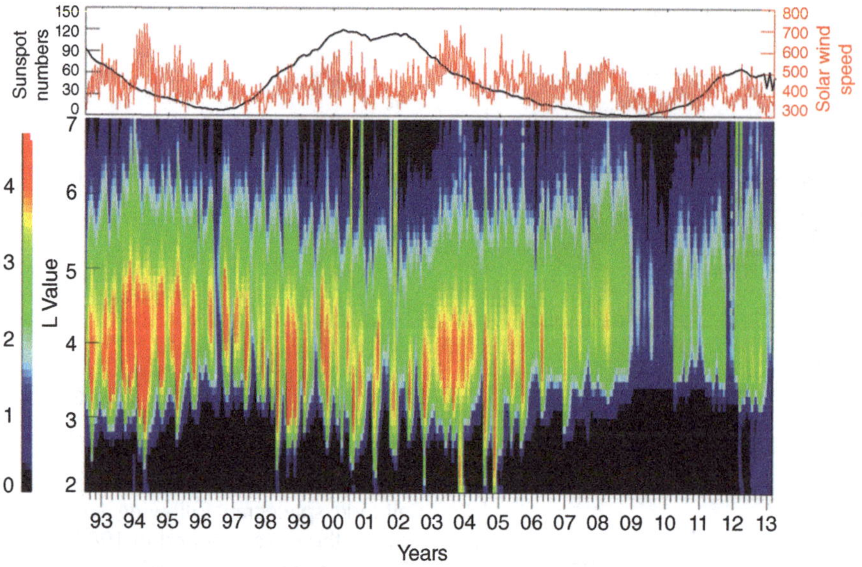

Figure 1.6 Long-term (1992–2013) record of $2 \leq E \leq 6$ MeV electrons measured primarily by the SAMPEX spacecraft at low altitude. The upper panel shows smoothed sunspot number (black) and solar wind speed (km/sec; red trace) for the same period of time (Source: X. Li.)

belt are strongly influenced by geomagnetic storm time processes. These lead both to significant electron loss and local acceleration deep within the inner magnetosphere. The outer Van Allen belt is highly variable on essentially all temporal scales and shows tremendous variability as well as significant variations in the outer belt spatial extent. This is well illustrated by Figure 1.6 adapted from Li et al. (2006). The main (lower) panel of the figure shows integral fluxes of $2 \leq E \leq 6$ MeV electrons measured by sensors on board SAMPEX. The data are color-coded representations of electron flux from L = 2.0 to L = 7.0 (vertical axis) versus time (horizontal axis) from 1992 to 2013. The color bar to the left is the log of electron flux (electrons/cm^2-s-sr). The upper panel of Figure 1.6 shows in black the smoothed sunspot number over the period 1992–2013, while the red trace shows the solar wind speed (km/sec) measured upstream of the Earth's magnetosphere.

As is evident from the figure, the outer Van Allen belt (as monitored from the ~500 km altitude of SAMPEX) typically extends from L ~ 3.0 to roughly L = 6.5. However, the belt's spatial extent was highly variable over the ~20-year SAMPEX lifetime. Also, the peak electron flux levels varied widely as a function of time. For example, in 1994, the peak >2 MeV electron fluxes were over 10^4 electrons/cm^2-s-sr, while in 2008–2009 the electron flux was often barely above background levels (~10^0 electrons/cm^2-s-sr). As is obvious from inspection of the figure, a main controller of the electron fluxes in the outer belt is the solar wind speed impacting Earth's magnetosphere (Baker et al., 1987). Indeed, solar wind speeds with V > 500 km/sec generally are associated with high outer radiation belt electron "content" (Baker, 2004). On the other hand, persistent solar wind speeds of $V \leq 300$ km/sec can lead to virtual disappearance of the outer belt (for electron energies $E \gtrsim 1$MeV). This occurred quite prominently in the profound solar minimum of 2008–2009.

The inner "edge" of the outer zone, as is obvious in Figure 1.6, shows considerable variability in spatial location. During particularly strong solar wind forcing conditions, the inward extent of the outer zone population can appear to reach down to L ~ 2.5. This is typically a rare and brief circumstance. During truly extreme forcing events such as the Halloween Storm events of late 2003 (see Baker et al., 2004), the slot region can be filled with multi-MeV electrons for weeks of time. An event of great strength was also seen in late 2004 (see Figure 1.6).

It has long been clear that large increases in multi-MeV electrons can cause significant spacecraft operational anomalies. For example, Figure 1.7 (taken from Baker, 1987) shows that star-tracker anomalies were observed on an operational geostationary Earth orbit spacecraft whenever the counting rate for E ~ 3 MeV electrons reached or exceeded 7 counts/sec. The vertical arrows show the precise timing relationship between the star tracker problems and the enhanced electron flux episodes during this 1980–1982 period. Many other electron-induced anomalies have been documented over the space age (see Baker, 2005, and references therein).

A final point to note is that newer spacecraft technologies can be quite susceptible to space radiation effects. An example is fiber optic cabling that is increasingly used on

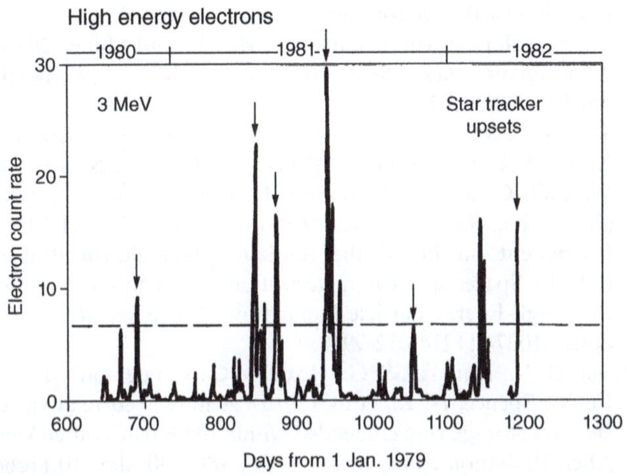

Figure 1.7 Temporal correspondence between star tracker anomalies onboard a geostationary orbit spacecraft (vertical arrows) and enhancements in E ~ 3 MeV electron fluxes measured by sensors on the same spacecraft. (Source: Baker, 1987.)

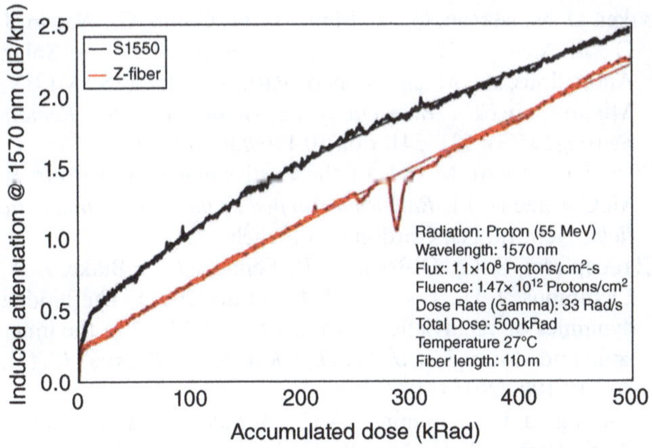

Figure 1.8 Induced attenuation of fiber optic cabling signals due to accumulated proton dose for two different types (S1500 and Z-fiber®) of material (Source: Alam, https://www.spiedigitallibrary.org/conference-proceedings-of-spie/10567/105672M/Performance-of-optical-fibers-in-space-radiation-environment/10.1117/12.2308184.full?SSO=1. Licensed under CC BY SA 4.0)

operational spacecraft. Different fiberglass materials and different designs can show substantially different induced attenuations of signals with increasing radiation dose exposure. An example is shown in Figure 1.8 from a study by Alam et al. (2006).

Figure 1.8 shows induced attenuation growth versus accumulated proton dose plots for the two fibers. S1550-HTA fiber with the higher NA (0.16) performs fairly well in comparison to Z-fiber®, which has relatively smaller NA (0.12). For both fibers, attenuation growth data can be represented well by the three-term "saturating exponentials" model (Alam et al., 2006). Solid lines in the figure represent best fit to the data. S1550-HTA shows a trend towards saturation, while Z-fiber® does not indicate any such behavior. Using the data acquired, radiation-induced attenuation of S1550-HTA fiber at room temperature and to a total accumulated dose of 50 kRad was predicted to be 1.33 dB/km and 0.76 dB/km in γ- and proton radiation environments, respectively.

1.4. DISCUSSION AND CONCLUSIONS

This chapter has reviewed evidence that galactic cosmic rays, solar energetic particles, trapped high energy particles, and magnetospheric electrons of moderate energies constitute a significant problem for the operation of spacecraft. These particle populations occur with different frequencies, and the nature of their damage mechanisms are also quite different. Certain types of spacecraft anomalies have been closely linked with environmental factors, while other classes of disruptions may or may not be related to space environmental conditions. More work needs to be done to clarify where the environment is (or is not) implicated. Also, more work is required to understand the physical nature of some disruption mechanisms in spacecraft systems.

There is an increasing physical understanding of the near-Earth space environment. This understanding is sufficient that there is even a significant predictive capability. The first line of defense for space environmental problems is good engineering design. However, systems often degrade over time, and sometimes unexpected sensitivities develop in space components. In such cases, a simple forecast of space environmental conditions may allow operators to change operational procedures, or at least be ready to recover quickly from operational problems should they occur. In yet other circumstances, it may be possible to turn off susceptible subsystems during periods of strong space environmental disturbance. Thus, we assert that accurate and reliable forecasts have great potential benefit to the operational community. In this way, space weather predictions and forecasts are a next logical step in our space environmental research efforts.

We have also shown in this chapter that long-term cumulative effects from space radiation can have a variety of severe effects. Total dose and displacement damage in electronic components and in solar power systems often spell the ultimate demise of space systems. This can be a cost driver over the entire course of a mission. For example, the possibility (or likelihood) of long-term degradation of solar array power output often forces designers to oversize the arrays. Similarly, degradation of thermal control surfaces often forces designers to oversize thermal radiator panels. This is done to be confident of end-of-life heat rejection, but then designers may have to add extra heaters which, in turn, require more solar array power. These compounding design aspects can drive the cost and complexity of space systems to extraordinary levels.

We also note that the degradation of semiconductor part performance often limits the effective operating lifetime of satellite subsystems in space. Having better knowledge of regional radiation environmental conditions would make a huge difference in accurately forecasting mission life and consequent program costs. Presumed knowledge of average space environmental conditions determine the operational regimes for spacecraft. These considerations dictate that certain orbits are not possible or cost effective to fly because of the severity of the radiation environments. We have shown in this chapter that the Van Allen Probes mission has revealed that for present, and recent, solar activity conditions, some parts of the near-Earth environment have been surprisingly benign and nonthreatening. Hence, getting the space system design right requires that we get the space environments right. This ultimately means that designers must be able to forecast with confidence what the expected fluences are for various particle populations over the course of the designed mission. Of course, there needs to be a reasonable sense of margins on these numbers because of the inherent variability of the solar-terrestrial system. This is where design and space weather climatology meet.

REFERENCES

Alam, M., Abramczyk, J., Manyam, U., Farroni, J., & Guertin, D. (2006). Performance of optical fibers in space radiation environment. *Proceedings, International Conference on Space Optics – ICSO 2006*, vol. 10567, 105672M. doi: 10.1117/12.2308184

Allen, J. H. (2002). Historical and recent solar activity and geomagnetic storms affecting spacecraft operations, *Proc. GOMAC*, Modern Space Syst. Issues, Monterey, CA.

Baker, D. N. (1987). Effects of the solar terrestrial environment on satellite operations, *Artificial Satellites*, *22*, 103.

Baker, D. N. (2002). How to cope with space weather. *Science*, *297*, 1486–1487. doi:10.1126/science.1074956

Baker, D. N. (2004). Specifying and forecasting space weather threats to human technology. In I. A. Daglis (Ed.), *Effects of space weather on technology infrastructure* (pp. 1–25). Kluwer.

Baker, D. N. (2005). Introduction to space weather. In K. Schere (Ed.), *Space weather: The physics behind a slogan* (pp. 3–20). Springer-Verlag.

Baker, D. N., Jaynes, A. N., Hoxie, V. C., Thorne, R. M., Foster, J. C., Li, X., et al. (2014). An impenetrable barrier to ultrarelativistic electrons in the Van Allen Radiation Belt. *Nature*, *515*, 531–534. doi:10.1038/nature13956

Baker, D. N., Jaynes, A. N., Kanekal, S. G., Foster, J. C., Erickson, P. J., Fennell, J. F., et al. (2016). Highly relativistic radiation belt electron acceleration, transport, and loss: Large solar storm events of March and June 2015. *J. Geophys. Res. Space Physics*, *121*(7), 6647–6660. doi:10.1002/2016JA022502

Baker, D.N., S.G. Kanekal, V.C. Hoxie, S. Batiste, M. Bolton, X. Li, S.R. Elkington, S. Monk, R. Reukauf, S. Steg, J. Westfall, C. Belting, B. Bolton, D. Braun, and B. Cervelli (2013). The Relativistic Electron-Proton Telescope (REPT) Instrument on Board the Radiation Belt Storm Probes (RBSP) Spacecraft: Characterization of Earth's Radiation Belt High-Energy Particle Populations, *Space Sci. Rev.*, *179*, doi:10.1007/s11214-012-9950.

Baker, D. N., Kanekal, S. G., Hoxie, V. C., Henderson, M. G., Li, X., Spence, H. E., et al. (2013). A long-lived relativistic electron storage ring embedded within the Earth's outer Van Allen Radiation Zone, *Science*, *340*, 186–190. doi: 10.1126/science.123351

Baker, D. N., Kanekal, S. G., Li, X., Monk, S. P., Goldstein, J., & Burch, J. L. (2004). An extreme distortion of the Van Allen belt arising from the 'Hallowe'en' solar storm in 2003, *Nature*, *432*, 878–881. doi:10.1038/nature03116

Baker, D. N., & Lanzerotti, L. J. (2016). Space weather. *American Journal of Physics*, *84*. doi.org/10.1119/1.4938403.

Baker, D. N., Mason, G. M., Figueroa, O., Colon, G., Watzin, J. G., & Aleman, R. M. (1993). An overview of the Solar, Anomalous, and Magnetospheric Particle explorer (SAMPEX) Mission. *IEEE Transactions on Geoscience and Remote Sensing*, *45*(3), 531–541. doi: 10.1109/36.225519

Cain, J. C. (1966). Models of the Earth's magnetic field. In B. McCormac (Ed.), *Radiation trapped in the Earth's magnetic field*. Newark, NJ: Gordon and Breach.

Claudepierre, S. G., O'Brien, T. P., Fennell, J. F., Blake, J. B., Clemmons, J. H., Looper, M. D., et al. (2017). The hidden dynamics of relativistic electrons (0.7–1.5 MeV) in the inner zone and slot region. *J. Geophys Res. Space Physics*, *122*(3). doi: 10.1002/2016JA023719R

Cummings, J. R., Cummings, A.C., Mewaldt, R. A., Selesnick, R. S., Stone, E. C., Von Rosenvinge, T. T., & Blake, J. B. (1993). SAMPEX measurements of heavy ions trapped in the magnetosphere. *IEEE Trans. Nucl. Sci.*, *40*(6).

Fennell, J. F., Blake, J. B., Friedel, R., & Kaneka, S. G. (2005). The energetic electron response to magnetic storms: HEO satellite observations. In T. I. Pulkkinen, N. A. Tsyganenko, & R.H.W. Friedel (Eds.), *The inner magnetosphere: Physics and modeling. Geophysical Monograph 155*. doi: 10.1029/155GM10

Fennell, J. F., Claudepierre, S. G., Blake, J. B., O'Brien, T. P., Clemmons, J. H., Baker, D. N., et al. (2015). Van Allen Probes show that the inner radiation zone contains no MeV electrons: ECT/MagEIS data. *Geophys. Res. Lett.*, *42*(5), 1283–1289. doi:10.1002/ 2014GL062874

Foster, J. C., Erickson, P. J., Baker, D. N., Jaynes, A. N., Mishin, E. V., Fennel, J. F., et al. (2016). Observations of the impenetrable barrier, the plasmapause, and the VLF bubble during the 17 March 2015 storm. *J. Geophys. Res.*, *121*, 5537–5548. doi:10.1002/2016JA022509

Gombosi, T. I., Baker, D. N., Balogh, A., Erickson, P. J., Huba, J. D., & Lanzerotti, L. J. (2017). Anthropogenic space weather. *Space Science Reviews*. doi:10.1007/s11214-017-0357-5

Klecker, B., McNab, M. C., Blake, J. B., Hamilton, D. C., Hovestadt, D., & Kaestle, H. (1995). Charge state of anomalous cosmic ray nitrogen, oxygen, and neon: SAMPEX observations. *Astrophys. J.*, *442*.

Li, X., Baker, D. N., O'Brien, T.P., Xie, L., & Zong, Q. G. (2006). Correlation between the inner edge of outer radiation belt electrons and the innermost plasmapause location. *Geophys. Res. Lett.*, *33*(14). doi: 10.1029/2006GL026294

Li, X., Selesnick R. S., Baker, D. N., Jaynes, A. N., Kanekal, S. G., Schiller Q., et al. (2015). Upper limit on the inner radiation belt MeV electron intensity. *J. Geophys. Res.*, *120*(2). 1215–1228. doi: 10.1002/2014JA020777

Lorentzen, K. R., Mazur, J. E., Looper, M. D., Fennell, J. F., & Blake, J. B. (2002). Multisatellite observations of MeV ion injections during storms. *J. Geophys. Res.*, *107*(A9), 1231. doi:10.1029/2001JA000276

Marvin, D. C., & Gorney, D. J. (1991). Solar proton events of 1989: Effects on spacecraft solar arrays. *J. Spacecraft and Rockets*, *28*, 713–719. doi:10.2514/3.26304

Mazur, J. E., et al. (2005). The creation of new ion radiation belts associated with solar energetic particle events and interplanetary shocks. In Gopalswamy, Mewaldt, & Torsti (Eds.), *Solar eruptions and energetic particles, Geophys. Monogr. Ser.*, *165* (p. 345), Washington, DC: AGU.

Selesnick, R. S., Baker, D. N., Jaynes, A. N., Li, X., Kanekal, S. G., Hudson, M. K., & Kress, B. T. (2014). Observations of the inner radiation belt: CRAND and trapped solar protons. *J. Geophysics. Res.*, *119*. doi:10.1002/2014JA020188

Selesnick, R. S., Looper, M. D., & Mewaldt, R. A. (2007). A theoretical model of the inner proton radiation belt. *Space Weather*, *5*, S04003. doi:10.1029/2006SW000275

Van Allen, J. A. (1966). Spatial distribution and time decay of the intensities of geomagnetically trapped electrons from the high altitude nuclear burst of July 1962. In B. M. McCormac (Ed.), *Radiation trapped in the Earth's magnetic field* (pp. 575–592), D. Reidel, Dordrecht-Holland.

Van Allen, J. A. (1983). *Origins of magnetospheric physics*. Washington, DC: *Smithsonian Institution Press*.

Van Allen, J. A., Ludwig, G. H., Ray, E. C., & McIlwain, C. E. (1958). Observation of high intensity radiation by satellites 1958 alpha and gamma. *Jet Propulsion*, *28*(9), 588–592.

Van Allen, J. A., McIlwain, C. E., & Ludwig G. H. (1959). Satellite observations of electrons artificially injected into the geomagnetic field. *J. Geophys. Res*, *64*, 877.

Vette, J. I. (1996). Models of the trapped radiation environment. *Vol. I: Inner Zone Protons and Electrons*, NASA SP-3024. Greenbelt, MD: Goddard Space Flight Center.

2

Effects of Space Radiation on Contemporary Space-Based Systems II: Spacecraft Internal and External Charging and Discharging Effects

Michael Bodeau[1] and Daniel N. Baker[2]

ABSTRACT

Satellites are a critical part of the technology infrastructure of modern society. This is made painfully obvious when major satellite anomalies and failures occur that reveal how dependent we are on the communications and other services that satellites provide. In this chapter, we review the history of spacecraft charging and the satellite anomalies it has caused. The first section covers surface charging, which was the first phenomenon to appear, and a later section reviews internal charging. It will be observed that rates of anomalies have risen with the introduction of new satellite technologies and fallen as the spacecraft engineering community improves their designs, only to rise again in newer generations of spacecraft. Space science (also called space weather) has played critical roles in establishing charging as the cause of the anomalies and by defining environments for engineers to use to analyze and test their satellite designs. The chapter concludes with a discussion of future collaborative efforts between the satellite engineering and space science communities that are needed to mature the understanding of space weather effects on satellites. Better predictions of space weather effects on satellites are needed to accurately assess the impacts of a one in 100-year or 300-year space weather event, which to date satellites have yet to experience, but which is conjectured to have dire to cataclysmic consequences to society (Nat. Res. Council, 2008), extrapolating from the anomalies that have been observed.

2.1. SURFACE CHARGING

In 1971, a pair of next-generation communication satellites built for the U.S. Air Force (flight 1, satellites 9341 and 9342) of the Defense Satellite Communication System II (DSCS II) were launched into geosynchronous orbit. Soon after, uncommanded unit state-changes on the spacecraft were reported (Fredericks & Kendall 1973; Fredericks & Scarf, 1973; Rosen, 1976; TRW Space Sciences Dept., 1972). Inquiries to other geosynchronous satellite programs revealed they, too, experienced anomalies. A list of 23 well-defined anomalies observed by several spacecraft was collected (Table 2.1). A surprising feature of the anomalies was that they mostly occurred at local times between very late evening to dawn local time (Figure 2.1), strongly suggesting an external environmental cause for the anomalies. (The anomalies shown by triangles and squares in Figure 2.1 are anomalies observed in real time while the satellites were transmitting data to the ground station. Arcs indicate time periods when the satellite was not transmitting, with the last normal state

[1] *Northrup Grumman Aerospace Systems, Redondo Beach, California, USA (ret.)*
[2] *Laboratory for Atmospheric and Space Physics, University of Colorado at Boulder, Boulder, Colorado, USA*

Space Physics and Aeronomy Collection Volume 5: Space Weather Effects and Applications, Geophysical Monograph 262, First Edition.
Edited by Anthea J. Coster, Philip J. Erickson, and Louis J. Lanzerotti.

DOI: 10.1002/9781119815600.ch2

Table 2.1 List of 23 satellite anomalies from 1971. (Reproduced with permission of TRW Space Sciences Dept., 1972.)

SATELLITE EVENTS, LIST OF							
TIME PERIOD			SATELLITE EVENTS	DATE*-TIME (ZULA)	ORBITAL POSITION	LOCAL SAT. TIME	ANGLE
P	1	Y	Star Sensor A Anomaly	7 May – 23:00	.04°S/97.9°E	05:39	85°
P	2	Y	OTU A and B Anomaly	18 May – 17:22	.08°S/95.1°E	23:43	356°
?	3		Receiver 3A Sensitivity	21 May – 15:00	.09°S/94.1°E	21:16	320°
P	4		Receiver 3A Lock-Up	13 Jun – 18:09	.10°S/87.4°E	23:57	359°
Y	5		False Flag (CCM)	21 Jul – 14:48	5.0°N/174.5°E	02:28	37°
Y	6		False Flag (CCM)	29 Jul – 23:05	3.3°N/51.5°E	02:32	38°
Y	7		Link 2 Data Quality	18 Aug – 01:23	5.3°N/15.7°W	00:23	6°
Y	8		False Flag (CCM)	26 Aug – 14:10	0.5°S/3.6°E	14:24	216°
?	9		False Flag (RIM)	2 Sep – 23:33	0.06°S/83.9°E	05:09	77°
P	10		False Flag (RIM)	5 Sep – 18:37,	0.37°S/83.6°E	00:11	3°
				5 Sep – 22:56, and	0.10°S/83.7°E	04:34	69°
				6 Sep – 00:22	0.01°N/83.8°E	05:58	89°
?	11	Y	False Flag (RIM)	14 Sep – 23:01	0.05°S/83.2°E	04:33	68°
P	12	Y	False Flag (RIM)	18 Sep – 17:27	0.33°S/82.9°E	22:59	345°
P	13	Y	False Flag (RIM)	19 Sep – 20:34	0.23°S/82.9°E	02:06	31°
Y	14		False Flag (CCM)	20 Sep – 14:17	5.3°S/152.5°W	04:05	61°
Y	15		False Flag (RIM)	20 Sep – 18:55	0.03°S/82.8°E	00:27	7°
Y	16		False Flag (CCM)	30 Sep – 11:26,	1.5°S/153.9°W	01:11	18°
				30 Sep – 11:37, and	1.8°S/154.1°W	01:22	21°
				30 Sep – 14:45	6.4°S/156.2°W	04:21	65°
Y	17		False Flag (CCM)	7 Oct – 22:43	0.8°S/23.8°E	00:18	5°
Y	18		RGA Reset	12 Nov – 06:10/11:30	103°W	23:20/04:40	350°/70°
Y	19		RGA Reset	21 Nov – 06:30/10:00, and	103°W	23:40/03:10	355°/48°
				21 Nov – 07:15/11:45	112°W	23:48/04:18	357°/65°
Y	20		RGA Reset	23 Nov – 07:30/16:30	112°W	00:03/09:03	1°/136°
Y	21		RGA Reset	24 Nov – 01:30/07:15, and	112°W	18:04/23:48	271°/357°
				24 Nov – 07:15/16:00	103°W	00:25/09:10	6°/137°
Y	22		RGA Reset	24 Nov – 07:00/15:00, and	103°W	00:10/08:10	3°/123°
				25 Nov – 08:00/15:55	112°W	00:32/08:28	8°/127°
Y	23		RGA Reset	26 Nov – 08:12	103°W	01:22	21°

*All dates are calendar year 1971

shown by the earlier local-time endpoint; the later local-time endpoint shows when transmissions were resumed and the anomaly was discovered. The actual time of the anomaly is unknown but is bracketed by the arc endpoints. Radial distance in the diagram is irrelevant and is only used to facilitate readability.)

It was suspected that the anomalies may have been induced by interactions of the spacecraft with energetic (up to 20 keV) electrons that are injected during magnetospheric substorms or magnetic storms preferentially into the local morning quadrant. A plot of the anomalies in Table 2.1 on the World Data Center A "musical scales" of the Kp index for 1971 (Figure 2.2) showed that, with the exception of anomaly 9, the anomalies are associated with Kp values $\geq 3_+$ coincident with or preceding the anomaly. A Kp $\geq 3_+$ indicates one to two substorms were likely within the 3-hour period, whereas a Kp of ≥ 7 indicates a worldwide magnetic storm was probable, which makes it possible for anomalous events at any local time. The temporal correlation was improved by using ground station magnetograms. Ground stations will only detect substorms near a satellite if: (1) the station and satellite are on approximately the same L-shell and (2) if the station and satellite are on nearly the same magnetic field line (magnetic longitude). The investigators found 19 "yes," 2 "probable," and 2 questionable "?" correlations of anomalies with ground magnetogram storm or substorm signatures. (The code to the left of the event number in Table 2.1 gives the correlation with U.S. ground station magnetograms, while the code to the right of the event number refers to correlations with magnetograms subsequently obtained from Swedish and Icelandic stations located closer to the satellite magnetic footprint.)

The strong correlation between substorms and anomalies does not explain how the former cause the latter. However, instruments on other spacecraft (e.g. ATS 1,

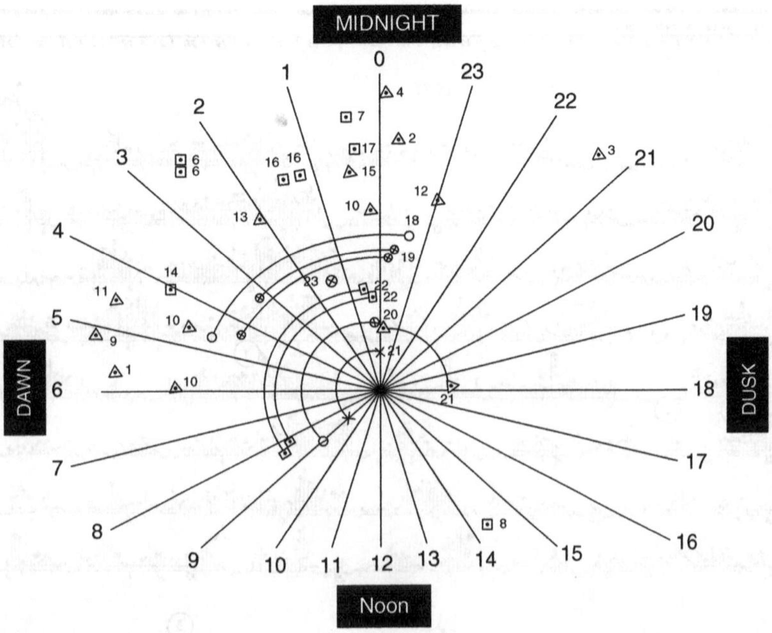

Figure 2.1 Local-time distribution of anomalies observed by several geostationary or near geostationary satellites. Numbers beside events are those tabulated in Table 2.1. (Reproduced with permission of TRW Space Sciences Dept., 1972.)

ATS 5, and OGO 5) provided clues to the causal relationship. Contemporaneous with the discovery of the anomalies, DeForest (1972) documented the charging of the ATS-5 spacecraft by energetic plasma sheet electrons (3 ≤ E_e ≤ 20 keV) injected into the local morning local-time sector at geosynchronous altitude. As an example, the substorm of October 16, 1969, injected electrons of mean energy ~5keV and protons of ~10keV, and number density ~1 cm^{-3} (Figure 2.3). After transition into eclipse, the entire spectra of electrons and protons are shifted by ~4.2 keV, with the electron energies retarded by this potential and the protons accelerated by it. An analysis of the spacecraft as a floating Langmuir probe indicated that illuminated spacecraft surfaces would be prevented from charging more than a few hundred volts negative relative to ambient plasma "ground" by photoemission, whereas a shadowed spacecraft surface would reach a potential nearly equal to the primary electrons, since the secondary electron emission is insufficient to prevent surface charging. The worst-case eclipse charging observed on ATS-5 was −9 kV. Similar corroborating evidence of surface charging was subsequently found in data obtained from OGO-5, IMP-6, and UK-4.

Such severe charging was a surprise. In a review of spacecraft charging, Garrett (1981) surveyed the early history of spacecraft charging. Early (1930s and 1940s) researchers predicted charging in interplanetary space between −3 and +10 V. Data obtained from rocket-borne sensors carried into the ionosphere in the 1950s showed charging of the rocket to −10 V to −60 V. Later observations by satellite-borne instruments in low altitudes were similar. Observations at higher altitudes reported from Vela 6, OGO 1, and OGO 3 were <100 V in magnitude. While these potentials are problematic for the measurement of ambient flux-energy spectra, they do not pose an engineering hazard to the function or health of spacecraft electronics. Charging to kV magnitudes could pose a hazard, as DSCS-II's experience demonstrated.

Spacecraft charging as a cause of the anomalies was initially met with skepticism in the satellite design community. Between 1960 and 1970, over 1200 satellites had been launched (Hiriart & Saleh, 2009). Many substorms had occurred in that decade, which meant that while *detection* of spacecraft charging to kV levels was new, spacecraft charging was not, and it did not appear to cause anomalies on these spacecrafts. If charging could cause anomalies, it should have been discovered and corrected long before DSCS II satellites were launched. Stronger evidence than an apparent temporal correlation between anomalies and substorms was required.

The satellite manufacturer TRW began ground tests on flight spare electronics and interconnecting cabling (Fredericks & Kendall, 1973; Fredericks & Scarf, 1973; Rosen, 1976; TRW Space Sciences Dept., 1972). First, static and low-frequency magnetic field susceptibility tests were performed. The field at the center of the coil was ~50 gauss. If the interconnecting cable was wrapped twice around the coil, anomalous behavior could be induced. Direct susceptibility of the circuits within the boxes was tested using AC magnetic fields between several

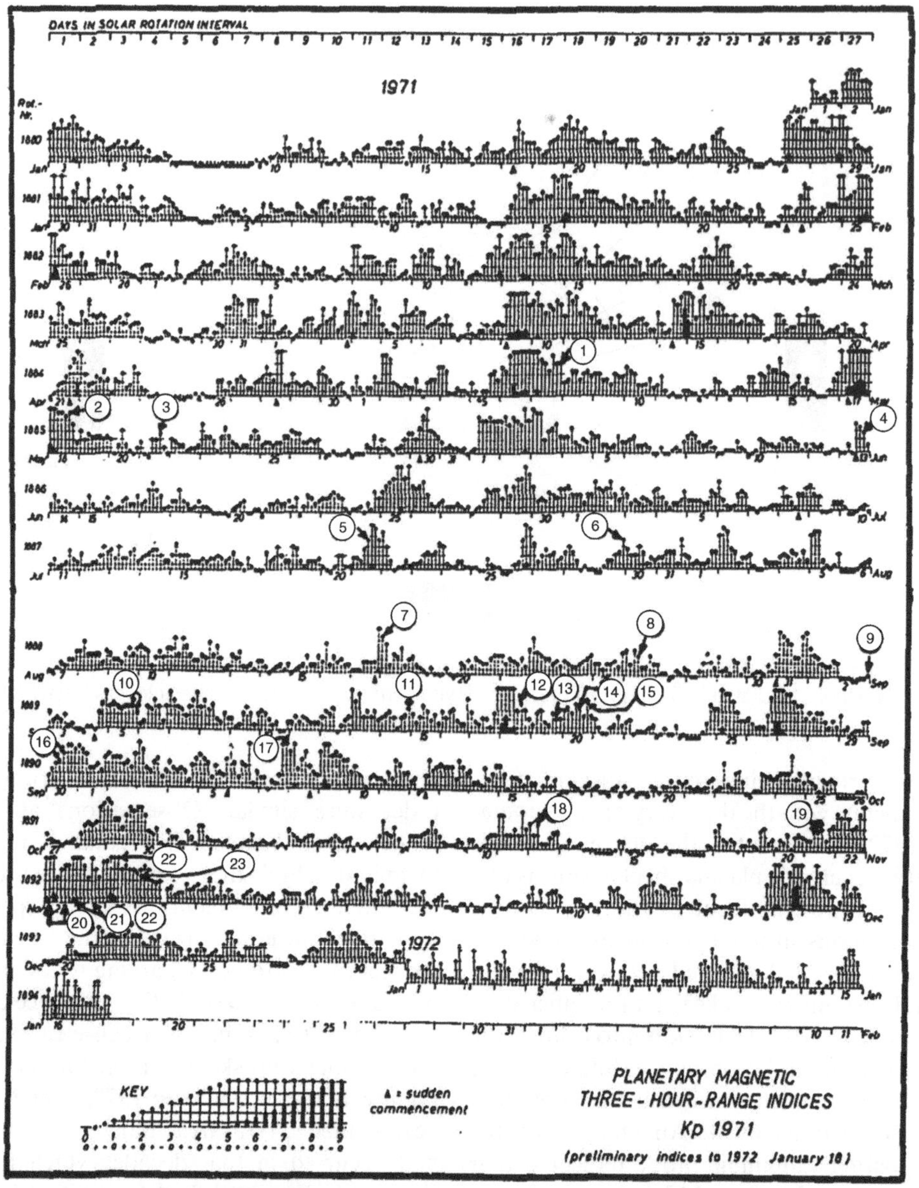

Figure 2.2 "Musical" diagram showing 3-hourly values of the planetary magnetic activity index Kp for 1971. Events listed in Table 2.1 are marked during the corresponding 3-hour period for which Kp is defined. (Reproduced with permission of TRW Space Sciences Dept., 1972.)

hundred Hz to 1 MHz. Susceptibility peaked at several 100 kHz with fields >50 gauss. Electric fields (E) up to 80 kV/m at frequencies up to 100 kHz were applied to the units, but no anomalies occurred. So direct exposure to large static or quasi-static fields of a magnitude expected from surface charging could not explain the upsets. However, on occasion arcs or corona discharges would occur in the test setup, and these produced anomalies.

A Tesla coil was used to produce high-frequency air arcs from the electrode to an electrically floating conductor across a 1 cm gap. The spacecraft electronics and cabling were electrically isolated from the Tesla coil power supply. Arcs from the Tesla coil produced anomalous behavior at distances up to 5 m, and at 0.7 m for corona discharges. When the units and interconnecting cables were shielded by aluminum foil, anomalies did not occur, even when the Tesla coil discharged adjacent to the foil. If just 15 cm of cabling was exposed, anomalies would occur. E-dot and B-dot probes (sensors that measure the rate of change of the E and B fields, and which require integrators to convert the measurements into E and B fields) measured the high-frequency transient E and B fields produced by the air arcs. The E field peaked at 950 V/m with a 20 nsec rise time at 0.5 m, and 400 V/m

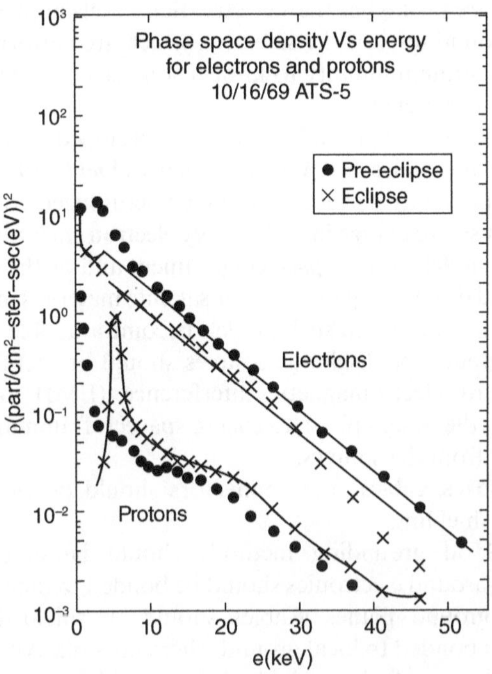

Figure 2.3 Phase space density of particles before and during eclipse (DeForest, 1972. Reproduced with permission of AGU.)

at 1.0 m, with approximately the same rise time. The magnetic field had a peak of ~0.01 gauss (0.8 A/m) and 20 nsec rise time. The hazard from the air arc was clearly due to its intense, fast rise-time E field.

It was not at all clear that electrostatic discharges from satellite surfaces would have the same characteristics as the high-frequency air arcs from a Tesla coil. The surfaces identified as the likely sources of discharges were the solar panels and multilayer thermal blankets. Ground tests of small samples of the solar array substrates, with and without solar cells, and of the thermal blankets were performed in a glass bell jar. Sources that produce electron beams of 1–30 keV energy that are in common use today to qualify spacecraft hardware were not available for these tests. Instead, high voltages (up to 20 kV) were applied across the samples to determine their breakdown voltage thresholds. A small, 35 cm brass rod antenna outside the bell jar detected the transient electric field radiated by the arc, and a standard electromagnetic compatibility (EMC) detection coil was used to measure the transient magnetic field.

The initial breakdown voltages were in the range of 2–9 kV (Table 2.2). These voltages are in the range observed on ATS 5 and expected, based on simple charging

Table 2.2 Initial breakdown voltages for solar panel and thermal blanket samples. (Reproduced with permission of TRW Space Sciences Dept., 1972.)

SPARKING POTENTALS			
	Breakdown Potential Diff. (V)	Configuration	Remarks
THERMAL BLANKET	~5000	Both terminals on top mylar 1ayer.	Breakdown around edges. Subsequent breakdowns require higher voltages. Test not continued past 15,000 volts. Aluminum vaporized on all layers of blanket, particularly along wrinkles of thin layers.
FIBERGLASS/ ALUMINUM	~6000	Across one sheet of fiberglass.	V decreases down to as low as 250 volts after repetitive sparking.
HONEYCOMB	~7500	Across two fiberglass sheets.	V decreases to as low as 700 volts after re-petitive sparking.
	~6000	Both terminals on same fiberglass surface.	V decreases to as low as 600 volts after re-petitive sparking.
SOLAR CELLS AND COVER GLASS ON SUBSTRATE	~7000	Both terminals on cover glass surface. Terminals cover space between cover glass.	V decreases down to ~1000 volts after repetitive arcing. Spark travels down cracks between cover glasses. No observable degradation of solar cells.
THERMAL COVER	~4000	Both flat terminals on teflon layer.	Breakdown around edge. Subsequent breakdowns require high voltages. Black paint flakes off. Small area cleared of silver.
	~2000	One flat terminal near edge of sample.	Same as above.
	~4000	One flat terminal in center of sample.	Spark seen in crack at 1000 V. Arc breaks through teflon at 4000 V. Some degradation as above.
	~9000	One smooth electrode in center of sample.	Arc becomes continuous at 13 kV. Hole burned in teflon layer of cover and webbing.

(2×10^{-5} torr.)

calculations, for DSCS-II (Figure 2.4). The electric fields radiated by arcs from the satellite materials (Table 2.3) were comparable to the fields radiated by the Tesla coil (in amplitude and fast rise time) that produced anomalous behavior in the electronics during the ground tests. The spacecraft electronics that were suffering in-orbit anomalies are closer to the solar panel surfaces and thermal blankets than the flight spares were during the ground test, so arcs from the spacecraft surfaces are credible causes of the in-orbit anomalies. Further, these field levels are far higher than electric field levels applied during ground tests of the satellites, which are performed to demonstrate immunity to known sources of electromagnetic interference.

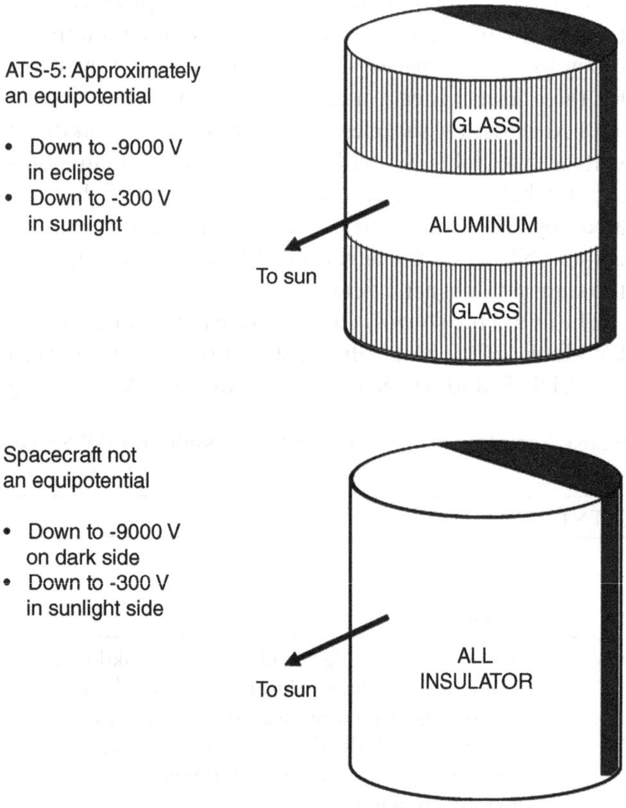

Figure 2.4 Analysis suggested that DSCS II dielectric surfaces could charge to kV levels even when in sunlight. (Reproduced with permission of TRW Space Sciences Dept., 1972.)

Multiple design guidelines were developed to prevent future anomalies (TRW Space Sciences Dept., 1972):

1. Spacecraft surfaces should be scrutinized for their propensity to charge in high energy electron environments, using models of the space environment and of the charging and discharging response of satellite materials in those environments, when such models become available.
2. Models of discharge sources should be included in standard electromagnetic interference (EMI) analysis during the design phase to ensure spacecraft immunity to pulses from the sources.
3. Boxes, cables, and connectors should provide adequate shielding.
4. Good grounding methods should be employed. Boxes around electronics should be bonded to their structural ground planes. Cables should be shielded with shields bonded to local ground where possible. All metallized layers of thermal blankets should be grounded. Aluminum honeycomb cores of solar panels should be electrically connected and grounded.
5. Electrical interface circuit design should provide filtering to reduce the circuit response bandwidth to the minimum required (and maximize the rise time to the maximum allowable). Protection should be provided against burnout of semiconductor devices such as MOSFETs, FETs, diodes, and transistors by large voltage transients induced by arcs.
6. Surface areas of exposed dielectrics or isolated conductors should be minimized. Transparent conductive coatings on solar cell cover glass and thermal blankets should be considered.
7. Monitoring instruments capable of detecting charging and pulses from ensuing arc discharges should be considered for flight. Electron emitter devices that "clamp" the spacecraft-plasma potential differences to safe, small values should be employed.

Table 2.3 Maximum E and B fields radiated by arcs on solar panel and thermal blanket samples compared to Tesla coil arcs. (Reproduced with permission of TRW Space Sciences Dept., 1972.)

	E			B	
	Amplitude (V/m)		Rise Time	Amplitude (γ)	Rise Time
	50 cm	100 cm	(ns)	30 cm	(ns)
TESLA COIL (in air)	~950	~400	~10	<.02	~15
THERMAL BLANKET	~700	~250	~15	<.02	~15
FIBERGLASS/ALUMINUM HONEVCOMB SUBSTRATE	~600	~325	~10	<.02	~15

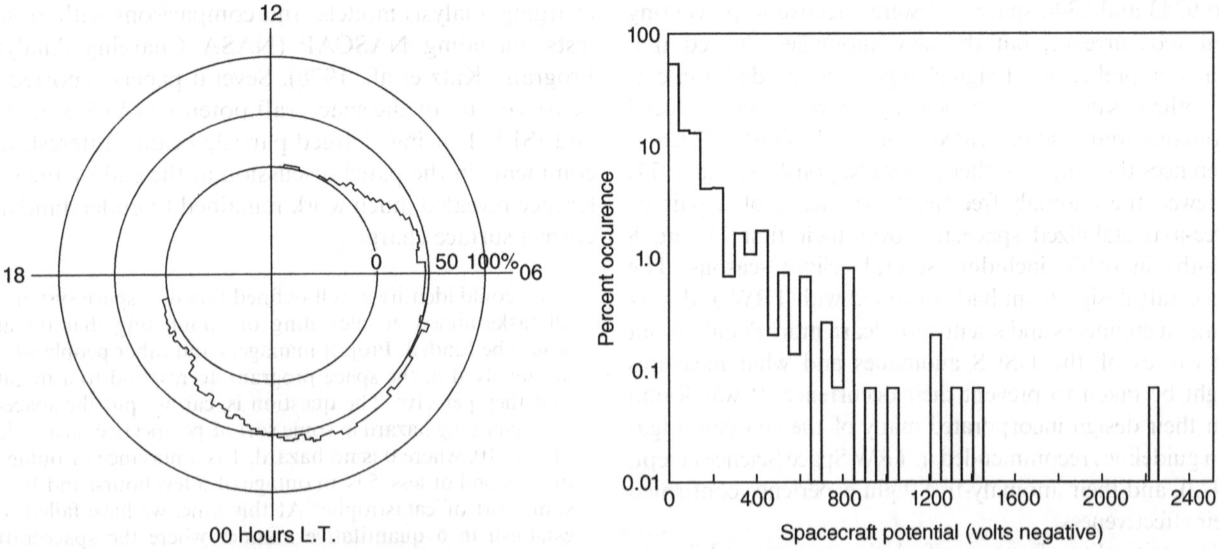

Figure 2.5 Distribution of ATS-6 charging events with local time and by voltage amplitude. (Shaw et al., 1976. Reproduced with permission of American Institute of Aeronautics and Astronautics, Inc.)

On June 2, 1973, satellite 9341 failed. Reviews of the space environment were undertaken (Pike & Bunn, 1976; Rosen, 1975). Ground station magnetograms showed that a series of magnetic substorms had occurred prior to and during the failure, and photographs of the southern hemisphere auroral band from a DMSP satellite confirmed an intense substorm (which saturated the camera) had been in progress. Measurements from ATS 5 also showed sudden changes in plasma properties in geostationary Earth orbit (GEO) just 12 minutes before the failure. This evidence strongly implicated spacecraft charging in the failure. Implementing the guidelines given above was now seen as necessary to not only prevent upsets but to protect the health of the spacecraft.

Intense research and study in the satellite and space environment communities ensued. The existence of spacecraft anomalies, and the later failure, induced by the environment was presented at several conferences, beginning in 1972 (Fredericks & Scarf, 1973) and continuing over subsequent years. In June 1975, a joint AIAA/AGU conference on spacecraft charging (Rosen, 1976) was held to consolidate all that was currently understood about the topic. A paper was presented (Shaw et al., 1976) documenting the detection of electrostatic discharges on a different geosynchronous satellite by a discharge monitor. The discharges fell into two categories. The first occurred near local midnight during magnetic substorms (determined by the presence of energetic electrons detected by the monitor and by ground-based geomagnetic indices). The second occurred during undisturbed geomagnetic conditions but was highly synchronized to the satellite spin angle. Most importantly, discharges were observed coincident with two types of engineering anomalies that occurred on the satellite carrying the monitor, providing the first direct evidence the electrostatic discharges could induce anomalies. Another paper (Reasoner et al., 1976) reported satellite charging data from 40 orbits by ATS-6. The local time distribution of charging events showed the now-familiar clustering in the midnight-to-dawn quadrant. The amplitude distribution of spacecraft structure ground voltages showed that ATS-6, a three-axis stabilized spacecraft, could charge to negative 1 to 2 kV in sunlight (Figure 2.5), much higher than observed on ATS-5, a spin-stabilized spacecraft.

In 1976, the U.S. Department of Defense and NASA cohosted the first Spacecraft Charging Technology Conference (Pike & Lovell, 1977). Fifty papers were presented dealing with subjects including (1) the geosynchronous plasma environment, (2) spacecraft modeling, (3) spacecraft materials characterization, (4) spacecraft materials development, and (5) satellite design and test. Further instances of environment-induced anomalies were reported at this conference.

The second launch of a pair of DSCS-II satellites (9343 and 9344) had occurred, and 100 anomalies of 6 different types occurred over their first 3 years. The case for a surface charging cause of these anomalies was presented (Inouye, 1977). Most of the 100 anomalies did not show as clear a correlation with substorms or midnight-to-dawn local time, but they all occurred in electronics that utilized bi-stable, flip-flop type logic which did upset when electric arcs were injected onto an engineering model spacecraft. There were no repeats of the RGA resets that had been observed on spacecrafts 9341 and 9342, confirming that changes quickly made to the RGA circuits prior to the launch of the next two 9343 and 9344 spacecraft were effective in preventing their reoccurrence, but the new anomalies showed that more comprehensive design changes were needed. Reports of other satellites experiencing environment-induced anomalies were also presented (Robbins & Short, 1977, and references therein). Another paper (Napoli & Seliga, 1977) reviewed the anomaly-free flight experience of a pair of three-axis stabilized spacecraft over their first 11 and 8 months in orbit, including several eclipse seasons. The spacecraft design team had consulted with TRW and government engineers and scientists to learn more details about the causes of the DSCS anomalies and what measures might be taken to prevent their occurrence. It was found that their design incorporated many of the charge-mitigation guidelines recommended in TRW Space Sciences Dept. (1972), and their anomaly-free flight experience confirmed their effectiveness.

The criticality of spacecraft design to susceptibility to spacecraft charging was further emphasized by comments given by R. Lovell and N. J. Stevens during the panel discussion at the end of the conference (recorded in Pike & Lovell, 1977, pp. 894–895):

> As a result of the tour that we took through the aerospace industry, it became evident to us, that the satellites that were experiencing anomalies were the ones that used computer level logic and stored commands. [The] satellite is configured such that any upset would start a whole series of commands. The NASA satellites, the ATS 5 and 6 and the SMS satellites were basically controlled from the ground. There was really very little that could get them into trouble. The SMS used latching relays. The ATS 6 depressed the logic levels so that it took something like a 15-volt spike to get the logic to change state. This is in addition, of course, to SMS and ATS 6's Faraday cage, and their shielded wire. Stu Bower's comment at the time was that operational satellites didn't have large weight margins and you probably couldn't do everything that was done for the NASA satellites. Exactly how much was really required we were never able to figure out.

The second Spacecraft Charging Technology Conference was held 2 years later, October 31 to November 2, 1978 (Finke & Pike, 1978). As pointed out in the preface to the proceedings, significant progress had been made since the first conference:

> The tenor of the presentations at this conference was gratifying. No longer are we working with first concepts and theories. We now have, in hand, space and ground test data that correlate well; analytical programs that can model the observed phenomena; and the imminent flight of the SCATHA spacecraft, which will extend our knowledge of the spacecraft charging environment. Although the technology is far from being complete, there was a feeling of confidence among the participants that we are beginning to develop a working knowledge of the problems and their potential remedies.

Many papers presented the results of charging and discharging tests of satellite materials, predictions of new charging analysis models, and comparisons with ground tests, including NASCAP (NASA Charging Analyzer Program; Katz et al., 1978). Several papers reported on active control of the spacecraft potential (ATS-5, ATS-6, and ISEE-1) using charged particle beams. Interestingly, comments in the panel discussion at the end of the conference revealed much work remained to understand and correct surface charging:

> If we could identify a well-defined threat to space systems, all tasks aimed at alleviating or eliminating that threat would be funded. Project managers and other people who are involved in the space program do respond to a threat that they perceive. The question is, can we put the spacecraft charging hazard in some sort of perspective on a scale of 0 to 10, where 0 is no hazard, 1 is a nuisance or outage of a second or less, 5 is an outage of a few hours, and 10 is some sort of catastrophe? At this time, we have failed to establish in a quantitative manner where the spacecraft charging hazard falls on this scale (Rosen, 1978).

Continuing skepticism regarding spacecraft charging and the lack of interest in flying environment sensors and discharge monitors by satellite program offices was reported.

> My introduction to the spacecraft charging problem was about 6 months ago when I was made project officer on the SCATHA program. Thus, I am the most junior member of this group. Although I was warned not to get involved with the SCATHA program and told it is a boondoggle, a WPA project for geophysicists, I do not hold this view. I believe it to be an important program and I think this is the perception of a large number of people. However, project personnel do not seem to consider spacecraft charging to be a hazard, and therefore nobody from the SAM0 Systems Program Offices attended (Kuck, 1978).

And

> However, we may be erring in attributing most spacecraft anomalies to charging. Some of the occurrences may be attributable to poor design. Currently, there is no system to identify the exact source of anomalies. We also do not know enough about the effects of ESD. That is, exactly what happens when there is a breakdown, what are the coupling mechanisms, what are the systems interactions, how does ESD couple into spacecraft systems? In short, there is a credibility gap in perceiving the actual hazard (Massaro, 1978).

The third joint DoD/NASA Spacecraft Charging Technology Conference was held in November 1980 (Stevens & Pike, 1981), continuing the 2-year cadence of these conferences. The SCATHA (P78-2) satellite had been launched in January 1979, and an entire session of the conference was dedicated to review the flight data obtained by the satellite since launch. SCATHA was the first satellite that flew instruments to (1) measure the electron and ion environment (and derive the satellite structure potential relative to ambient plasma "ground"; (2) measure the potential of dielectric material samples relative to spacecraft structure ground as they respond to the

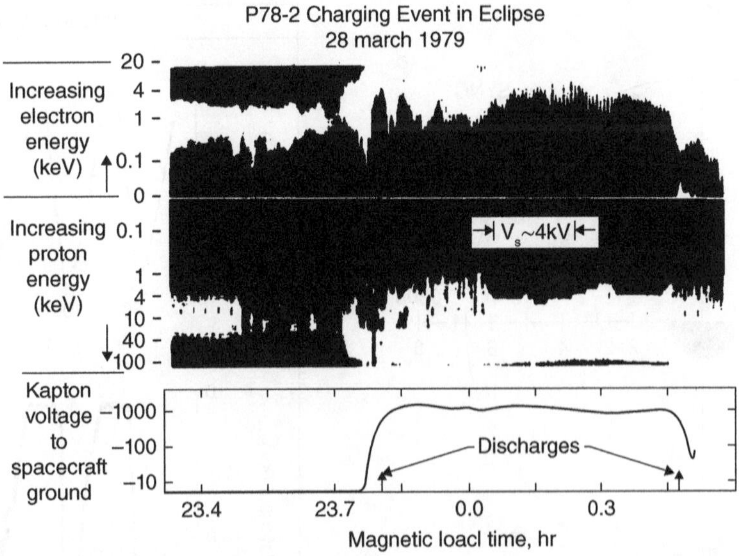

Figure 2.6 Discharges detected during eclipse charging event. (Koons, 1981. Reproduced with permission of NASA).

environment; and (3) detect electrostatic discharge (ESD) events when they occur by their RF emissions outside the spacecraft and by transient pulses induced on wires inside the spacecraft.

Electron and ion spectrograms recorded during several charging events were similar to those observed by ATS-5 and ATS-6, showing a clear shift in ion energy as the spacecraft charged negatively during the events (Fennel et al., 1981). However, the flux energy spectrum was notably asymmetric during these events, with electron flux higher in directions perpendicular to the local magnetic field lines. It was also noted that the ion composition during these events was dominated by O+ (Fennel et al., 1981; Johnson, 1981; Johnson et al., 1981). Both features differ from the usual analysis assumptions of uniform isotropic flux-spectra with ions dominated by H+. The Satellite Surface Potential Monitor (SSPM) experiment provided detailed differential charging histories during two notable charging events for dielectric film samples, as well as longer term statistical information on charging probabilities versus local time and altitude (L-shell; Mizera & Boyd, 1981). The electrostatic discharge monitors (Damron et al., 1981; Koons, 1981) did detect ESD events on the satellite exterior. On average, discharges occurred two to three times per day but typically occurred in spurts over a couple of hours, followed by several days of inactivity.

An eclipse charging event on 28 March 28 is shown in Figure 2.6. The electron flux average energy increases from ~1 keV to 20 keV, and the ion spectrum shows the satellite frame potential dropped to as much as −8 kV and maintained ~−4 kV until the satellite entered sunlight. Coincident with the jump in electron energy, a Kapton SSPM sample charged to ~−1 kV relative to

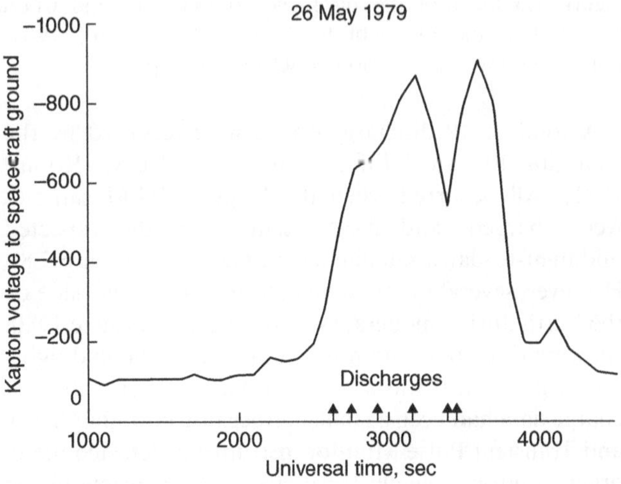

Figure 2.7 Multiple discharges occurred during sunlit (differential) charging event. (Koons, 1981. Reproduced with permission of Aerospace Corp.)

spacecraft ground. A discharge occurred 5 minutes later, and another was observed during the transition into sunlight (but not from the SSPM samples).

Six discharges were detected on May 26, 1979, while the vehicle was in sunlight (Koons, 1981). A Kapton sample (on the shaded end of the vehicle) charged from a background level of ~ −100 V to nearly −1 kV, and the occurrences of discharges are noted (Figure 2.7). These six events all occurred at the same satellite spin angle relative to the sun, suggesting it was a single area of the spacecraft that was discharging. This had been previously observed on another spacecraft (Shaw et al., 1976) but did not repeat for the remaining discharge events observed on SCATHA.

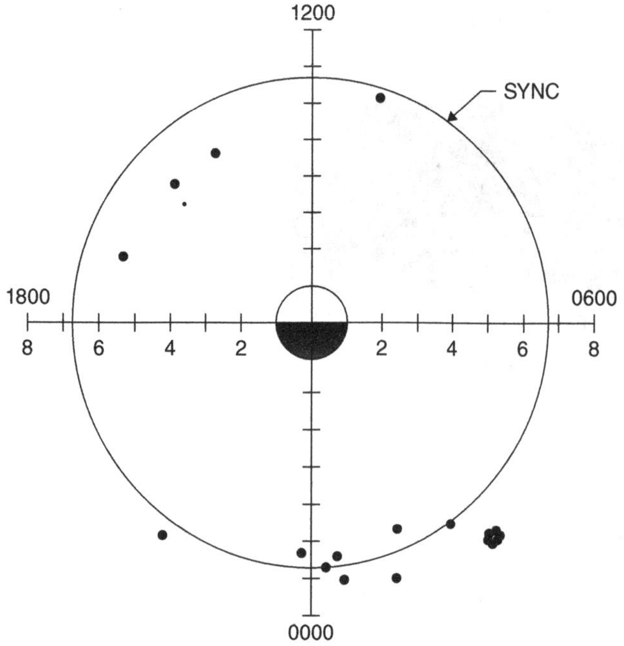

Figure 2.8 Most of the discharges detected by the CEEA occurred in the midnight to dawn sector. (Koons, 1981. Reproduced with permission of Aerospace Corp.).

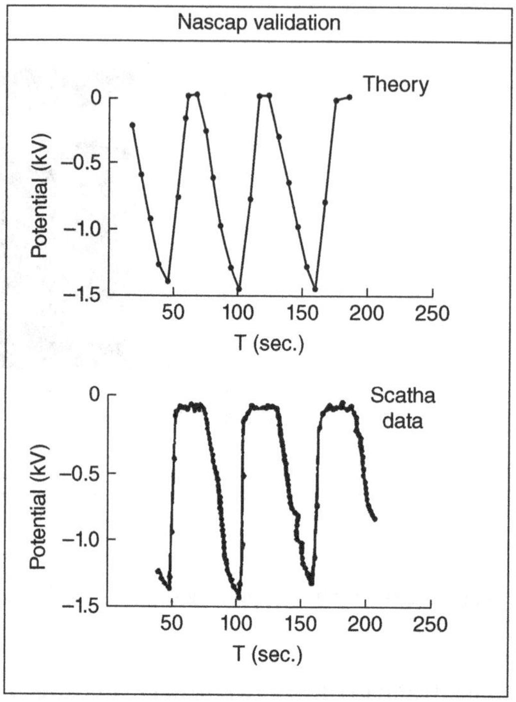

Figure 2.9 Comparison of NASCAP predictions and in-orbit observations of Kapton charging on P78-2.(Saflekos et al., 1981. Reproduced with permission of NASA.)

A total of 19 discharge events were recorded by the Charging Electrical Effects Analyzer (CEEA) (Koons, 1981). All occurred when the Kapton SSPM samples were charged, and most occurred in the expected midnight-to-dawn satellite local time sector (Figure 2.8). However, several events were detected on the day side of the Earth during moderate substorm events in June 1980, showing discharges are not just limited to the midnight-to-dawn sector. It was also noted that processing the discharge data had been a challenge because both the CEEA and Transient Pulse Monitor instruments detected many other nondischarge electrical events (2538 events in the case of CEEA), most related to commands and equipment turn-on or turn-off transients.

The multitude of papers presented during the Material Characterization and Material Response sessions gave further insights into spacecraft charging. Most material charging tests had been performed (and are still performed) using mono-energetic electron beams. Tests using multiple energy beams demonstrated that surface charging can be suppressed when a low flux of very high-energy (100s of KeV) penetrating electrons is added to the usual high flux of nonpenetrating (10s of keV) electrons, because of the enhanced radiation–induced conduction in the bulk material. Some results indicated that discharges would not occur (Adamo & Nanevicz, 1981), or that discharges could occur during multibeam exposure but are smaller (Treadaway et al., 1981), while yet other results (Balmain & Hirt, 1981) indicated that a multienergy beam produces larger ESD events but with a longer wait time between discharges (i.e. at a much-reduced rate of occurrence). Obviously, more work would be required to establish the proper and necessary methods of achieving higher environmental fidelity. Several papers reported on the secondary electron emission yields of materials, a key property that limits surface charging for high-resistance materials. Tests demonstrating the ability of indium oxide (IO) and indium-tin-oxide (ITO) semi-conducting coatings to prevent charging of dielectric films and glass substrates were also presented.

Many papers were presented comparing analytical model predictions of charging to measured or observed charging. NASCAP predictions of material sample charging and charging of the P78-2 spacecraft were compared to observed charging, using analytical (single or double Maxwellian) fits to the measured environments as inputs. Figure 2.9 (Saflekos et al., 1981) and Figure 2.10 (Schnuelle et al., 1981) provide examples showing good agreement between predictions and observations of material differential charging and spacecraft frame absolute charging, respectively.

Charging-induced anomalies on board the METEOSAT-1 spacecraft were also reported (Hoge, 1981). About 150 anomalies, mostly spurious status changes, were observed over 3 years. There was some correlation with local ground station magnetometer data as well as the planetary Ap index, suggesting that charging and discharging might be the cause. A review of the satellite design showed that ~80% of the surface is nonconductive. The large exposed metallized surfaces of the forward, aft, and antenna thermal shields (blankets) were not grounded. Second surface mirrors of aluminized FEP Teflon were also ungrounded. The shields, mirrors, and solar cell panels were deemed likely sources of ESD. A review of the command circuits that were upsetting showed they utilized low noise-immunity amplifiers and TTL type logic. Other circuits that utilized high noise immunity amplifiers and latching relays did not upset (similar to a finding that N. J. Stevens shared at the second charging conference [Finke & Pike, 1978]). Electron beam tests of sample coupons of the thermal shield, the second surface mirrors, and the ungrounded radiometer mirror confirmed they would charge and discharge, but the breakdown voltages were lower, and discharge current levels were substantially smaller than expected. A pulsed current injection test was performed on an engineering model spacecraft that was able to reproduce some of the in-orbit upsets, but only when the discharges were large and the injection points were located close to the equipment that upset. The injected pulses were much larger than the currents generated during the coupon charging tests, which raised doubts that charging and discharging was the cause of the anomalies.

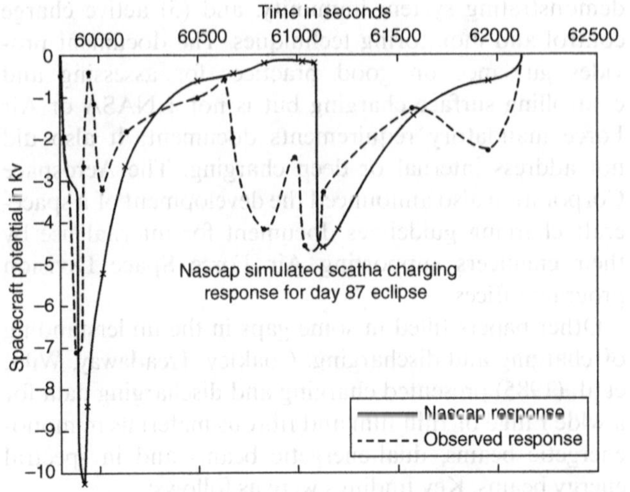

Figure 2.10 Comparison of NASCAP predicted spacecraft frame charging with observed response of SCATHA. (Schunelle et al., 1981. Reproduced with permission of NASA.)

A spacecraft-level charging test was performed to resolve those doubts (Reddy, 1981). The surface potential observed during the test showed large differentials around the satellite circumference and increased in magnitude rapidly at beam energies above 15 keV (Figure 2.11).

Discharges occurred at rates of 0.1 to 1 per second on METEOSAT-1 (Hoge, 1981). The discharges produced E fields typically ranging between 500 and 1500 V/m, and occasionally up to 5 kV/m. Other researchers have measured E field peaks up to ~10 kV/m at 30 cm from charged dielectric samples, and have shown an approximate R^{-3} falloff of field level with distance (Nanevicz et al., 1978; Nanevicz et al., 1981). These field levels exceed levels typically used during EMC tests of a satellite, demonstrating that a satellite could pass its EMC qualification and yet still be susceptible to spacecraft charging. ESD immunity testing requires different levels (and methods) of testing.

Changes were adopted for the second follow-on spacecraft (METEOSAT-2). Grounding of the thermal barrier

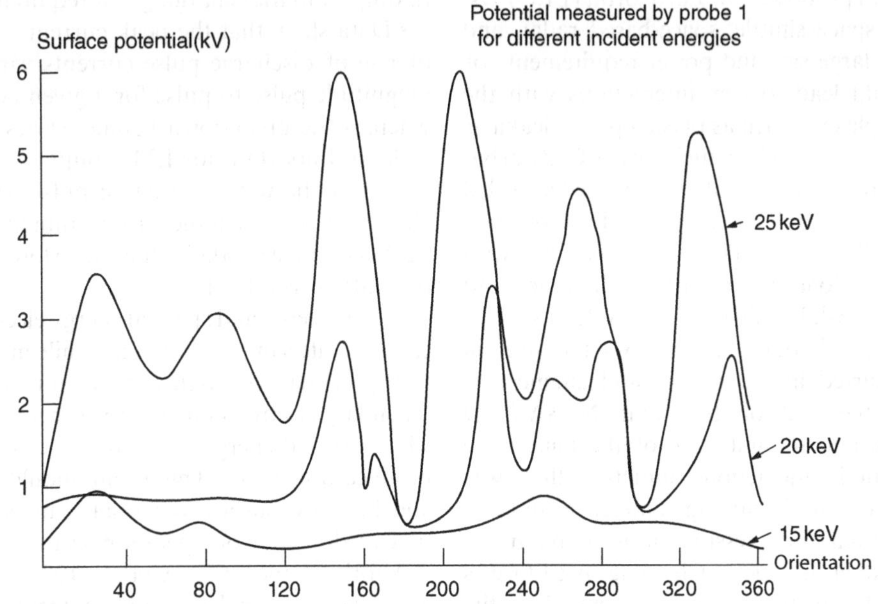

Figure 2.11 Large differential potentials were induced around the METEOSAT-1 spacecraft, which increased with beam energy. (Hoge, 1981. Reproduced with permission of NASA.)

metallization was planned to reduce discharge amplitudes. Critical commands were switched from the susceptible TTL level logic back to more noise immune relay logic, (noncritical commands continued to use the sensitive TTL circuits, providing a baseline reference with flight 1). An electrostatic analyzer to monitor the ambient particle environment and an ESD event detector were added to the payload to acquire in-orbit data to determine if charging and ESD are in fact associated with anomalies or not. METEOSAT had performed the recommended tests, made some of the recommended design upgrades, and planned to fly environment monitors and ESD event detectors—all consistent with recommendations from the DSCS-II anomaly investigation.

A draft MIL standard (Holman, 1981) and a charging analysis and design guideline handbook covering analyses and tests for spacecraft charging (Stevens, 1981) were presented. Charging analyses were to be performed using standardized worst-case environments, drawn from events observed by ATS-5, ATS-6 and SCATHA. Capacitive and arc discharge tests of spacecraft and analyses that couple the structural surface currents and radiated E and H fields to harnessing were proposed. The peak currents and fields and their rise and fall times were TBD, as were the coupling analysis methods. It was pointed out that a criterion for predicting when disruptive discharges can occur was still lacking at this time, hampering efforts by designers to quantitatively determine the environmental risk faced by a design. Guidance was given on methods to minimize those risks.

The last session of the conference was dedicated to the announcement of a new joint NASA/DoD initiative on Spacecraft-Environment Interactions. The focus of the new initiative was on proposed low Earth orbit (LEO) systems, such as the space shuttle, space-based radar, and space station. The large size and power requirements of these systems would lead to new interactions with the dense low altitude plasma, such as plasma power leakage, distinct from the charging and arcing issues of GEO orbit.

A fourth conference was held 3 years later, titled "Spacecraft Environmental Interactions Technology—1983" (Purvis & Pike, 1985). Three sessions dealt with LEO plasma interactions, particularly with large- and high-voltage systems, while only one session dealt specifically with spacecraft charging. This shows the shift in emphasis that occurred in the science and engineering community with the end of the joint NASA/DoD SCATHA program and the initiation of the joint LEO Space Environmental Interactions initiative that was announced during the third charging conference, and discussed in a paper given in the session on future programs and directions (Pike et al., 1985). Charging in LEO was discussed, especially in the auroral regions, but the observed potentials of roughly −14 V on DMSP Flight 6 in a 9.5 keV electron flux (Besse et al., 1985) and ~+28 V during pulsed electron gun operations in the shuttle (STS-3 mission; Williamson et al., 1985) suggested that hazardous frame potentials do not appear to occur in LEO.

A paper in the spacecraft charging session announced the release of "Design Guidelines for Assessing and Controlling Spacecraft Charging Effects," NASA TP-2362 (Purvis et al., 1985). The document is divided into five sections: (1) spacecraft charging concepts of importance to the designer, (2) the modeling techniques to be used to assess whether the design is adequate for environmental immunity, (3) specific guidelines for protecting systems and subsystems, (4) test procedures for demonstrating system immunity, and (5) active charge control and monitoring techniques. The document provides guidance on good practices for assessing and controlling surface charging but is not a NASA or Air Force mandatory requirements document. It also did not address internal or deep charging. The Aerospace Corporation also announced the development of a spacecraft charging guidelines document for internal use by their engineers supporting Air Force Space Division program offices.

Other papers filled in some gaps in the understanding of charging and discharging. Coakley, Treadaway, Wild, et al. (1985) presented charging and discharging data for a wide range of thin film and fibrous materials in monoenergetic beams, dual-energetic beams and in spectral energy beams. Key findings were as follows:

- For worst-case testing of satellite dielectrics, monoenergetic 25 keV electron beams should be sufficient to bound the amplitude and pulse width of discharges without significant over- or understress (Figure 2.12). This finding simplifies charging test requirements.
- Data show that the peak current amplitude and rate of rise of discharge pulse currents varies 1–2 orders of magnitude pulse to pulse for a given beam environment, whereas the breakdown voltage varies much less (~30% or less). Important for EMI coupling assessments, some materials showed a discharge pulse with a fast rate of change of the discharge current ranging between 0.1 and ~2 A/ns, whereas others showed a slower and less hazardous 0.01 A/ns rate of rise.
- Dual energy and spectral energy electron environments can generate sample discharges while maintaining low surface potentials. Low-energy electrons can cause enough secondary electron emission to keep surface potentials low while the mid-energy electrons deposit enough charge to produce discharges. This result should help explain why and how low surface potentials can be measured at the time of discharge on operational satellites.

Another paper (Chaky et al., 1985) demonstrated that discharges occur at lower voltages (some below 1 kV) when the charge polarity is positive-dielectric/negative-metal

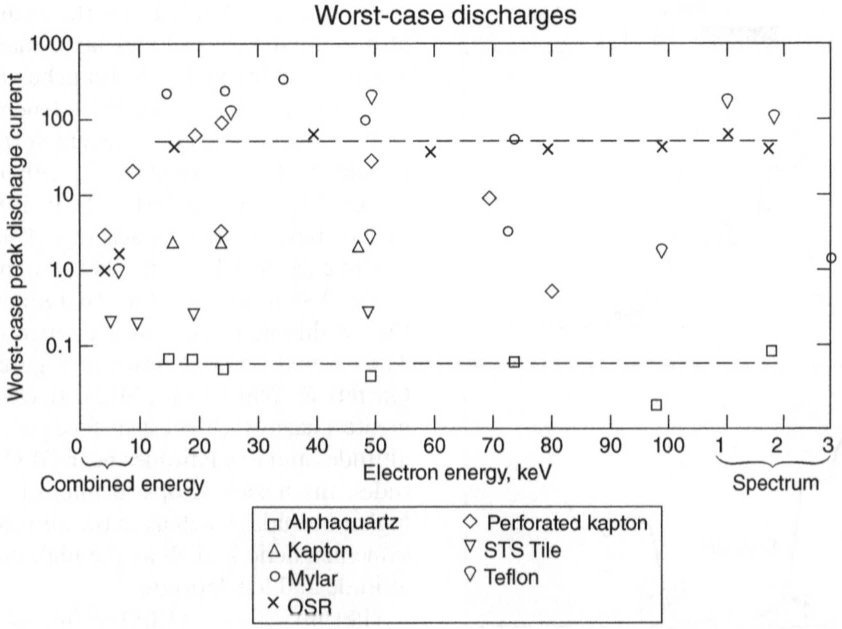

Figure 2.12 Worst-case discharges are insensitive to mono-, dual-, and spectral beam energies but do differ from one material to another. (Coakley, Treadaway, Wild, et al., 1985. Reproduced with permission of NASA.)

than when the dielectric is charged negative relative to the underlying conductor. The latter polarity was the one most commonly used in laboratory tests (hence it was termed "normal gradient," whereas the positive-dielectric/negative-metal polarity is termed "inverted gradient"). Positive-dielectric/negative-metal is expected to occur for sunlit dielectrics on a spacecraft whose frame is negatively charged, such as solar arrays, whereas negative-dielectric/positive-metal would be expected for shadowed dielectrics (Figure 2.13). An apparent consequence of the lower breakdown threshold is much slower rise times and peak currents that would reduce the EMI hazard from the inverted-gradient polarity of discharge. A case of possible positive- dielectric/negative-metal charging coincident with discharges on SCATHA was examined (Figure 2.14).

A major review paper (Frederickson, 1985) provided details about the initiation and propagation of discharges in dielectrics, drawing upon a rich literature from the high-voltage insulation community. He asserts that two phenomena explain typical laboratory results:

1. Radiation produces large electrostatic fields inside insulators via the trapping of a net space charge density.

2. The electrostatic fields initiate discharge streamer plasmas similar to those investigated in high-voltage electrical insulation materials; those streamer plasmas generate the pulsing phenomena which have been seen by many workers.

Breakdown streamers typically initiate under E field stress of 10^5 V/cm, and occasionally as low as 10^4 V/cm, and propagate at speeds of 10^5–10^6 m/s. The streamer is a tube of highly ionized plasma with a diameter of 0.1–10 microns, which will continue to propagate as long as the E field at the streamer tip exceeds the threshold. If the streamer approaches the vacuum surface of the dielectric, the pressure will burst through the surface, releasing the electron-rich plasma and gas to interact with the external electrostatic fields and produce significant transient

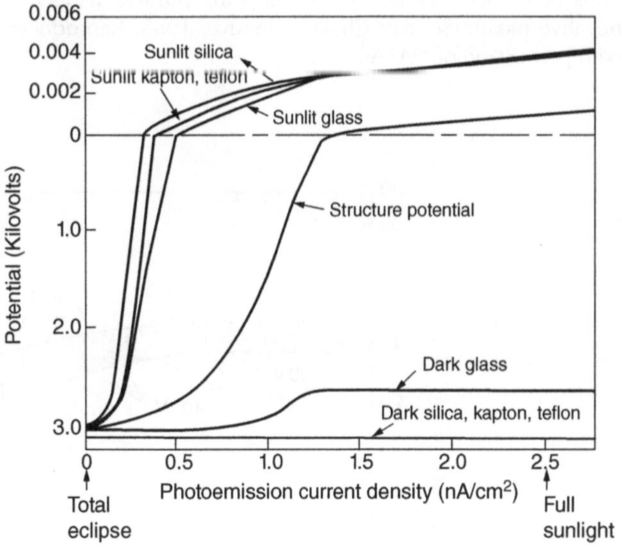

Figure 2.13 Effect of sunlight (UV) intensity on charging of dielectrics and structure. (Source: Chaky & Inouye, 1985. Reproduced with permission of NASA.)

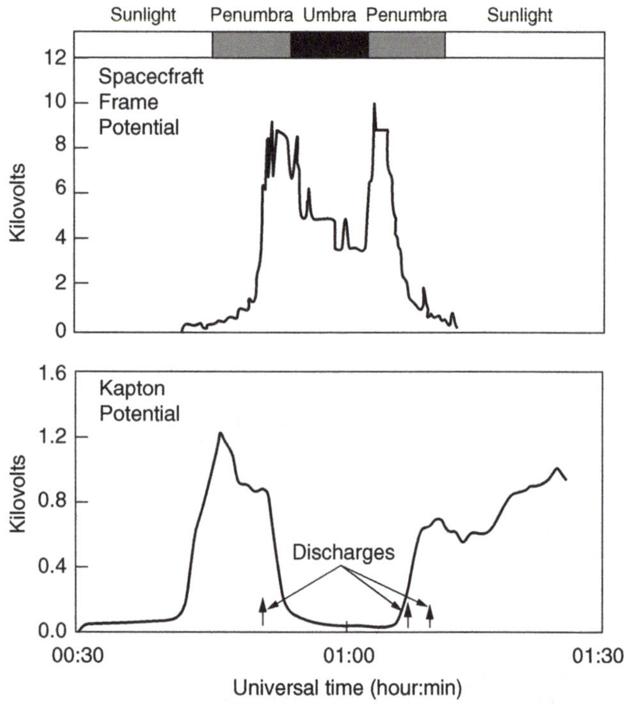

Figure 2.14 Example of discharge occurrences during conditions of reduced sunlit intensity favoring positive dielectric/negative metal. (Source: Chaky & Inouye, 1985. Reproduced with permission of NASA.)

electromagnetic fields and surface currents. Neutral gas is also emitted that includes large mass fragments of the polymer chains and side branches (Kendal, 1985) that can modify surface properties (such as secondary electron emission yield) of adjacent surfaces and likely contributes to the electrostatically enhanced contamination reported by others (Clark & Hall, 1985).

The next (fifth) Spacecraft Charging Technology Conference was held in 1989, a gap of 6 years (Olsen, 1993). A spacecraft surface charging map, now known as the "wishbone map," was presented (Evans et al., 1993). The colored version shown in Figure 2.15 below is from Garrett & Whittlesey (2012). It clearly shows that the greatest hazards (highest surface potentials) occur at high altitudes and low latitudes near GEO orbit. At lower altitudes, the hazard drops in intensity and only occurs at higher latitudes (such as in the auroral arc) that share the same magnetic L-shell as the high hazard region at high altitude and low latitude.

The shift in focus to LEO environment effects announced at the prior (fourth) charging conference produced new results reported at this conference. New analysis models and tools to predict charging effects in LEO orbit (particularly polar orbits that transit the hazard zone in the auroral arc) were presented. Analysis predictions using the POLAR code showed good agreement with observations

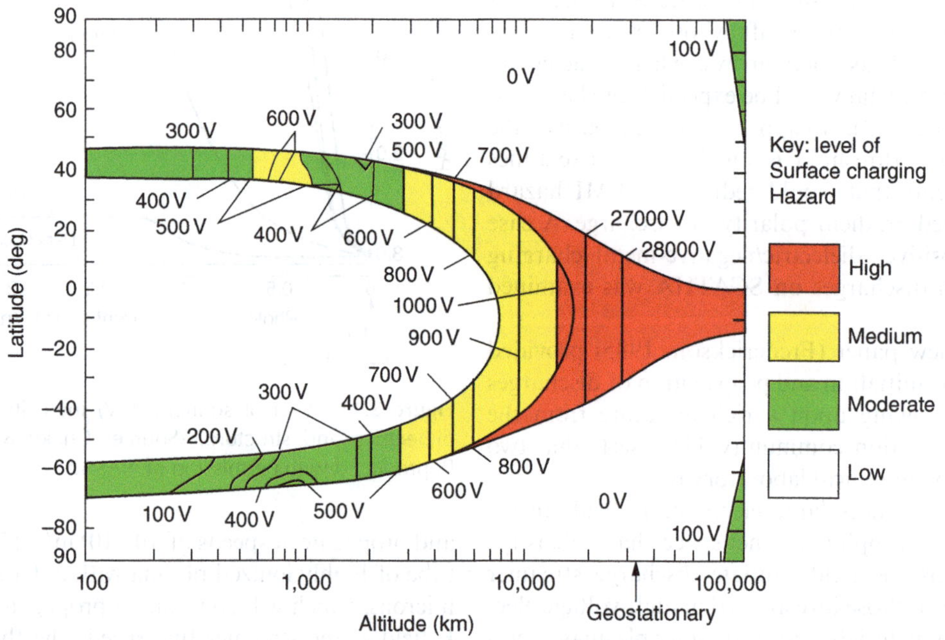

Figure 2.15 "Wishbone map" of surface charging hazard risks versus altitude and longitude (Evans et al., 1993). (Source: Garrett & Whittlesey, 2012. Reproduced with permission of John Wiley & Sons.)

from DMSP F6 and F7 for several typical conditions but failed to model more severe charging that occurred during conditions of low plasma density (i.e. larger sheath dimensions; Cooke et al., 1993). Most significantly, a DMSP frame potential of −1.5 kV was reported. While this is less than the typical eclipse results in GEO, it exceeds the low-altitude charging levels shown in the wishbone map. In addition, the analysis suggested there could be spacecraft-size scaling effects on charging (i.e. more severe charging for larger size spacecraft).

In LEO orbit, the satellite velocity exceeds the thermal velocity of the ambient ions. This results in a high ion flux on the ram facing surfaces (in the velocity vector direction) and a depletion in antivelocity vector direction in the spacecraft wake. This creates the possibility of severe differential charging of dielectrics and electrically isolated conductors on the wake side of the spacecraft. Several analytical studies of wake effects were reported (Laframboise & Luc, 1993; Soubeyran et al., 1993), as were plans for orbital experiments and ground tests to explore the LEO interactions (Ferguson, 1993; Martin & Mallott, 1993). These and subsequent efforts ultimately led to new guideline documents tailored for high-voltage-system operations in the unique LEO environment (NASA, 2007).

Wilkinson (1993), from the National Oceanic and Atmospheric Administration (NOAA), presented a review of trends in satellite anomalies in GEO and other orbits. NOAA collects space weather data from its geosynchronous orbiting GOES satellites and its low-Earth orbiting NOAA satellites, and operates a system of data centers that collect, archive, analyze, and disseminate solar-terrestrial environment data. NOAA's National Geophysical Data Center had also collected anonymous satellite anomaly reports: date, time, location, and other pertinent information. The events ranged in severity from minor operational problems to permanent spacecraft failures. The database contained over 3000 anomalies collected between 1971 and 1989, with contributions from seven countries: Australia, Canada, Germany, India, Japan, United Kingdom, and the United States.

This database was examined for correlations with potential environmental causes. The anomaly count for geosynchronous satellites was found to rise significantly for local times between ~23:00 and ~07:00 over a background average count of ~80 anomalies per 1 hr local time bin (Figure 2.16). This midnight-to-dawn jump in anomalies is a signature for surface charging that has been shown previously for specific satellites. Statistical tests showed that the observed local time distribution had a 0.00035 probability of being produced from random causes. At the same time, there was little to no local time

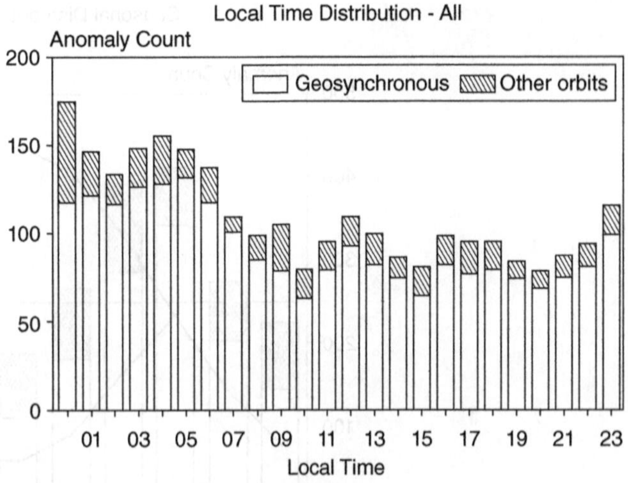

Figure 2.16 Local time distribution of all anomalies. (Source: Wilkinson, 1993. Reproduced with permission of US Air Force Geophysics Lab.)

dependence for the smaller number of anomalies occurring on (the smaller number of) satellites in other orbits.

The GEO spacecraft anomalies also rose around equinox seasons (Figure 2.17). Around equinox, conditions are more favorable for charging: more surface area of the spacecraft has little or no sun exposure, which reduces photoemission from more surface area. Less sun exposure also lowers surface temperatures, which in turn lowers electrical conductivity and charge leakage. Magnetic substorms and energetic plasma injections also occur more frequently around equinox due to greater geo-effectiveness of solar wind interactions with Earth's magnetosphere during equinox season (Russel-McPheron effect).

The local time and seasonal effects show up more clearly when examining a subset of anomalies that occurred on NOAA's own GOES 4, 5, and 6 satellites (Figure 2.18). The phantom commands were attributed to surface charging, whereas the "other" anomalies were predominantly telemetry errors attributed to single event upsets that have no seasonal dependence. The local time distribution of phantom commands had a 0.0000022 probability of being random, whereas the other anomalies had a high 0.94 probability of being random. The phantom commands and number of major magnetic storms both show a strong seasonal dependence that is anticorrelated with solar declination. This is also a pattern that had been shown previously to be signature of surface charging.

The NOAA database shows that surface charging anomalies occur more frequently, and on more satellites, than the few cases documented in the literature. As a consequence of the strong correlations of anomalies with

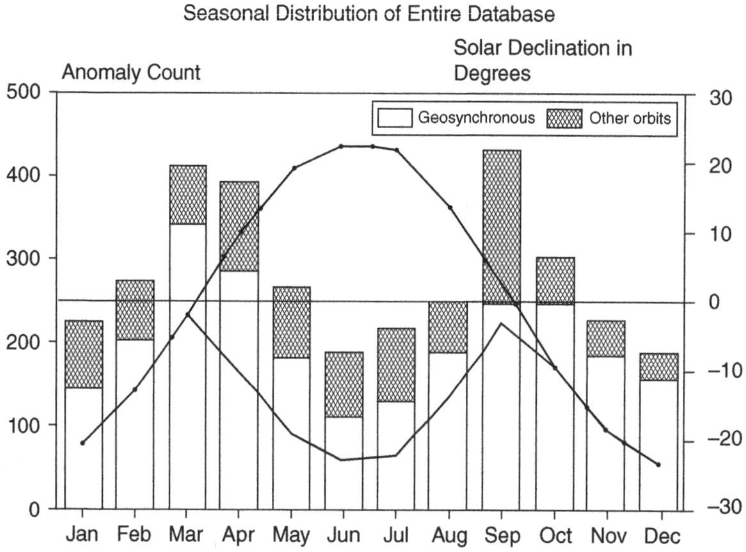

Figure 2.17 Seasonal distribution of all anomalies. (Source: Wilkinson, 1993. Reproduced with permission of US Air Force Geophysics Lab.)

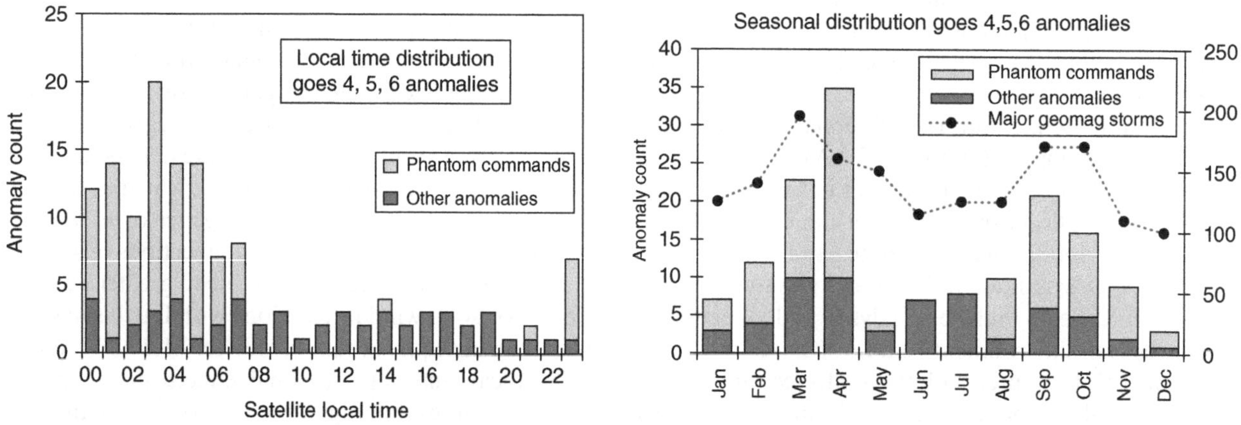

Figure 2.18 Local time and seasonal distribution of anomalies on GOES 4, 5, and 6. (Source: Wilkinson, 1993. Reproduced with permission of US Air Force Geophysics Lab.)

midnight-to-dawn sector local times and equinox seasons, it became common practice to presume that any anomaly occurring at these local times and seasons was due to surface charging, and events outside these local times or seasons were presumed due to other causes. However, other anomaly cases show this practice is not always valid.

Occurrences of triggered surface charging–induced anomalies outside of the local midnight-to-dawn sector have been reported (some contemporaneously with the fifth charging conference and others in the subsequent years). NASA's first Tracking and Data Relay Satellite (TDRS-1) suffered a number of outages triggered by uncommanded deactivation of a command verification unit (CVU) (Bodeau et al., 1989; Daughtridge et al., 1991). When the CVU is commanded offline, ground commands to the spacecraft are no longer processed. All other satellite functions continued normally, but after a 20-minute absence of processed commands from the ground, a watchdog timer would cause the satellite to go into a safe-hold mode. In this mode the satellite tracked the sun to maintain power from the solar arrays, but the high gain antenna pointing controls were suspended, resulting in loss of high gain antenna services.

Initially, it was suspected that single event upsets were to blame, based on experiences of SEU (single-event upset) events in solid-state random-access memory devices that were recently introduced into the control systems of TDRS-1 and other, contemporaneous spacecraft (Shoga et al., 1987; Wilkinson et al., 1991). However,

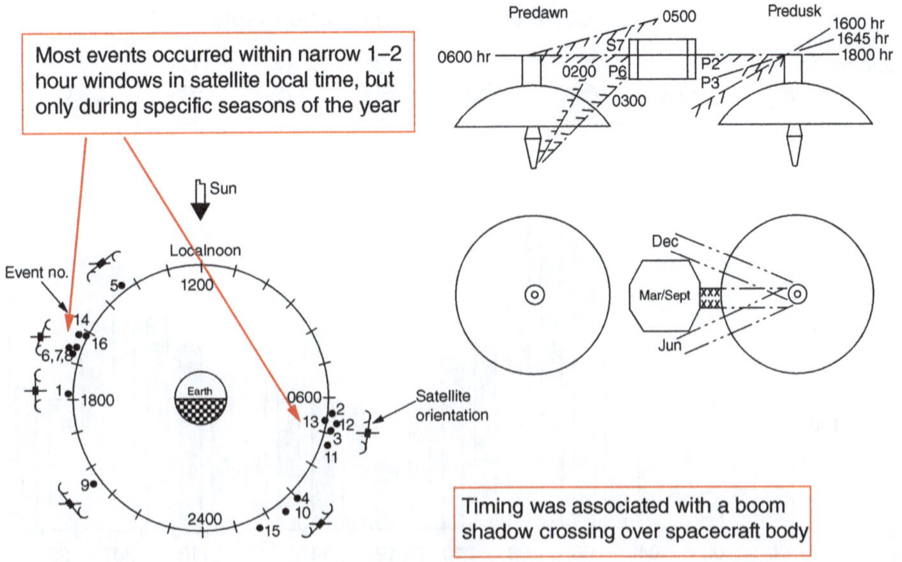

Figure 2.19 TDRS-1 CVU anomalies associated with very specific local time and seasonal dependence. (Source: Bodeau et al., 1989. Reproduced with permission of IEEE.)

such memory devices were not part of the CVU circuitry. As the CVU anomaly counts accumulated, an unusual diurnal and seasonal pattern emerged (Figure 2.19). Anomalies occurred only within a narrow range of dates around equinox and not at other times of year. The local time distribution of events clustered within two narrow bands in local time, one between midnight and dawn and the other near dusk. The former is typical of surface charging, but the cluster near dusk is not. The combined seasonal and local time distributions were associated with shadows on the dielectric surfaces of the spacecraft body cast by the booms that support the high gain antennas. Presumably, charging was occurring in the shadows and discharges triggered by subsequent exposure to sunlight, reminiscent of the discharges that repeatedly occurred at specific sun-spacecraft angles on spin-stabilized spacecraft (Damron et al., 1981; Shaw et al., 1976).

The investigation team established that, even though TDRS-1 had a reasonably good Faraday cage construction (all aluminum honeycomb body panels), and the harnessing leading to the CVU was shielded, the CVU-disable interface circuits were unique in that they employed flip-flop–type devices without filtering, the same type of circuitry that suffered upsets on DSCS-II. Analysis showed that the low-frequency style terminations for the harness shields allowed enough coupling from the high-frequency components of ESD transients to trip the flip-flop circuits. Because the CVU disable circuit was intended for use only during ground test and not for use on-orbit, it was not designed to the same standards as flight (which prohibited use of flip-flops at unit interfaces, a lesson learned from DSCS-II). The traces were cut to the flip-flop circuits of the CVUs destined for the next TDRS flights to prevent spurious activation. No repeats of CVU disables occurred on those subsequent flights.

Other reports of surface discharge–caused anomalies at atypical local times emerged. Telecom 1A showed a very atypical local time pattern (Catani, 2001) that appeared to be triggered by shadowing and subsequent sun exposure on the Earth-facing top of the spacecraft. Ground tests motivated by this experience later confirmed that discharges of charged dielectrics could be triggered by sunlight, meteoroid impact plasma, and gas pulses (Levy et al., 1997).

The most definitive case of triggered ESD occurring outside the typical midnight-to-dawn sector actually occurred in the early 1980s (but was reported decades later in Bodeau, 2007b). A satellite in GEO experienced 3760 events over 1713 days, for an average rate of 2.2 events per day. The daily rate ranged from 0 to 16 events. A plot of the local time distribution for the events is shown in Figure 2.20. Over half of the events occurred within two narrow local time windows that coincided with times when one of the antenna blankets emerged from shadow into sunlight. There was an additional small cluster of events near local midnight that only occurred during winter, caused by a shadow cast by the antenna feeds on the front of the otherwise sunlit antenna blanket when the sun declination was low. Roughly a dozen design changes were defined and quickly incorporated in the next spacecraft builds (which were consistent with the yet to be released NASA guidelines of Purvis et al., 1985). The changes eliminated the anomallies on those satellites.

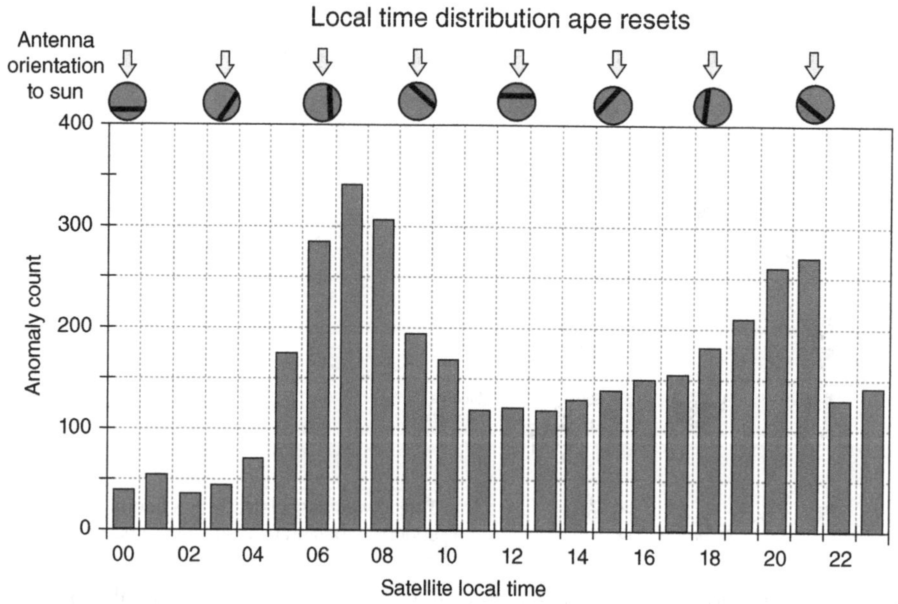

Figure 2.20 Local time distribution of antenna positioner electronic upsets (Source: Bodeau, 2007a. Reproduced with permission of European Space Agency.)

In summary, satellite designers and manufacturers responded to the early surface charging–induced anomalies by applying engineering judgment and risk tolerance to guide their selection of which protection methods to use, and subsequent flight history detremined if their judgment was correct or not. The long gaps between spacecraft charging conferences after the third conference, and the drop in new work dealing with surface charging in GEO, show that the satellite industry believed they had adopted enough mitigations into their designs that surface charging was no longer affecting satellite performance or reliability.

The satellite industry experienced rapid growth in the 1990s, responding to the increasing demand for satellite communication services. Satellites grew in both size and power to accommodate the much larger numbers of RF receivers and transmitters necessary to fill that demand. During 1997, Space Systems/Loral launched five next-generation high-powered spacecraft. In 1998, they reported (Hoeber et al., 1998) that two satellites with GaAs cells had lost a combined 18 circuits over the first year of their life, and a third spacecraft with a Si cell array had lost a circuit. The failure symptoms were consistent with low impedance shorts between cells at different points within a string, and shorts between high voltage cells and the array electrical ground. All of the anomalies occurred when other, instrumented, satellites measured a charging environment characteristic of a solar substorm, suggesting that surface charging was involved (Katz et al., 1998). However, surface charging and ESD on solar arrays was nothing new, and the satellite industry had been operating solar arrays with ESD for decades without loss of power. So, these initial reports of solar array failures due to arcing were met with confusion and skepticism, despite the concurrence with surface charging environments.

The investigation teams performed ground tests that replicated the failures and provided insight to the role that surface charging played (Figure 2.21). The surface of the solar cell cover glass charges relative to the solar cells (and power system and structure electrical ground). When the differential voltage reaches a threshold, a discharge occurs that releases a small amount of plasma. These discharges most commonly occur along cover glass edges, where the sides of the underlying solar cells and other conductors were exposed. If the voltage difference between the adjacent cells or between cells and ground conductors is high enough, current will flow through the plasma, triggering a vacuum arc sustained by the solar array power. The arrays with failures were operating at 75 V and 80 V, higher than the 30 V and 50 V used on older generation satellites. And three to five strings of solar cells shared blocking diodes, whereas older designs typically used a single diode per string. These new array designs could deliver much more power to the triggering plasma than older designs. It is the large amount of energy dissipated by the vacuum arc in the small gap that ultimately causes a permanent failure. In the case of these arrays, the Kapton insulation under and between the cells was pyrolized into a graphite-like char. As the char accumulated, it became more conductive and drew current away from the vacuum arc, causing spontaneous

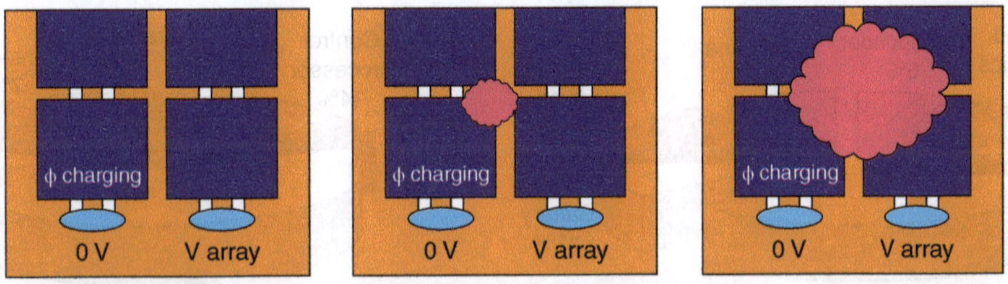

Figure 2.21 Surface charging initiated sustained vacuum arcs begin with conventional charging of the cover glass (left figure), followed by discharges (middle figure). If the array operating voltage and current is high enough, a much more energetic secondary vacuum arc is initiated between adjacent cells (right figure) that damages the panel insulation (Source: Katz, Davis & Snyder, 1998. Reproduced with permission of American Institute of Aeronautics and Astronautics, Inc.)

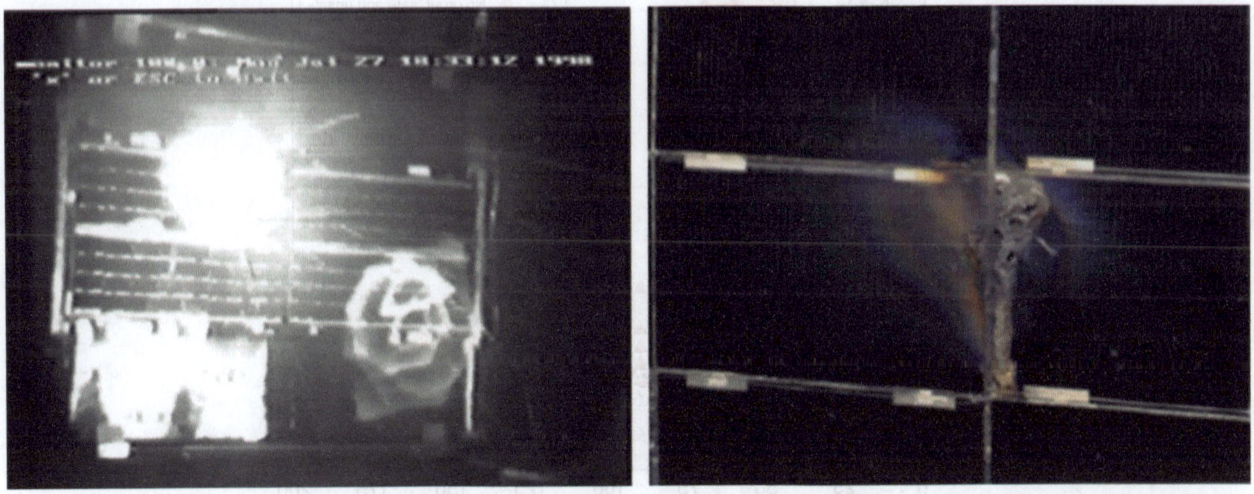

Figure 2.22 Sustained arc and permanent damage that occurred during EOS-AM ground test. (Lai, 2011, reprinted by permission of the American Institute of Aeronautics and Astronautics, Inc.; Ferguson & Hillard, 2011)

interruption of the arc. However, the persistent low resistance shunt path between the strings remained, and the string current was unavailable to the higher impedance spacecraft loads.

Several mitigation steps were proposed. The first was to add silicone adhesive "grout" lines between adjacent cells to isolate them from discharge plasma (gaps on the arrays with failures were exposed). Ground test of a solar cell coupon with grout that generated over 1000 ESD events was unable to produce a sustained arc or array damage. This method was to be applied to arrays already manufactured to minimize delays in satellite deliveries. For future arrays, additional measures were recommended: reroute the solar cell string layout to reduce the maximum cell-to-cell voltage difference to 50 V or less (based on the failure-free history of 50 V arrays), and reduce the maximum available current to the short-circuit output of a single string of solar cells by placing blocking diodes at the end of each string of solar cells.

Apparently, a previously unknown sustained arc damage threshold had been crossed by the new designs. The satellite industry, with the support of government and academic experts, responded by testing their next-generation high-voltage solar array designs to determine if they were susceptible to the same failure mode (see Bodeau, 2012, 2014, and the large number of references contained therein). As a result, other susceptible arrays were discovered by ground tests and fixes applied before launch (e.g. the EOS-AM 127V array suffered a severe sustained arc in test, Figure 2.22, and had to be redesigned, per Lai., 2011; Fergusson & Hillard, 2011). Unfortunately, not all susceptible designs were caught, and solar array failures continued to plague the satellite industry, with most occurring between 1998 and 2001 (Figure 2.23; Frost & Sullivan, 2004).

Initially, there was uncertainty and controversy over how to perform these tests, and there was strong suspicion that the methods used in the early tests may have

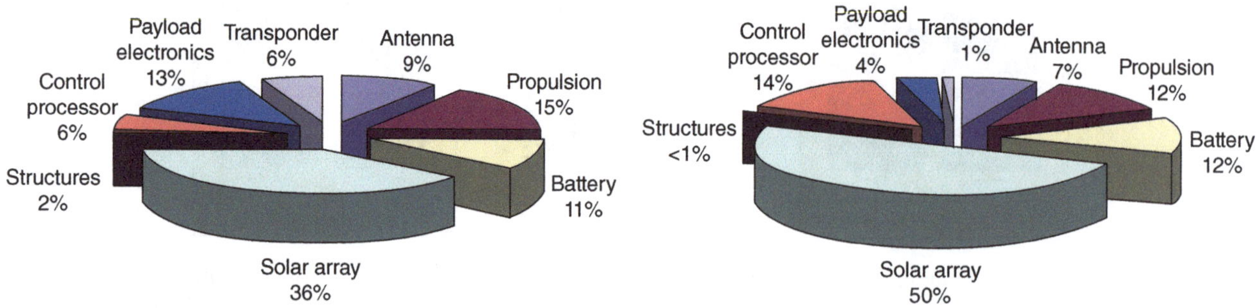

Figure 2.23 Source of commercial satellite anomalies by number (left) and amount (right) of insurance claims (Source: Frost & Sullivan, 2004. Reproduced with permission.)

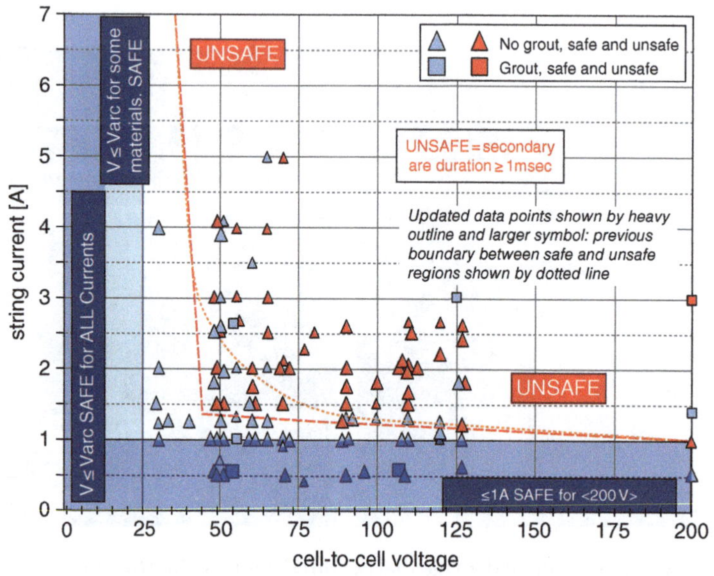

Figure 2.24 Safe voltage and current limits that prevent sustained arcs derived from solar array coupon tests and other vacuum arc tests. (Source: Bodeau, 2014. Reproduced with permission of IEEE.)

produced false alarms. Over the ensuing decade, significant analytical and experimental efforts were spent on refining the test methods. New industry solar array testing standards were released that required ESD testing of solar cells (AIAA-S-111, 2005) and solar panels (AIAA-S-112, 2005). Ultimately, an international solar array test standard, ISO 11221 (2011), was released in August 2011 that defined best practices for qualifying solar array designs.

In addition, meta-reviews of the test results accumulated by the satellite industry (e.g. Bodeau, 2012, 2014), augmented by the extensive body of work developed by the power industry that defined the characteristics of metal vapor vacuum arcs (Lafferty, 1980), led to the definition of rough voltage and current thresholds for self-sustaining vacuum arcs (Figure 2.24).

It was now understood that older array designs did not suffer string failures primarily because the maximum current available from a string of solar cells was well under 1A, and the use of blocking diodes at the end of each string prevented current sharing among strings. As a result, industry plans to use much larger solar cells with maximum currents of several amps were dropped. The voltage threshold for a sustained arc was less clear. It appeared to be between 30 V and 50 V for GaAs and Si solar cells (but is as low as ~8–10 V for solder and 12–20 V for Cu and Ag per Lafferty, 1980). As the operating voltage increases above the threshold, the likelihood of triggering a vacuum arc increases. Guidance to change solar cell string layouts to reduce the cell-to-cell voltage difference to 50 V or less was given; however, limiting the string current was the primary defense.

Unfortunately, currents from multiple strings of solar cells must be combined and routed to the spacecraft on a much smaller number of conductors, due to the limited number of slip rings or cable wrap assembly conductors available to transfer power from the rotating solar arrays to the spacecraft. So, sustained arcs are possible from the array backside into the spacecraft. Sustained arcs have been observed in ground tests (e.g. Bodeau, 2013; Leung et al., 2010; Sevoz et al., 2012) and in orbit (e.g. Bodeau, 2015a; Leggett, 1993; Maejima et al., 2004; Raine, 1978) for diode boards, power harnesses, slip ring assemblies, and the primary power input circuits of internal spacecraft electronics, though other trigger mechanisms besides surface discharges (such as meteoroid impacts, tin whisker fusing or other conductive debris, or intermittent shorts though defective or worn insulation) are suspected for many of these. The common element is that plasma bridges exposed conductors with sufficient voltage difference and source current capability to trigger a long-duration, damaging vacuum arc. However, ground tests showed that the probability of a surface discharge (Qu et al., 2005) or meteoroid impact (Akahoshi et al., 2008) triggering a sustained arc on the array backside is low, which explains the paucity of backside failures observed in orbit (relative to the large number of satellites). As a result, there has been no effort made to expand ISO 11221 and AIAA S-112 to cover testing of this other hardware.

While the possibility that surface charging and discharging could initiate array failures was sufficiently demonstrated by ground test that new design standards and qualification tests were adopted, the case that surface charging actually initiated the in-orbit failures was weak. Bodeau (2015b) reviewed several proxies for the surface charging environment at the times of 12 of the string failures and found that Kp, the most widely available index for charging conditions, was at best useless and at worst misleading, as an indicator that surface charging was the root cause. Much stronger evidence was found using ground station magnetometer data and auroral images (in the X-ray photon energy range) around the magnetic footprint of the satellite (see Figure 2.25). Examples of each are shown in Figure 2.26. Magnetograms and auroral images near a satellite's footprint had been used during the DSCS-II anomaly investigation discussed earlier. Lanzerotti et al. (1998) demonstrated a clear correlation between large ground station magnetogram excursions and actual surface charging of charge plate detectors carried by TELSTAR-4. The severe charging of the charge-plate sensor was in turn used as a proxy for surface charging preceding the failure of Telstar 401, which was only 8 degrees longitude away. This proximity is critical since Garrett et al. (1980) showed that the electron energy spectrum from sensors on spacecraft 1976-059A and ATS-6 closely matched when the satellites were within ±7 degrees of longitude of each other, and the electron flux-energy spectrum from 1976-059A could reasonably predict the charging of ATS-6. Hones et al.

		Magnetogram from Groundstation Near Satellite magnetic Footprint				POLAR
Event	Kp	Station 1	Station 2	Station 3	Station 4	Pixie
1	1+to 2–	+	NA	NA	–	NA
2	2–to 3+	+	–	+	+	–
3	3–	+++	+	+++	+++	NA
4	3–	+++	+	+++	+++	NA
5	3–	+++	+	+++	+	+
6	4–	+++	+++	+++	+++	+
7	4–	+++	+++	+++	+++	+
8	4–	+++	+++	+++	+++	+
9	4	+++	+	+++	+++	+++
10	4+ to 5	+++	+	+	+++	NA
11	4+ to 5–	+++	+	+++	+++	+++
12	7–	+++	NA	NA	+++	+++

Figure 2.25 Evidence of substorm activity from local magnetograms and auroral X-ray images for 12 solar string failure events. (Source: Bodeau, 2015b. Reproduced with permission of IEEE.)

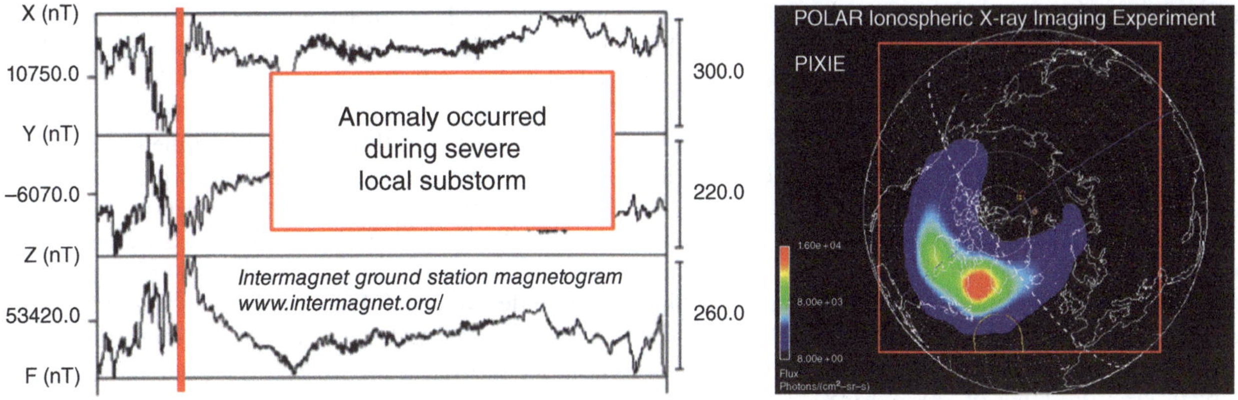

Figure 2.26 Example ground magnetogram and auroral X-ray image near satellite magnetic footprint. (Source: Bodeau, 2015b. Reproduced with permission of IEEE.)

(1996) showed that the particle energy spectra measured by satellite detectors in GEO and LEO matched when the LEO satellite was near the magnetic footprint of the GEO spacecraft. So, using data from other nearby satellites or sensors located near the magnetic footprint provides credible evidence for or against surface charging as root cause. Bodeau (2015b) made strong appeals for making such proxy data available to the satellite industry.

In summary, this review has shown that spacecraft surface charging has been the source of satellite anomalies and failures since the early 1970s. Establishing that connection required expertise and data from both the space science and satellite engineering communities. Engineering solutions were typically developed and implemented before a detailed mechanistic understanding of the nature of the charging and discharging processes and their interactions with satellites were developed, and before clear climatological understanding of the likelihood and duration of severe surface environments were developed. However, new susceptibilities appeared over time as technology advanced, and consequently, the collaboration between space science and industry has continued over the ensuing decades. A road map for future efforts has been developed and articulated as a result of that continuing dialog between space scientists and satellite engineers, which is discussed at the end of this chapter.

2.2. INTERNAL (DEEP) CHARGING

Significant progress was made in the 1970s and 1980s, first in detecting spacecraft surface charging and measuring the hot plasma environment (1–100 keV) that causes it, and then in developing laboratory techniques and analytical tools to predict the responses of spacecraft to those environments. Recurring anomalies on some spacecraft were clearly correlated to significant surface charging environments occurring in local times roughly between midnight and dawn during disturbed geomagnetic conditions. Ground tests demonstrated that severe electromagnetic transients would result from those discharges, and engineering guidelines were developed that reduced the incidence of charging and discharging or reduced the coupling of the electromagnetic transients into spacecraft harnesses and circuitry. As these guidelines were implemented in new spacecraft designs, the rates of anomalies dropped.

However, anomalies were still occurring, and studies showed many of these anomalies had no correlation with surface charging environments (substorms and high fluxes of electrons in the 1–100 keV range). Instead, the anomalies correlated well with periods of high flux of energetic electrons (100s of keV to MeV), which can easily charge thick dielectrics and isolated conductors below the surfaces of a spacecraft. Charging of insulators and isolated conductors below the surfaces of the spacecraft were termed "deep charging" or "internal charging."

In 1986, two reports were pubished documenting anomalies that were potentially caused by deep charging. The first, Leung et al. (1986), reported on a set of 42 power-on-reset (POR) anomalies that occurred on Voyager 1 during its passage through the Jupiter radiation belts. Leung et al. described the power on reset as follows:

> The Voyager FDS is an on-board computer system containing a volatile memory system. It was feared that power-line undervoltage transients could cause malfunctions of the memory and computer operations with no warning of the malfunctions. To avoid the problem, a POR system is incorporated in the FDS.... The key element is the undervoltage sensor that continually monitors the power supply voltage. If an undervoltage occurs, the sensor sends a digital signal to the delay logic electronics which stop processing, stop the internal clock, reinitialize the computations as needed, wait a period of time, and restart computer activity if the undervoltage condition has ceased. The minimum period of time lost is about 175 ms. These 175-ms outages can be identified by gaps in the data stream and by 175-ms discrepancies in the timing of spacecraft activities.

By identifying when the 175 ms delays occurred, it was shown that the PORs occurred mostly between 8R_J down to ~4.5 R_J and back out to 8 R_J, with some lag (R_J = one Jupiter radius). The accumulation of PORs were best correlated with the accumulation of high energy electron flux (i.e. fluence), as shown in Figure 2.27. The typical electron integral flux-energy spectrum in that region is shown in Figure 2.28. The average equivalent shielding of the Voyager primary structure was 0.2 cm (81 mils), whereas there was only about 0.013 cm (5 mils aluminum) of shielding provided by thermal blankets to exterior cables. The electron flux with energy >150 keV, which is able to penetrate those blankets, is about 4x higher than the electron flux ≥1 MeV that is able to penetrate the primary structure. This led the investigsators to suspect external harness charging.

The POR circuit is in a different circuit card than the undervoltage detection circuit, and the connecting cable is bundled with other wiring, some of which is routed outside the spacecraft. It was suspected that these other cables that exited the shielding of the spacecraft were exposed to the higher external fluences and were the sources of the ESD. This hypothesis was supported by ground tests that exposed wire bundles to electron fluences expected at Jupiter. Discharges were detected, but when all of the wire center condutors were terminated to 50 ohms, the peak amplitudes were ~0.8 V with pulse widths of <50 ns, well under normal signal levels. However, if some of the center conductors were unterminated, the peak voltages in the adjacent 50 ohm terminated conductors rose to ~100 V with a pulse width of ~500 ns, more than sufficient to cause a POR. Leung et al. (1986) pointed out that it was

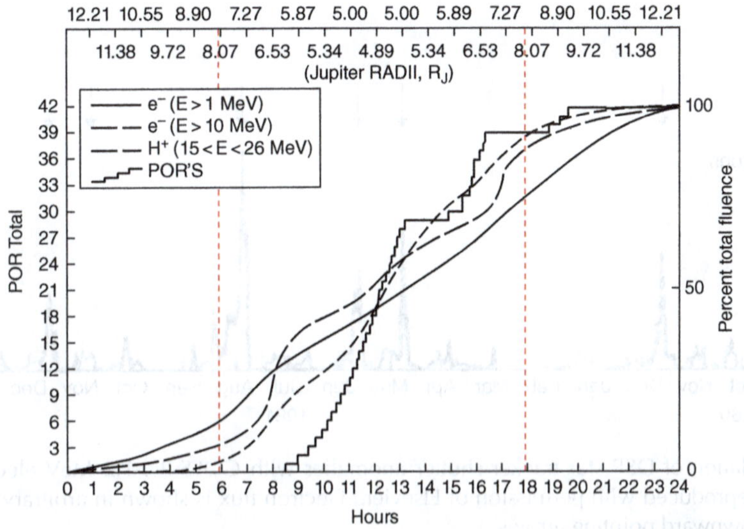

Figure 2.27 Correlation of PORs with high energy electron fluence. (Source: Leung et al., 1986. Reproduced with permission of American Institute of Aeronautics and Astronautics, Inc.)

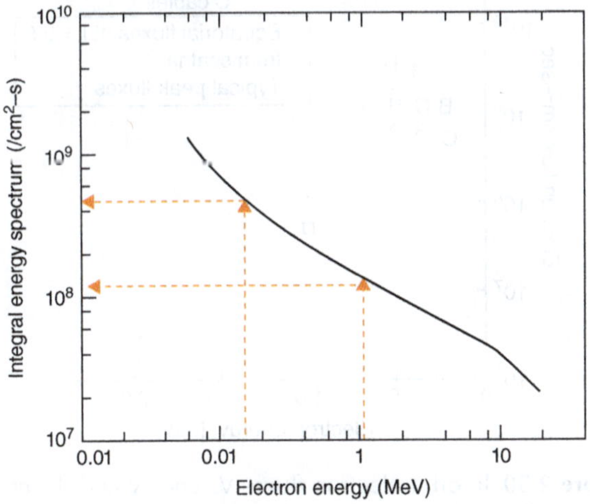

Figure 2.28 Typical electron flux-energy spectrum at 4.5 R_J. (Source: Leung et al., 1986. Reproduced with permission of American Institute of Aeronautics and Astronautics, Inc.)

common practice at this time to build spacecraft harnesses with extra spare wires to facilitate late design changes or repairs. These extra wires were not terminated if not utilized. It was highly likely that such wires were present in the Voyager harnesses and were the sources of discharges responsible for the PORs. JPL subsequently issued a design guideline that prohibited abandoned, ungrounded wires in any flight harness. The authors also pointed out that the environment at Jupiter was ~100 times more severe than Earth's radiation belts. So it was not clear if similar anomalies could occur in Earth orbit, or if the guideline to ground isolated conductors in harnesses and inside the spaccaft was necessary.

Surface charging was shown to be unlikely. The other science instrument flown on Voyager showed that the external plasma environment that could have caused surface charging was too cold and dense to allow charging to potentially hazardous levels. Instead, they indicated that surface charging was limited to ~100 V or less, well below discharge thresholds. This conclusion was also supported by the extensive analysis effort that went into the design of Voyager to preclude surface charging problems and the test program that verified that the designed-in immunity would work (Whittlesey, 1978).

The second report in 1986 was Vampola (1986). He reported previously undisclosed anomalies that had been occurring on US Air Force satellites beginning in 1977. In one case, a spurious signal was occuring in an exposed power-control cable which appeared to cause a subsystem failure. Environmental data from other satellites showed surface charging conditions were not present. Instead, the energetic electron flux was markedly enhanced, which led to a tentative assignment of an electrostatic discharge caused by thick dielectric charging in the exposed cable. After the launch of NTS-2 (a demonstration satellite for the Global Positioning System, GPS) in 1978, a series of clock anomalies occurred that were correlated with high fluxes of energetic electrons (~10^{12} e-/cm^2-day) as estimated from a dosimeter instrument on the spacecraft. Spurious pulses in a cable irradiated by these energetic electrons were thought have caused a bit shift in a tuning register.

Vampola (1986) also reported that the US Air Force's Defense Support Program (DSP) satellites had been experiencing a series of star sensor shutter anomalies. The star sensor is used for precision spacecraft attitude

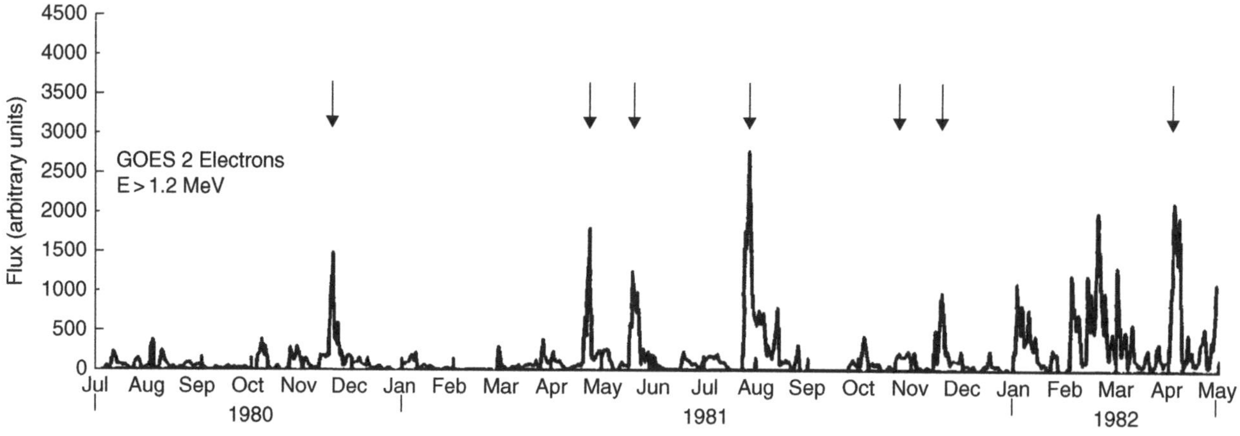

Figure 2.29 Correlation of DSP star tracker shutter anomalies with GOES-2 >1.2 MeV electron flux. (Source: Vampola, 1986. Reproduced with permission of Elsevier.) Electron flux is shown in arbitrary units, occurrences of anomalies by downward pointing arrows.

determination. The sensor has a conical sunshade to prevent the intrusion of sunlight into the star sensor field of view. A sun sensor is mounted on the exterior surface of the sunshade to sense the imminent intrusion of sunlight, which occurs as the spacecraft orbits the Earth despite the sunshade, and to command a shutter to close to protect the star sensor. The anomalies were uncommanded shutter closings. DSP satellites had been in orbit since 1969. The first shutter anomaly occurred in 1976, following a magnetic storm, and a second occurred in 1978. During the geomagnetically active period from 1980 to 1982, a series of shutter anomalies occurred. Figure 2.29 shows a clear corelation of these anomalies with elevated fluxes of ≥1.2 MeV electrons measured by environment sensors on the GOES-2 spacecraft during this 2-year period. A cable connects the sun sensor to electronics inside the spacecraft that control the shutter, and this cable exits the spacecraft and is routed outside along the star-sensor shade, where it is highly exposed to the energetic electron flux. The probable cause for the anomaly was deemed to be charging and discharging within the exposed cable section. (The references did not indicate if supporting ground tests had been performed that replicated the anomaly.)

A similar anomaly occurred on GPS: an uncommanded switch from autotrack to manual track mode in the solar panel orientation control system. Environment monitors on SCATHA, which was in the vicinity, showed that *surface* charging could not have been the cause, whereas the flux of energetic electrons was exceptionally high. And within an hour of this GPS anomaly, SCATHA suffered its first environmentally induced anomaly, an uncommanded mode change in its magnetometer, which is connected to electronics inside the spacecraft by a control cable that is exposed for its 7 m run along a boom. Cables

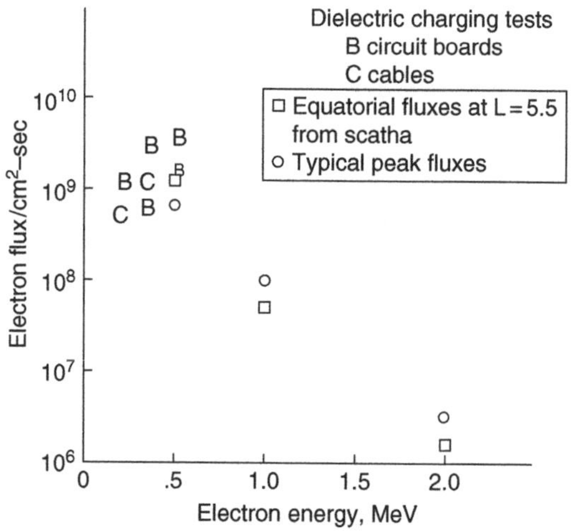

Figure 2.30 In-orbit electron fluxes Vs energy and fluxes that produced ESD in ground tests. (Source: Vampola, 1986, and references therein. Reproduced with permission of Elsevier.)

and circuit boards had been tested for charging and discharging (Vampola, 1986, cited Mallon et al., 1979, and Wenaas et al., 1979). Comparisons of the flux levels used in the ground tests to the typical and peak flux levels expected in the GPS orbit (Figure 2.30) indicated that the flux at energies of 1MeV or higher (which could penetrate the spacecraft structure) were too low to cause discharges of cables or circuit boards inside. Hence, charging and discharging inside the spacecraft was deemed implausible.

Based on this review of anomalies, Vampola (1986) offered the following design guidelines:

1. The electron fluence required to produce discharges is on the order of 10^{12} e-/cm², which must be accumulated on a time scale that is short compared to the time it takes

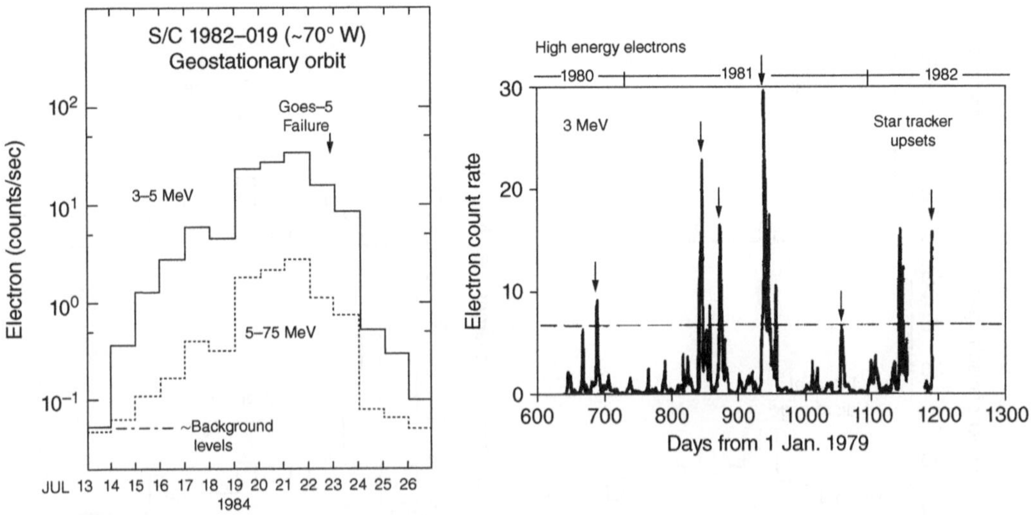

Figure 2.31 Correlation of two anomalies to high fluxes of extremely energetic electron flux. (Source: Baker et al., 1987. Reproduced with permission of Elsevier.)

charge to bleed off (believed to be a few hours). The corresponding threshold flux is a few times 10^8 e-/cm^2-sec. Fluxes below this threshold should not produce discharges.

2. Based on the flux energy spectra observed, eliminating electrons of energy below 1 MeV should prevent thick dielectric discharges, which requires shielding equivalent to 60 mils (1.5 mm) of aluminum.

3. Because circuit boards are contained within boxes that in turn are shielded by the spacecraft structure, special precautions for circuit boards need not be taken.

4. Cables exposed to the environment require shielding, and the digital circuits that use signals from exposed cables should be designed to survive and not respond to thick dielectric discharges.

A review of deep charging by Baker et al. (1987) implicated (Figure 2.31) high fluxes of extremely energetic electrons (3–10 MeV) in the hardware failure on GOES-5 and also correlated flux peaks in this energy range to the star tracker upsets previously reported by Vampola (1986; Figure 2.29), suggesting the location of the discharges might be deeper inside the spacecraft. It was also noted that the peaks in this energy range cluster heavily during the decline from solar max to solar min and showed a 27-day periodicity corresponding to arrivals of high-speed solar streams originating from near equatorial corona holes.

The discharge detector data from SCATHA was analyzed to determine the number of events due to internal charging and their distribution in altitude and local time in Koons and Gorney (1992) and again in Koons et al. (2000). If the surface charge monitors (Kapton, Teflon. and quartz) were charged >100 V (negative) at the time a transient event was detected, it was deemed to be due to surface charging, and if the charge monitors showed <100 V, then the event was deemed to be due to deep or internal charging. The local time and radial distributions of the discharge events shown in Figure 2.32 support this division. The charging events classified as surface charging show the expected concentration at local times between midnight and dawn, whereas the events classified as internal charging show a broader distribution with a peak near local noon, consistent with the known asymmetry in the local time distribution of energetic electron flux (Lanzerotti et al., 1967). The number of discharge events increased rapidly when SCATHA approached its perigee at ~5.5 Re, (Re = earth mean radius) because the peak of the energetic electron flux is between L = 4 Re and 5 Re. The number of events when SCATHA flew at geosynchronous altitude (6.6–7.0 Re) is much smaller.

Koons et al. (2000) found that the deep charging–induced anomalies appear to have a threshold of ~1000 e-/cm^2-s-sr for the ≥1.4 MeV electron flux detected by LANL (Los Alamos National Lab) sensors on a GEO spacecraft (Figure 2.32). Assuming the maximum possible 4π steradian exposure, the fluence accumulated would be ~10^9 per day and it would take 1000 days to accumulate the 10^{12} e-/cm^2 required to cause discharges in ground tests. However, the electrical charge-decay time constant for the spacecraft materials was believed to be a few hours, so the threshold fluence for a discharge could not be accumulated over the 1000 days. Since the flux at L = 5.5–6.0 Re near SCATHA's perigee is higher than at 6.6–7.0 Re (GEO), the 1.4MeV flux at the altitudes where most of the events occurred would be higher than the flux LANL detected, and the threshold flux for anomalies would be higher than Figure 2.33 suggests. Additionally, if the source of the ESD was lightly shielded, the much

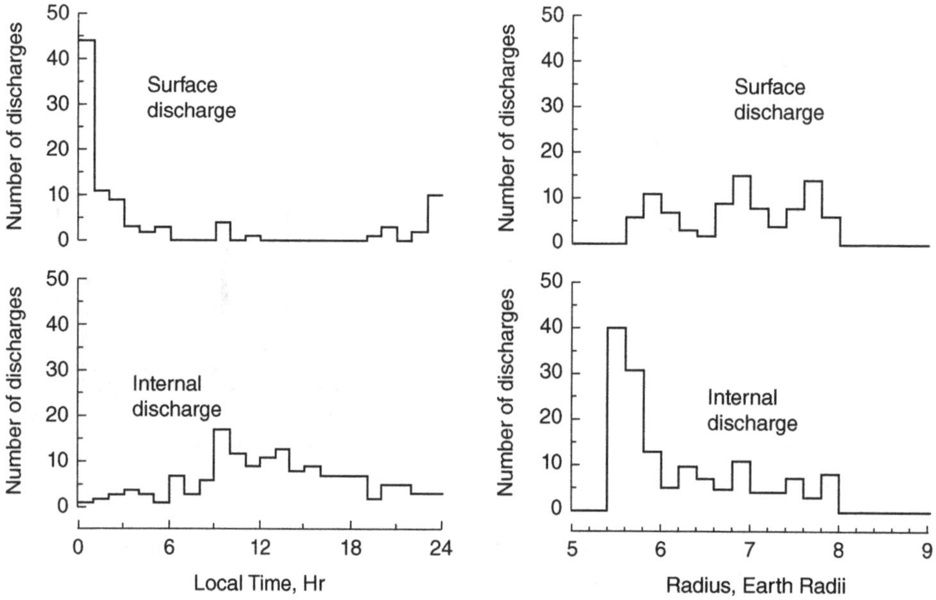

Figure 2.32 Designation of surface versus internal discharges is supported by their local time and radial distance distributions. (Source: Koons & Gorney, 1992. Reproduced with permission of American Institute of Aeronautics and Astronautics, Inc.)

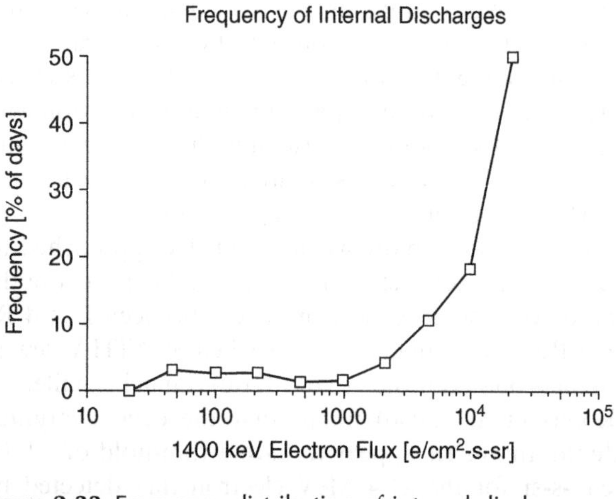

Figure 2.33 Frequency distribution of internal discharges versus 1.4 MeV electron flux in GEO. (Source: Koons et al., 2000. Reproduced with permission of US Air Force Geophysics Lab.)

higher integral flux at lower energies of a few hundred kiloelectrons might be able to overwhelm the charge decay rate and reach breakdown levels.

Finally, Koons and Gorney (1992) noted (Figure 2.34) that the surface discharges produced larger signals in the pulse detectors than the internal discharges and the large number of normal switching events detected by the same sensors. The larger peak voltages could easily upset digital logic circuits but are unlikely to cause permanent damage.

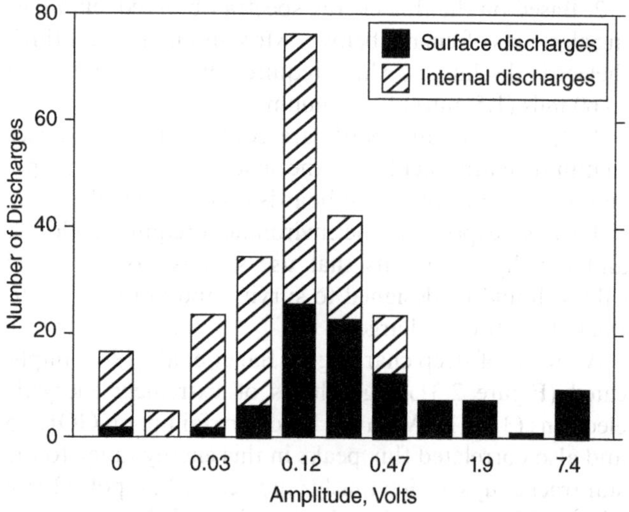

Figure 2.34 Number and amplitude of surface versus internal discharges. (Source: Koons & Gorney, 1992. Reproduced with permission of American Institute of Aeronautics and Astronautics, Inc.)

New data from the METEOSAT series of spacecraft provided additional evidence of deep dielectric charging. Recall that surface charging was suspected as the cause of anomalies on METEOSAT F1 because of correlation with disturbed geomagnetic conditions but could not be conclusively proven. METEOSAT F2 was equipped with an electron spectrometer (50 eV–20 keV) in an attempt to demonstrate a direct link between anomalies and surface charging, and multiple design changes were made to reduce the occurrence of anomalies (Coates et al., 1989). Nevertheless, anomalies continued to occur, especially radiometer stoppage or position jumps. Differential surface charging was observed, but only 1 of >100 anomalies on F2 occurred at any time near an observed differential charging event, and in all cases, there was no evidence that the spacecraft was charged to more than a few volts. Similar to F1, the anomalies showed a dependence on geophysical parameters: local time, solar aspect angle, and geomagnetic activity. Since the anomalies correlated with sustained periods of high geomagnetic activity, it was suspected that an elevated flux of energetic electrons might be causing the anomalies, and a new electron spectrometer that monitors the flux in the 30 keV to 300 keV range was added to the third flight (called METEOSAT-P2 by Coates et al., 1989, and METEOSAT-3 by others).

Coates et al. (1989) reported that METEOSAT P2 experienced anomalies at a rate 4 times higher than observed on F2, but that the anomalies on both flights show the same dependence on local time and season and therefore likely have the same cause. The P2 anomalies did show a good correlation with electron fluxes in the 43–300 keV range, which makes deep charging a plausible mechanism. However, the strong seasonal dependence (peaking around equinox) indicated a second causal factor was involved.

Later investigations using the total number of anomalies that occurred over the METEOSAT-3 mission life confirmed a strong correlation of some anomalies with the energetic electron flux. Hilgers et al. (2000) focused on a set of ~500 anomalies involving the radiometer. They found a poor correlation with the 24-hour average flux of electrons of energy 200–300 keV but found good correlation with the build-up of charge over a 4- to 5-day period. Andersson et al. (2000) looked at the full set of 724 anomalies of 18 types that occurred on METEOSAT-3 (an average rate of 1 every 5 days), plus a set of 192 anomalies reported for a GEO communications satellite, Tele-X (an average rate of <1 in 10 days). They found the rates for some anomalies on both spacecraft correlated with elevated fluxes of energetic electrons. It was found that none of the five energy channels of the METEOSAT-3 was preferred for the prediction of anomalies. Surprisingly, the best prediction of anomalies was made using the integral >2 MeV electron flux from GOES. Taken together, the two studies suggest that the source of the anomalies may be due to charging and discharging from a more heavily shielded source from charge accumulated over multiple days. These implications are clearly at odds with earlier studies that claimed the charge would leak out after several hours so only the much higher flux at much lower energies (few 100 keV) incident on thick dielectrics outside the spacecraft could cause ESD.

A new USAF satellite, CRRES (Combined Release and Radiation Effects Satellite) was launched on 25 July 1990 into a geosynchronous transfer orbit to investigate fields, plasmas, and energetic particles inside the Earth's magnetosphere. On this spacecraft was an instrument, the Internal Discharge Monitor (IDM), that was designed to detect the occurrence of ESD from small samples of cables and printed circuit board materials (Figure 2.35 and Figure 2.36). Prior to launch, an engineering model

Figure 2.35 Photograph of IDM experiment tray, with sixteen 5 cm x 5 cm planar PCB sample trays and four 20 cm long cable trays. Not shown is the thin 8 mil (0.2 mm) aluminum cover. (Source: Coakley, Treadaway, & Robinson, 1985. Reproduced with permission of IEEE.)

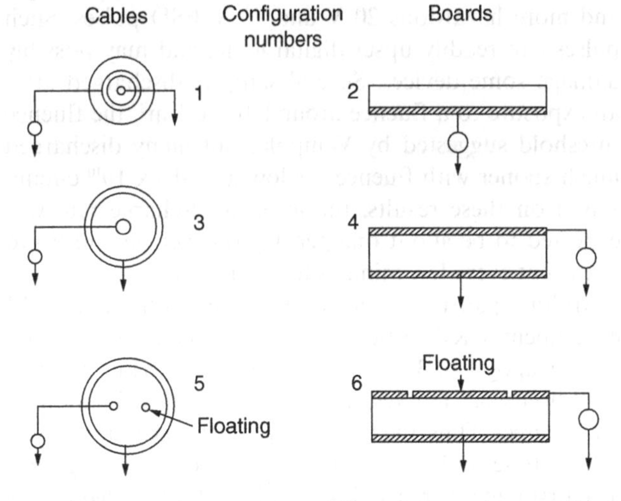

Figure 2.36 Configurations used for IDM samples. Small circles indicate connections to an oscilloscope and arrows show DC grounds. Oscilloscope connections had a 25 ohm impedance (50 ohms to ground at the experiment exit in parallel with a 50 ohm coax transmission line and 50 ohm scope input). Note that configurations 5 and 6 include DC isolated conductors. (Source: Coakley, Treadaway, & Robinson, 1985. Reproduced with permission of IEEE.)

of the instrument was exposed to high energy electron flux to assess if the samples would discharge and the ability of the instrument to detect those discharges (Coakley, Treadaway, & Robinson, 1985).

The samples were exposed to an electron spectral source that matched the predicted CRRES orbit flux-energy spectrum at a flux of 4.5 pA/cm^2 for 18 continuous hours. After that period, the flux was increased to 70 pA/cm^2 for an additional 87 minutes, then to 140 pA/cm^2 for another 44 minutes, and then back to 4.5 pA/cm^2 for a final 105 minutes. The changes to higher flux provided additional discharge data within a shorter time period. The high rates are within the bounds of storm level fluxes expected during the CRRES flight. The beam energy was set to 2.5MeV, but after traversing multiple thicknesses of metallic plates, the experimental dose profile and stopped-charge density profile in a 100 mil (2.5 mm) fiberglass circuit board closely matched the profiles expected from the on-orbit environment.

The results from the test are summarized in Table 2.4 and Figure 2.37. Generally, all of the circuit board coupons discharged, with higher rates of discharge during the higher flux exposure. Board coupons with configurations 2 (only backside ground plane) or 4 (ground planes on top and bottom) typically produced discharges <0.5 V with narrow pulse widths (~10 ns or less), which may be too small to upset digital logic circuits. The thickest (125 mil = 3.18 mm) fiberglass boards produced larger pulses with 1–2 V peaks that may be large enough to upset logic circuits. Samples that had electrically isolated conductors on the surface (configuration 6) produced much larger and more hazardous 20 V and 75 V ESD pulses. Such pulses can readily upset digital logic, and may possibly damage some devices. Several samples discharged after an exposure to a fluence around 10^{12} e-/cm^2, the fluence threshold suggested by Vampola, but many discharged much sooner with fluences as low as 3–10 x 10^{10} e-/cm^2. Based on these results, the in-orbit discharge rate was expected to be about one per day on average and up to one event every few minutes following a storm.

Multiple papers reported the flight results from the IDM experiment. CRRES flew in a geostationary transfer orbit with an apogee of 33,500 km, a perigee of 348 km, a period of 9 hours 52 minutes, and an equatorial inclination of 18.2 degrees. Thus, the IDM is exposed to the inner belt for a short time, and to the outer electron belt for a long time. The first paper (Frederickson et al., 1991) reported results from the first 10 months, while Frederickson et al. (1992, 1993) reported on the full 14-month mission. The IDM experiment detected 4300 pulses from the 16 samples (combined area of about 300 cm^2) over the 14-month mission. As a group, the number of pulses per 10 hr orbit was strongly influenced by the electron fluence per orbit (Figure 2.38). However, the behavior of individual samples varied markedly (Figure 2.39). Some pulsed frequently early in the mission, while others had few or no discharges and then discharged frequently in the latter part of the mission. Overall, the discharge rate after orbit 600 was nearly 6 times higher than earlier in the mission. Note that, unlike the ground test setup, the IDM instrument was only designed to count discharges, not measure the peak voltage, pulse width, or pulse energy.

A composite plot of the discharge count per orbit vs electron fluence per 10 hr orbit is shown in Figure 2.40. It shows the connection between discharge rate and fluence per orbit in a different format than Figure 2.38. Significantly, it was also noted that there were no discharge pulses during orbits with a fluence per orbit $\leq 2 \times 10^{10}$ e-/cm^2 (average flux of 5.5×10^5 e-/cm^2-s or current density of 0.09 pA/cm^2). This finding was later codified as the 0.1 pA/cm^2, "No-ESD Safe-Flux" limit in NASA HDBK 4002 (NASA, 1999), which is discussed later. However, Frederickson et al. (1992) expressed reservations with this interpretation:

> However, during the last seven months when the samples were most likely to pulse, there were no orbits with fluence below 3.5×10^{10}. Thus, it may be that no pulses were experienced in orbits with fluence below 2×10^{10} simply because no samples were sufficiently charged to produce pulses when such orbits occurred. We don't believe our data is sufficient to provide a firm guideline. One might wish to add a safety factor and assume that only fluxes below 1×10^5 e-/cm^2-s (0.016 pA/cm^2) are safe.

In preflight ground EMC tests, a MIL-STD-1541 pulser (MIL-STD-1541A, 1987) was used to apply arcs to the IDM chassis and the shield around its cables. These external pulses would trigger pulse counts in all 16 channels simultaneously, which showed that the IDM could detect discharges external to the IDM but within the CRRES spacecraft. A total of 206 such events were detected, and they correlated with 674 anomalies of 8 types that occurred elsewhere on CRRES (Figure 2.41; Violet & Frederickson, 1993). Most of these anomalies were of the "phantom command" type involving the science instruments, and none resulted in any permanent damage. The CRRES spacecraft anomalies and IDM coupon discharge rates correlated with exposure to high flux levels of energetic (>300 keV) electrons (Figure 2.42) and showed little correlation with proton flux, SEU rates in other detectors, or with surface charging conditions. There were instances where multiple instruments were upset simultaneously, like the IDM experiment, suggesting that interconnecting wiring was either the source of the anomalies or was highly coupled to the source. Frederickson et al. (1992) also noted that CRRES had several hundred feet of multiwire cables, overwrapped with metalized polymer foil, which by itself provides very little radiation shielding. Some of that cable length was

Table 2.4 Summary of CRRES IDM preflight deep charging ground test results. (Source: Coakley, Treadaway, et al., 1985. Reproduced with permission of IEEE.)

Planar Samples	FEP 60-mil	FEP 90-mil	FEP 90-mil	FEP 90-mil	FBG 47-mil	FBG 125-mil	FBG 125-mil	FBG 125-mil	PTFE 93-mil	MYLAR 10-mil	ALUM. 40-mil	ALUM. IRT
Configuration # (Fig. 2.36)	#2	#2	#4	#6	#2	#2	#4	#6	#2	#6	#6	#6
No. of events In low flux	0	0	5	3	2	11	2	8	4	7	1?	0
No. of events in high flux	5	5	2	51	8	19	14	6	3	1	4	1
No. of events in second low flux	0	0	2	0	0	0	8	4	1	0	0	0
Peak in low flux	–	–	0.08 V 10 ns	>>–2 V >20 ns	0.15 V 5 ns	2 V 60 ns	1 V 15 ns	>>±4 V –	0.1 V 10 ns	0.05 V 10 ns	? ±0.02 V 10 ns	– –
Peak in high flux	–0.45 V 2 ns	+0.2 V –0.1 V 50 ns	0.15 V 5 ns	+75 V 10 ns	0.25 V 10 ns	2 V 60 ns	1 V 15 ns	>+20 V 50 ns	0.2 V 10 ns	0.05 V 10 ns	>±4 V >20 ns	>2 V –

? Means that the event may have been noise.

		Cables			
		Aljac	LWC	Kapton	RG316
	Low flux	2	3	0	0
	High flux	1	0	I	0
	Second low flux	0	1	0	1
Low flux peak		0.3 V	+0.4 V –15 ns	– –	– –
High flux peak		–0.05 V 5 ns	– –	>>0.02 v –	0.12 V 10 ns

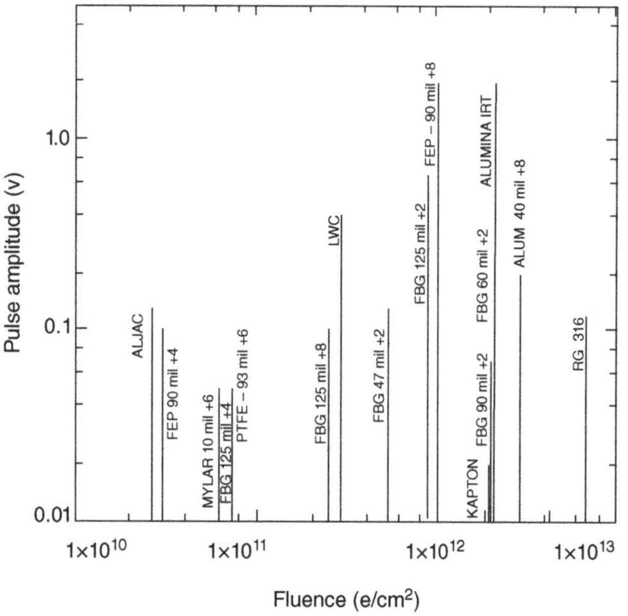

Figure 2.37 Incident fluence to first discharge during low flux exposures (Source: Coakley, Treadaway, & Robinson, 1985. Reproduced with permission of IEEE.)

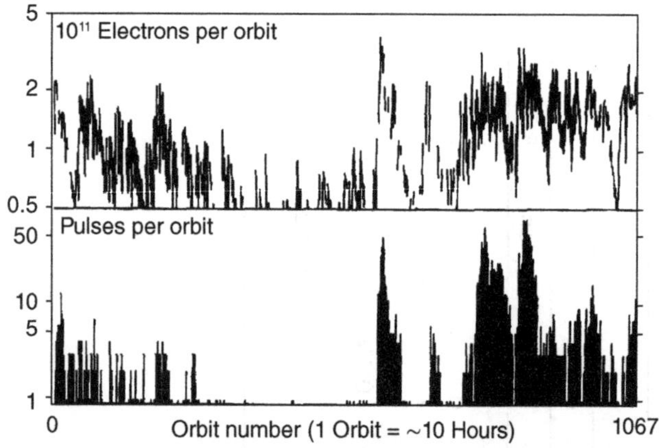

Figure 2.38 Number of discharges per orbit (lower panel) is driven by the incident internal electron fluence per orbit (upper panel). (Adapted from Frederickson et al., 1992.)

installed near the surface of CRRES and therefore was highly exposed, suggesting that cables were the likely cause of anomalies on CRRES.

Finally, Figure 2.42 shows that most of the IDM pulses and CRRES anomalies occurred during transit of the peak of the outer electron belt at L values between 3.5 and 5 Re, while few anomalies occurred around GEO (~6.6 Re) where the electron flux is lower. So the applicability of these discharge rates and anomaly rates to the majority of satellites that operate in GEO was uncertain. However, that uncertainty was subsequently dispelled when the Anik E1 and E2 satellites suffered failures in January 1994.

There are multiple press accounts contemporaneous with the Anik failures (e.g. Baker et al., 1994; Hanlon, 1994; INTELSAT, 1994), and several later, more-detailed discussions of the failures and the space weather that caused them (e.g. Lam et al., 2012; Odenwald, 2001), from which the following account is based. Anik E1 and E2 were launched into geosynchronous orbit in 1991 and were expected to operate through 2003. On 20 January 1994, the Intelsat-K satellite, which used the same basic satellite design as Anik E1 and E2, briefly began to wobble, resulting in a temporary service outage. Two hours later, Anik E1 began to roll end-over-end uncontrollably, interrupting telephone service to 40 northern Canadian communities and preventing the Canadian press from delivering news to over 100 newspapers and 450 radio stations. After 7 hours, Telesat Canada engineers were able to correct Anik E1's pointing problems by switching to a spare momentum wheel system. (Momentum wheels act like gyroscopes that keep the satellite orientation stable and the antennas pointed to the intended service area.) Seventy minutes later, Anik E2's momentum wheel system failed. Unfortunately, its backup system had also failed (it was not known if the backup system had failed during this space weather event or had failed sometime earlier). While the satellite was alive and otherwise healthy, it continued to spin slowly, rendering it useless. After 5 months of effort, engineers had developed and implemented a new ground-based control system to monitor the satellite's orientation (by the drift of its antenna pattern) and to command other systems (magnetic torquers and chemical thrusters) to correct the satellite's orientation, allowing service to be restored.

All three satellites shared a common design (however, INTELSAT, 1994, alludes to some unspecified enhancements to its satellite that made it more resilient to space weather). Having three satellites suffer similar anomalies within a 24 hour period pointed to some common external and nonrandom cause. A review of space weather was made, and deep charging caused by the elevated flux of energetic (>2 MeV) electron flux observed by the GOES satellites was implicated as the probable cause (Figure 2.43). It was noted that a large near-equatorial coronal hole had passed across the sun from 8 to 13 January and was the source of high-speed solar wind streams that indirectly caused the enhanced electron flux preceding the Anik failures (Lam et al., 2012).

A review of the GOES >2 MeV electron flux since Anik E1 and E2 were launched in 1991 shows (Figure 2.44) multiple episodes where the energetic electron flux was as high or higher than the flux on 20 January 1994 but did not produce failures. It is possible that the characteristics of the electron storm preceding 20 January (such as its duration or perhaps the energy spectrum) made it particularly hazardous. It is also possible, based on the

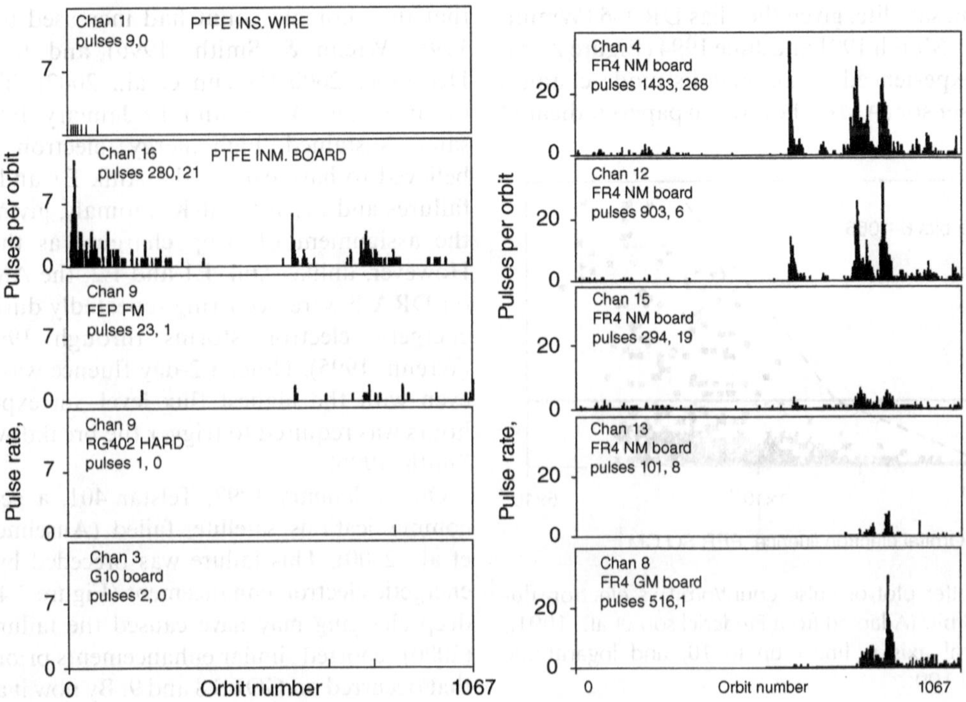

CHA	SAMPLE DESCRIPTION	VM	GEOM	PULSES
1	SC18 WIRE TYPE ET, 7 mil PTFE	1	1	9, V < 16
2	TS TRIAX CABLE, RAYCHEM 44/2421	5	5	0
3	MEP G10 SOLITHANE COATED ONLY	50	7	2, V < 70
4	FR4 EPOXY FIBERGLASS, 0.317 cm	5	2	1433, V < 40 16, 25<V<60 252, V>60
5	RG 316 CABLE, BELDEN 83284	0.5	3	0
6	SOLID AL JACKET RG402 CABLE	1	3	1, V < 30
7	ALUMINA, 0.102 CM, Cu electrodes	40	6	0
8	FR4 EPOXY FIBERGLASS. 0.317 cm	1	4	516, V < 45 1, V<60
9	FEP TEFLON, 0.229 cm, Al electr.	100	6	23, V < 70 1, V<70
10	FEP TEFLON, 0.229 cm, Al electr.	0.2	4	0
11	PTFE FIBERGLASS, .229 cm, 3M "250"	1	4	0
12	FR4 EPOXY FIBERGLASS, 0.317 cm	5	2	903, V < 40 2, 40<V<60 4, V>60
13	FR4 EPOXY FIBERGLASS, 0.317 cm	100	6	101, V < 60 8, V<60
14	MEP G10 SOLITHANE, LEAKY PAINT	<1	8	0
15	FR4 EPOXY FIBERGLASS, 0.119 cm	0.25	2	294, V < 60 19, V<60
16	PTFE FIBERGLASS, .229 cm, 3M "250"	0.2	2	280, V < 30 2, 30<V<60 19, V>60

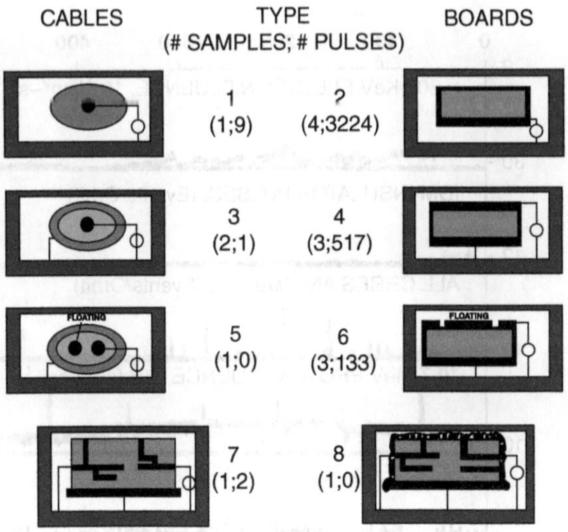

Figure 2.39 Pulses/orbit by sample types with geometries listed in table. (Adapted from Frederickson et al., 1991.) VM is the max voltage observed during preflight ground tests (see Table 2.4), while the PULSES column shows the number of pulses and peak voltage. Where multiple peak voltages are listed, the lowest peak value is based on the channel range and is certain, while higher thresholds are based on the extent of cross-talk to adjacent channels and are uncertain.

CRRES flight history, that it took a few years before the aging of the dielectrics increased the resistivity sufficiently that charging and discharging could have occurred, although Intelsat-K, launched in June 1992, was roughly a year younger than E1 and E2. This shows the challenge of predicting discharges and anomalies from external space weather data alone.

Additional corroborating evidence for deep charging as the cause of the Anik failures was provided by a set of phantom command anomalies on a geosynchronous communication satellite, given the alias DRA-δ (Wrenn, 1995). Between March 1991 and June 1994 (Figure 2.45), this satellite experienced 57 anomalous mode changes of an attitude sensor (AME). Follow-on papers indicated that the anomaly count had increased to 130 through 1995 (Wrenn & Smith, 1996) and to 214 through December 2000 (Wrenn et al., 2002). Three of these occurred on 14, 16, and 19 January 1994 during the same sustained high-energy electron enhancement believed to have caused the Anik E1 and E2 hardware failures and the Intelsat-K anomaly, giving credence to the assignment of deep charging as the root cause. However, unlike Anik E1 and E2, the AME anomalies on DRA-δ were occurring repeatedly during the earlier energetic electron storms through 1992 and 1993 (Wrenn, 1995). Note, a 2-day fluence was used because even with the highest flux level, an exposure of >30 hours was required to trigger the breakdowns (Wrenn & Smith, 1996).

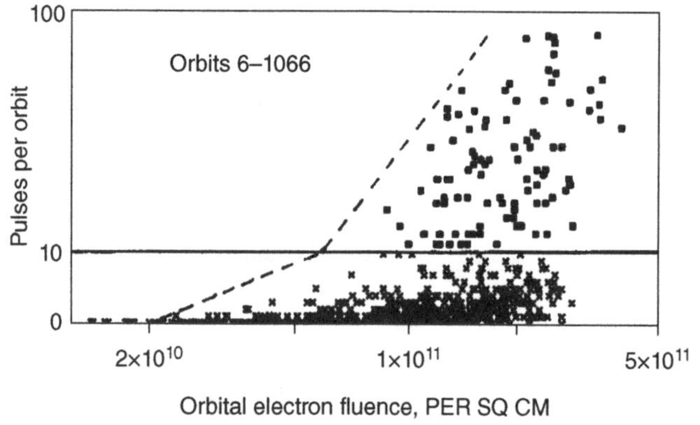

Figure 2.40 Scatter plot of pulse count/orbit vs electron fluence for each orbit. (Adapted from Frederickson et al., 1991.) Note the vertical axis is linear up to 10, and logarithmic between 10 and 100.

On 11 January 1997, Telstar 401, a geosynchronous communications satellite, failed (Anselmo, 1997; Love et al., 2000). This failure was preceded by a significant energetic electron enhancement (Figure 2.46), suggesting deep charging may have caused the failure. Love et al. (2000) reported similar enhancements prior to anomalies that occurred on GOES 8 and 9. By now it appeared clear

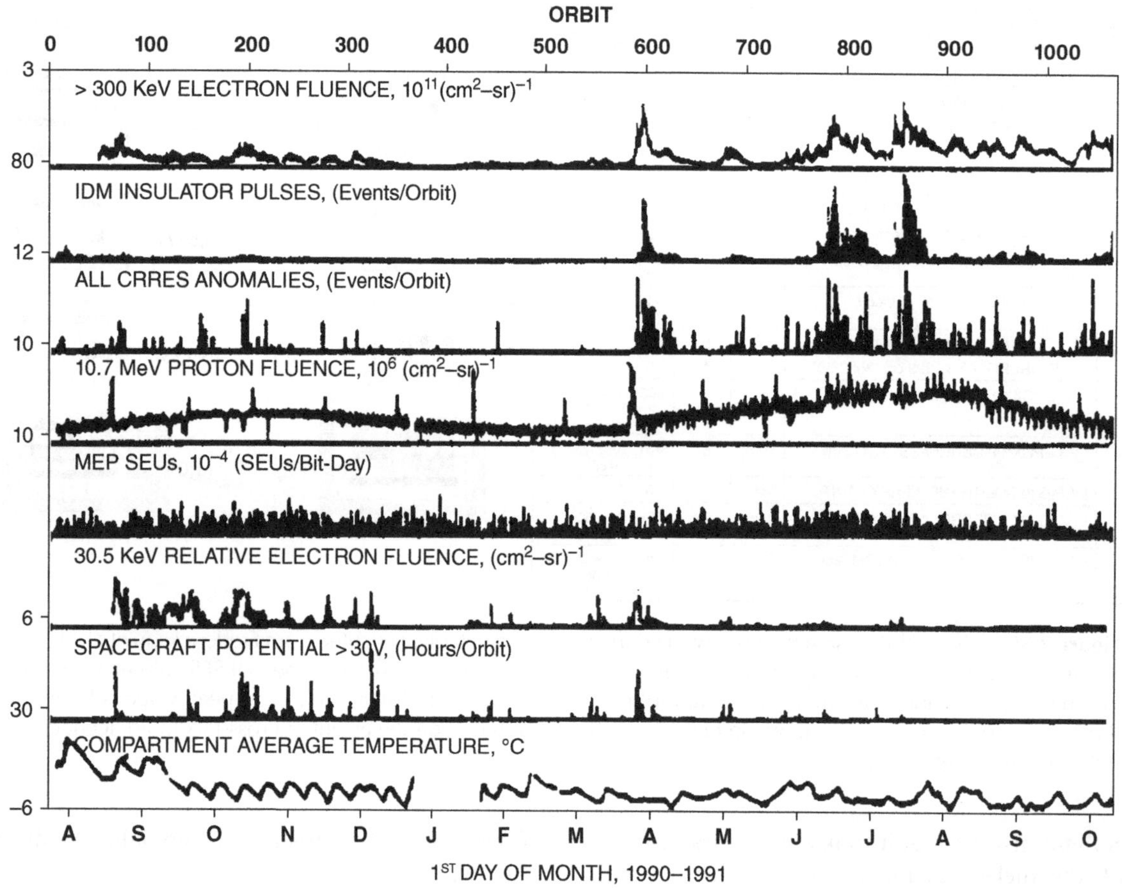

Figure 2.41 The IDM insulator pulses (second panel) and CRRES anomalies (third panel) correlate with the >300 keV electron fluence (first panel), but not with energetic proton fluence (fourth panel), SEU counts (fifth panel) or surface charging (sixth and seventh panels). (Source: Violet & Frederickson, 1993. Reproduced with permission of IEEE.)

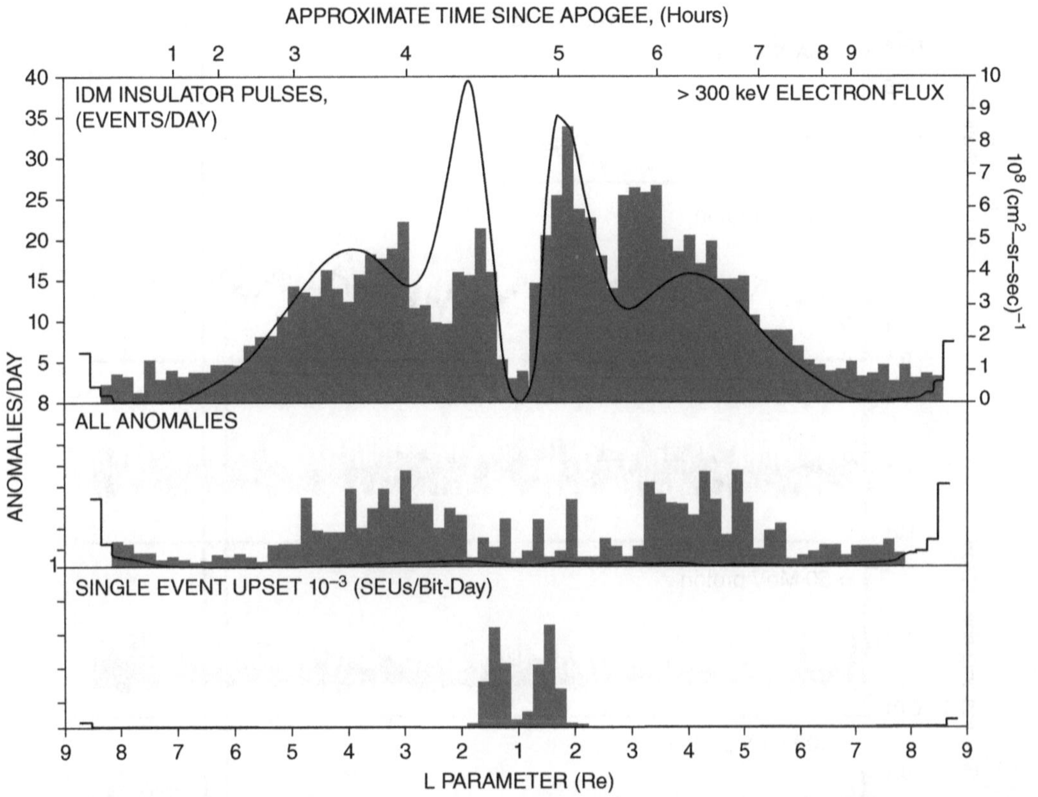

Figure 2.42 CRRES IDM pulse rate (bars, top panel), spacecraft anomaly rate (bars middle panel), and SEU rate (bars bottom panel) organized by geomagnetic L-shell shows the majority of anomalies are related to the high flux of energetic electrons (solid line top panel). (Source: Violet & Frederickson, 1993. Reproduced with permission of IEEE.)

that high fluxes of energetic electrons were hazardous to the function and health of satellites.

On 19 May 1998, Galaxy 4 lost attitude control, interrupting service, due to damage to its primary spacecraft computer. The failure was traced to a random short circuit, a failure mechanism that had been identified and designed out of subsequent spacecraft. Attempts to switch over to the redundant computer showed the backup unit was already dead. Satellite telemetry showed power to the backup unit was open circuit (blown fuse). This failure was traced back to a power transient that had occurred in June 1997 (engineering model tests showed that a short circuit within the computer replicated the bus transient) (Bodeau, 1999). This satellite failure interrupted service to pagers across North America, (because ~80% of pager service utilized the same satellite).

Given the prior history of Anik and Telstar, space weather was suspected as the cause. The anomaly investigation team at Hughes, which manufactured G4, and researchers in the space weather community (Baker et al., 1998) noted that an extended period of elevated energetic electron flux that began in late April and continued into to May (Figure 2.47). Baker et al. (1998) reported that other satellite anomalies occurred in this time frame (Equator-S failed on 1 May, and anomalies occurred on Polar, on 1 May, and GMS, on 4–7 May). The occurrence of multiple satellite anomalies following an extended exposure to energetic electron flux echoed the failures on Anik E1 and E2 and the anomaly on Intelsat-K, giving greater credence to deep charging and ESD as the cause of the failure.

Other evidence, however, cast doubt on deep charging as the root cause. G4 was built using the mature HS601 spacecraft design (Bodeau, 2007b). At the time of the failure, there were 36 other HS601 and 601HP spacecraft in orbit, several within a few degrees of longitude of G4, and none experienced failures. In addition, over half of these satellites, including G4, had been launched early enough to experience all or some of the severe energetic electron flux levels that occurred in 1993–1995. As shown in Figure 2.48, the energetic electron flux persisted for months at levels orders of magnitude higher than the flux preceding the G4 computer failures. If deep charging was the cause of the G4 failures, the fleet of 601 spacecraft should have experienced multiple failures during this more severe space weather, but no failures occurred. The

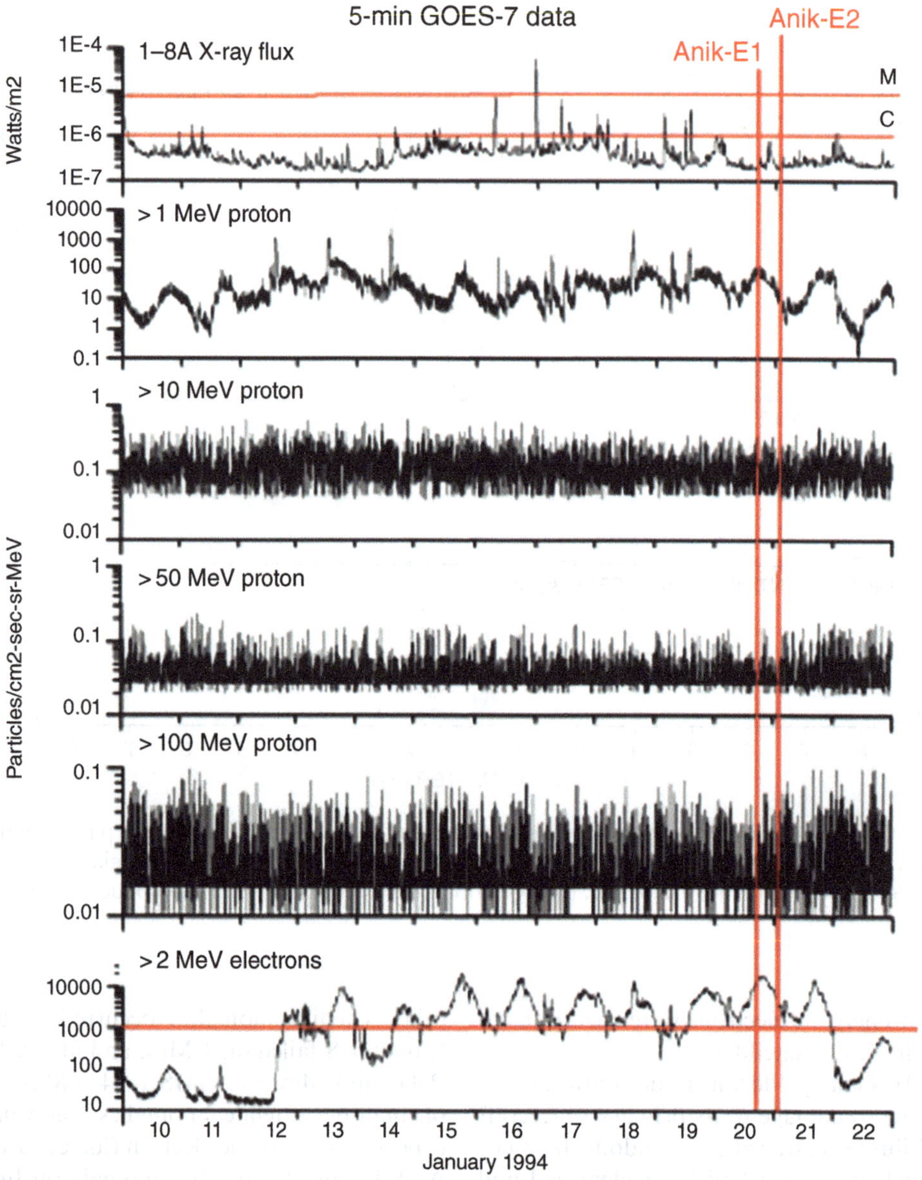

Figure 2.43 Review of space weather prior to Anik failures implicated deep charging from the week-long sustained high flux of energetic electrons (bottom panel) preceeding the events. (Source: Lam et al., 2012. Reproduced with permission of AGU) Note that the horizontal red line in the bottom panel indicates the flux threshold used by NOAA to determine when to issue alerts of adverse deep charging conditions.

investigation team was forced to look for a different cause, ultimately determining (on the basis of analyses, engineering tests, and the G4 telemetry) that a short circuit induced by a tin whisker was the cause of the computer failure on June 1997. Since the first two failures, seven additional computer failures (from a total population of 44 computers on HS601 spacecraft) had occurred (Bodeau, 2007b). Almost all of the failures had nearly identical telemetry signatures, and all occurred in computers susceptible to tin whiskers (except for the single random short circuit on May 1998). This experience showed that environment data alone is insufficient to prove root cause, and that the flight experience of similar spacecraft is more relevant than the contemporaneous failures of satellites with different designs.

Given the strong evidence of deep charging induced operational anomalies and the potential for satellite hardware failure, others began to look for additional evidence of deep charging impacts on satellites. Leach and Alexander (1995) and Baker (2000) reviewed multiple reports of anomalies, some of which were attributed to deep charging. Pilipenko et al. (2006) found significant

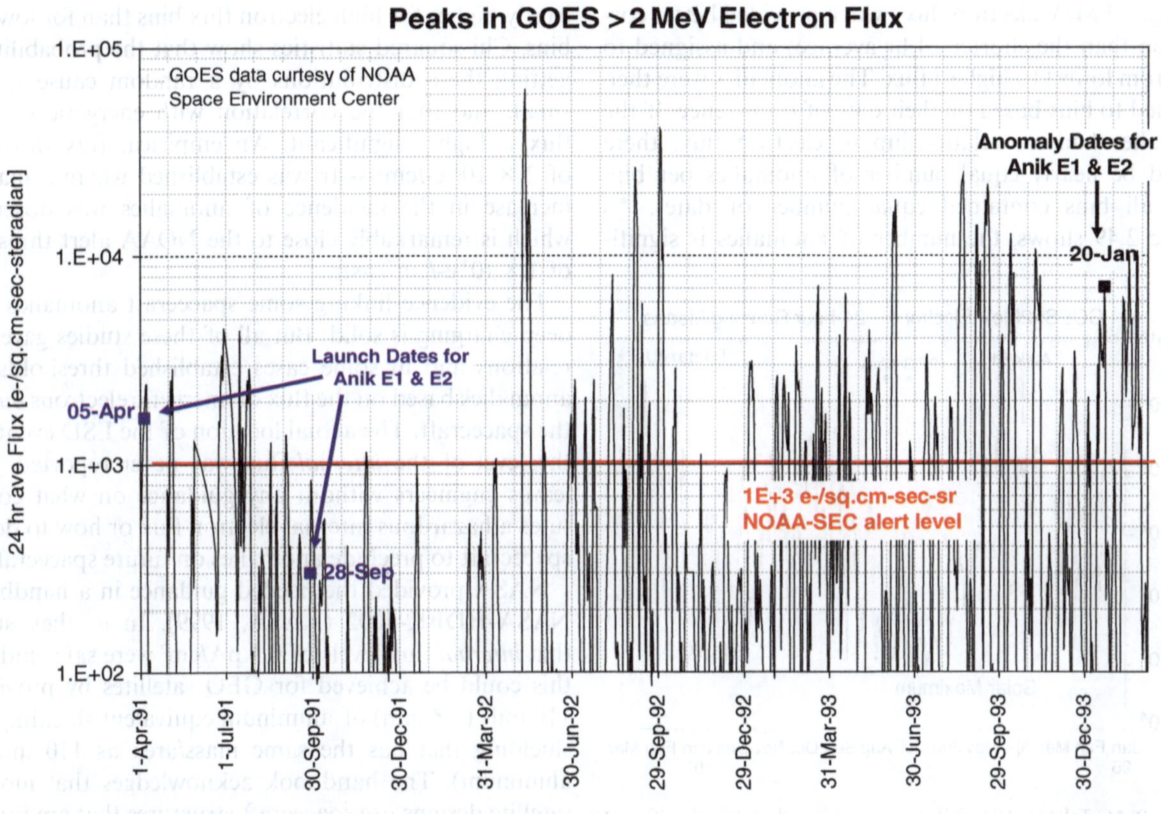

Figure 2.44 Energetic electron flux history from the launch of Anik E1 and E2 until the failures on 20 January 1994. Intelsat-K was launched in June 1992, roughly a year after E1 and E2.

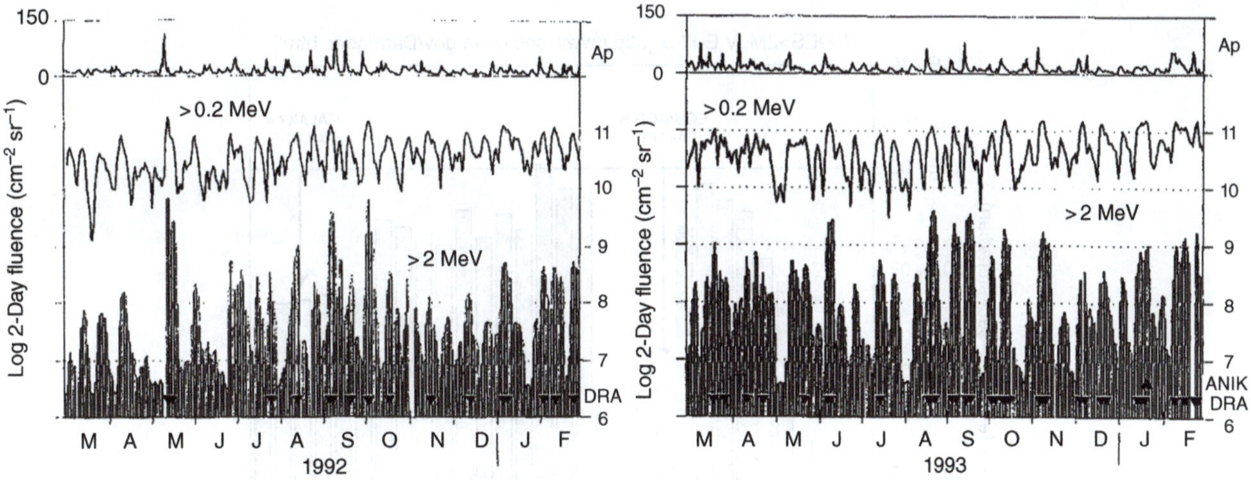

Figure 2.45 Correlation of AME switching anomalies on DRA-δ with peaks in 48 hr electron fluence of energy >0.2 MeV (METEOSAT-3) and >2 MeV (GOES-7). (Source: Wrenn, 1995. Reproduced with permission of American Institute of Aeronautics and Astronautics, Inc.)

statistical correlation between energetic electron flux and the occurrence of anomalies contained in the NOAA anomaly database.

At the sixth Spacecraft Charging Technology Conference, Balcewicz et al. (2000) reported results of an investigation into anomalies observed in the fleet of GEO communication satellites designed and manufactured by Hughes. (The results of this study were available to and used by the G4 failure investigation team.) All of the dates covered by the study were ranked by the 7-day average >2 MeV electron flux (which provided better correlation than the shorter 24 hr average) and assigned to bins from lowest to highest flux. The anomalies were then assigned to bins based on their date of occurrence. If the anomalies had no relationship to electron flux, there should be nearly equal number of anomalies per bin, since all bins contained equal number of dates. As Figure 2.49 shows, the number of anomalies is significantly higher for high electron flux bins than for low flux bins. Chi-squared statistics show that the probability of getting these distributions by a random cause is very small, and that the correlation with energetic electron flux is highly significant. An empirical flux threshold of 2×10^3 e-/cm^2-s-str was established where a marked increase in the incidence of anomalies was observed, which is remarkably close to the NOAA alert threshold of 1×10^3 e-/cm^2-s-str.

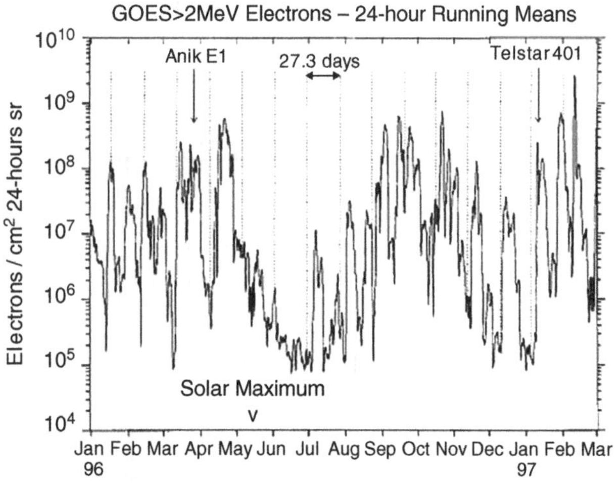

Figure 2.46 Telstar 401 failure was preceded by significant energetic electron enhancement, similar to Anik. (Source: Love et al., 2000. Reproduced with permission of IEEE.)

The evidence linking some spacecraft anomalies with deep charging is solid. But all of these studies gave correlations and in some cases established thresholds, for anomalies based on the flux of energetic electrons *outside* the spacecraft. The actual location of the ESD event and the level of the *internal* flux was never reported. This leaves engineers without any guidance on what constitutes a hazardous internal electron flux or how to design spacecraft to preclude anomalies on future spacecraft.

NASA provided the needed guidance in a handbook, NASA-HDBK-4002 (NASA, 1999). In it, they stated that *internal* flux levels of 0.1 pA/cm^2 were safe, and that this could be achieved for GEO satellites by providing 110 mils (2.8 mm) of aluminum equivalent shielding (i.e. shielding that has the same mass/area as 110 mils of aluminum). The handbook acknowledges that modern satellite designs use spacecraft structures that are thinner than this, and some use thermal blanket close-outs with far lower mass, and that adding the extra mass at the

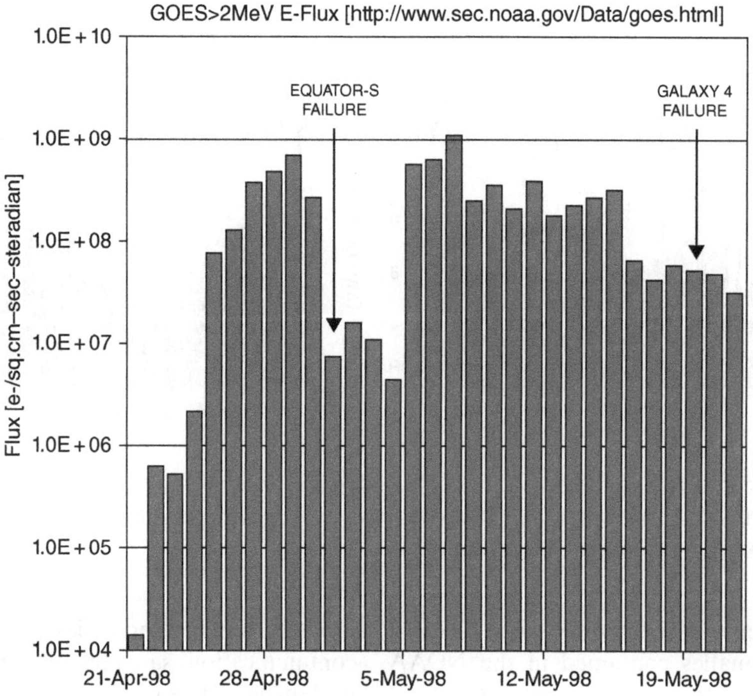

Figure 2.47 Episode of elevated energetic electron flux preceding the G4 and Equator-S failure. (Source: Baker et al., 1998. Reproduced with permission of EOS.)

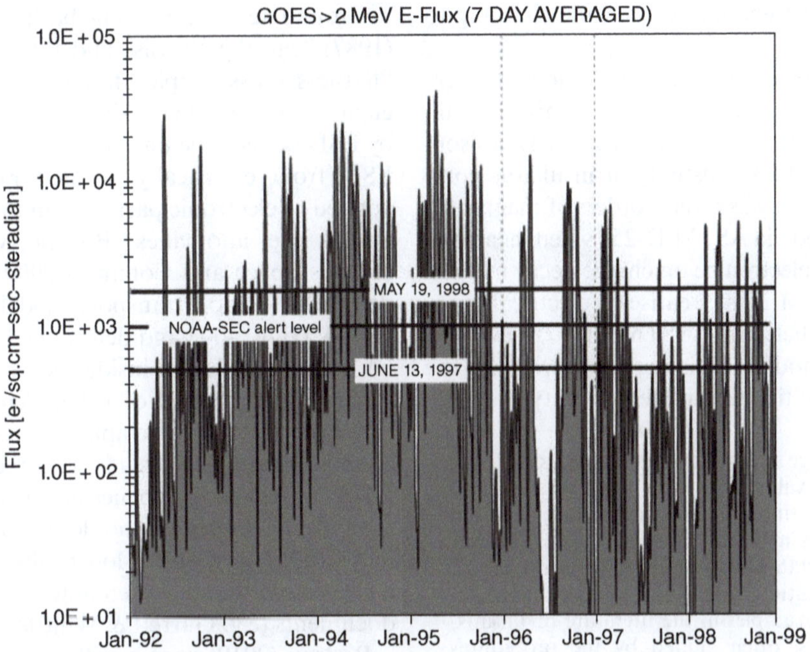

Figure 2.48 G4 and other similar spacecraft had been exposed to much higher electron flux levels without failures. (Source: Bodeau, 2007b. Reproduced with permission of AGU.)

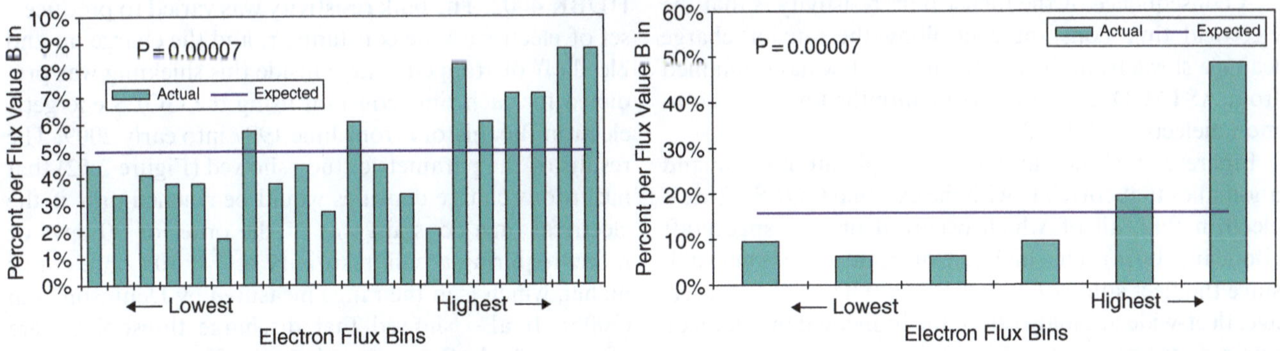

Figure 2.49 Relationship of anomalies to energetic electron flux (7-day average GOES >2 MeV flux) for the entire HS601 fleet (left-hand plot) and for a specific unit type (right-hand plot) (Source: Balcewicz et al., 2000. Reproduced with permission of US Air Force Research Lab.)

spacecraft level is not practical. But they recommend localized shielding. Charging analysis methods were also presented (details in Appendix D of HDBK-4002, NASA, 1999), based on a worst-case external electron flux (from 11 May 1992) and insulator resistivity and dielectric constants, with recommended values for these properties collected from engineering handbooks and displayed in Table I in section 6.2 of NASA, 1999. The table also listed the electrical time constant (which is given by the product of dielectric constant and bulk resistivity and which gives the rate of charge leakage out of the insulator), which was <6 minutes for printed circuit board material, and about 2–4 days for thin film polymers. Caution was given that the handbooks may have only reported minimum resistivity values and that the actual resistivity may be higher. Recommendations were given to perform resistivity measurements on the materials being used in spacecraft hardware, using standard test method ASTM D-257 (ASTM, 1991) or an electron beam charge decay method. In practice, the same radiation transport tools used to compute the total dose expected for electronics over a satellite mission life (using models of the shielding mass distribution around the electronics) were used to compute the internal flux. If the internal flux was shown to be less than 0.1 pA/cm^2, the design was declared safe. If the flux was higher than this "sure-safe" threshold, analyses using resistivity values derived from ASTM D-257 methods were performed to show that charge leakage was fast enough to preclude charging to hazardous levels.

However, evidence mounted that these guidelines were not sufficient. A series of studies (e.g. Dennison et al., 2006; Frederickson et al., 2003; Frederickson & Dennison, 2003) showed that the bulk resistivity of insulators commonly used on spacecraft was several orders of magnitude higher than determined via ASTM D-257 when measured in a vacuum using an electron beam charge decay method (Table 2.5). Even ASTM, in its treatise on electrical properties of engineering dielectrics, ASTM STP 728 (Wintle, 1983), states that method D-257 is not appropriate for the determination of the ultimate volume resistivity:

> After applying a voltage step, the current decays to a steady state or quasi-steady value. How long this takes depends upon the temperature, time, material and electric field, but delays of a few hours may well be needed to establish a semblance of steady-state conditions at room temperature. While adequate precautions are usually observed in scientific work, there are examples in the literature of unsatisfactory measurements, often guided by the procedures outlined in the national standards of various countries. These tests, frequently based on a 60-s delay, are quite inappropriate as a measure of conductivity.

A consequence of the higher bulk resistivity is that the electrical time constant controlling the rate of charge leakage slows from the few minutes to few days obtained from ASTM D-257 to weeks or months for many common dielectrics (Table 2.5).

Figure 2.50 shows another set of phantom-command anomalies that correlate with the external GOES >2MeV electron flux, all of which occurred on one spacecraft (Bodeau, 2007c). This high rate of occurrence is unusual, since the fleet study of Balcewicz (2000) showed an average, fleet-wide anomaly rate at the individual unit level of one per century.

Ground tests employing both the MIL-STD 1541A (1987) "zap the chassis" and ISO-14302 (BS ISO, 2003) "harness cross-couple" test methods performed on an engineering model unit showed the upset was not caused by ESD outside the unit, whereas other tests showed that ESD from electrically isolated radiation spot shields applied to electronic parts within the control circuits replicated the anomalies. Boeing (which had acquired Hughes Space and Comm in 2000) then performed the detailed radiation transport modeling recommended in NASA HDK 4002 and determined that the flux incident on these spot shields inside the units was far below the "safe flux" threshold of 0.1 pA/cm^2 (Figure 2.51). The threshold flux level also appeared to drop over time (consistent with aging trends observed by CRRES and DRA-δ). As a stop-gap measure, it was suggested that the "safe-flux" threshold should be dropped an order of magnitude based on the low peaks shown in Figure 2.51, a suggestion that has been noted in subsequent guidance documents (e.g. Garrett & Whittlesey, 2012).

Bodeau (2010) analyzed the rate of charge accumulation and E-field build-up inside thick dielectrics and electrically isolated conductors shielded by the 110 mils of aluminum equivalent shielding recommended by NASA HDBK 4002. The bulk resistivity was varied to produce a set of electrical time constants, τ, and the charge up and bleed off of trapped charge inside this shielding was predicted for each time constant using the GOES energetic electron flux history from June 1987 into early 2009. The results of the parametric study showed (Figure 2.52) that hazardous charge densities would be reached only if the electrical time constant was of the order of 30 days or more, requiring a bulk resistivity of 1 × 10^{19} ohm-cm or higher, which is in the range measured by Dennison et al (2006). It also showed that discharge thresholds were often reached after accumulating sufficient charge from

Table 2.5 Charge decay tests in vacuum yield resistivities and electrical time constants that are order(s) of magnitude higher than ASTM method. (Source: Dennison et al., 2006. Reproduced with permission of IEEE).

COMPARISON OF DARK CURRENT AND POLARIZATION DECAY CONSTANTS [10]

Material	Dark Current Resistivity (Ω-cm) ρ_{DC}	Decay Times (days) Measured τ_{DC}	Measured τ_P	Tabulated τ_{ASTM}	Ratio of Decay Times τ_{DC}/τ_P	τ_{DC}/τ_{ASTM}
Kapton (KA001 Polyimide)	5.0×10^{19}	75	–	3.5	–	4×10^1
Teflon (PTFE)	4.3×10^{19}	73.4	0.13	2.0	6×10^2	4×10^1
Teflon (FEP)	3.5×10^{19}	71.0	0.083	2.0	9×10^2	3×10^1
Teflon (PFA)	2.6×10^{19}	51.2	0.11	2.0	5×10^2	3×10^1
Tefzel (ETFE)	3.1×10^{19}	68.0	0.24	0.27	3×10^2	3×10^2
Urethane Potting Compound (Conothane EN-11)	1.6×10^{18}	2.2	0.81	0.015	3×10^0	4×10^2
FR4 Printed Circuit Board	1.1×10^{18}	5.2	21.5	0.015	3×10^0	5×10^2
PTFE Composite	3.0×10^{20}	341	0.75	2.1	4×10^2	2×10^2
Alumina (Al_2O_3)	2.9×10^{17}	21.4	0.26	0.001	8×10^1	3×10^3

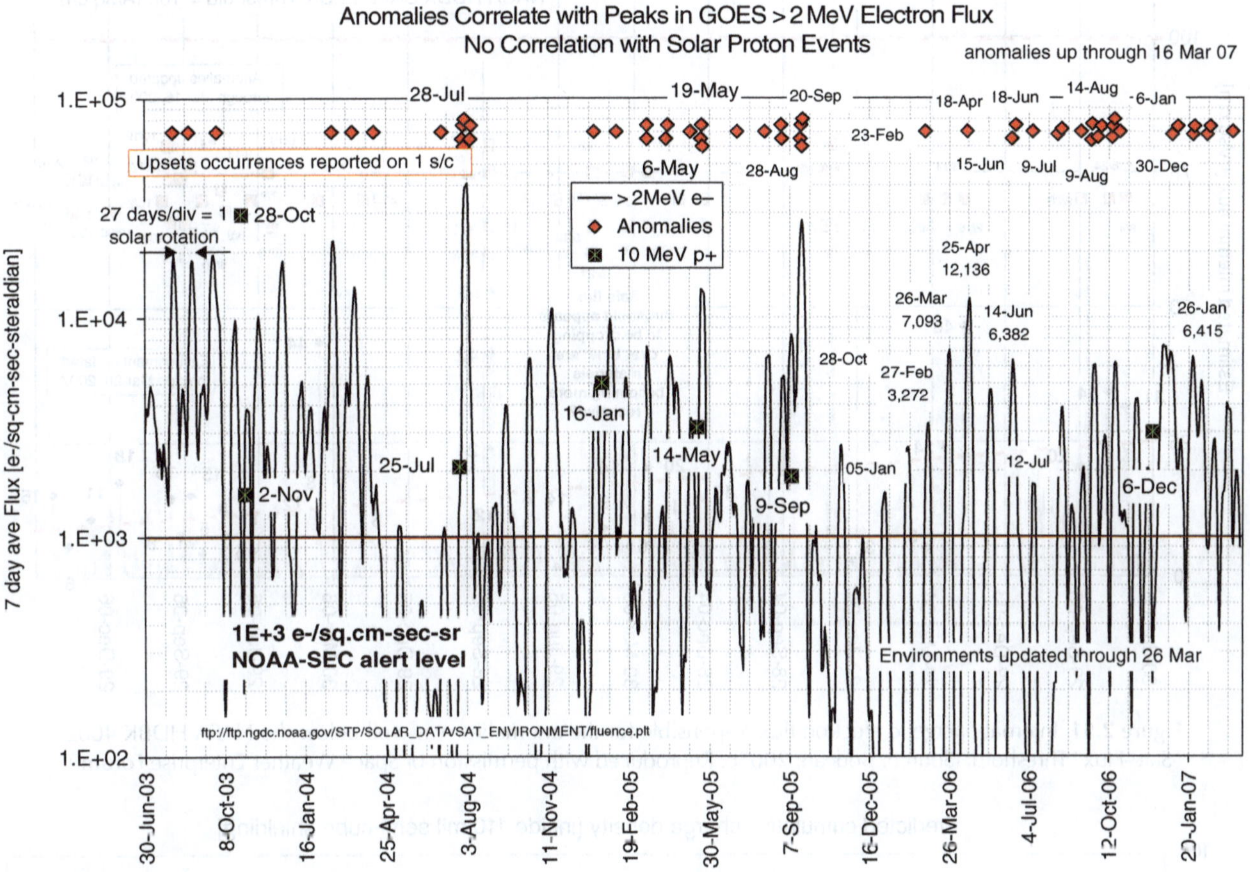

Figure 2.50 Example of 54 anomalies on one spacecraft correlated with the external energetic electron flux. (Source: Bodeau, 2007c. Reproduced with permission of Space Weather Enterprise Forum.)

several energetic electron storms, explaining why anomalies may not occur during every event (e.g. in Figure 2.51).

In summary, the unambiguous correlation of these anomalies with the energetic electron flux, combined with ground tests demonstrating that ESD from ungrounded spot shields could reproduce the anomaly (while ESD outside the unit could not), showed that deep charging and ESD inside units was the correct root cause. This conclusion was further supported by analyses using resistivity values in the range measured by new advanced test methods that showed hazardous levels of charging would occur, despite the internal flux level being below NASA HDBK 4002's safe-flux threshold. (Past anomaly investigations that assumed that ESD within units was implausible and assigned the likely ESD location to more exposed cables should be reexamined.)

However, there were still doubts that internal charging was the root cause, because this set of evidence did not explain why the anomaly rate on this particular spacecraft was so much higher than observed on three other, very similar spacecraft. The command control circuits used on the four spacecraft utilize common elements (Figure 2.53). The first element is a large input filter, used to buffer signals coming in on the command line. As a consequence of the large number of phantom command anomalies that occurred in the 1970s that were caused by surface charging–generated ESD, new unit designs adopted the guidance given in NASA TP-2361 in 1984 to desensitize command circuits to fast transient pulses. A large R-C filter was put at the input that requires hundreds of milliseconds for a low-current command-line input step pulse to charge up. After those hundreds of milliseconds, the filter output voltage reaches the reference voltage, and the comparator then sends a pulse to the downstream circuitry that executes the function being commanded. (It is this robust design that prevented any unit upsets by the external discharges applied during the anomaly investigation.)

Of the four similar spacecraft, the two oldest were anomaly free. A comparison of the unit design used by the two older spacecraft with the unit design of the third, newer spacecraft that had the anomalies revealed key design differences. The units on the third spacecraft had a reference voltage that was about one third that of the units used on the two older spacecraft. Those older units also incorporated a small capacitor located between the

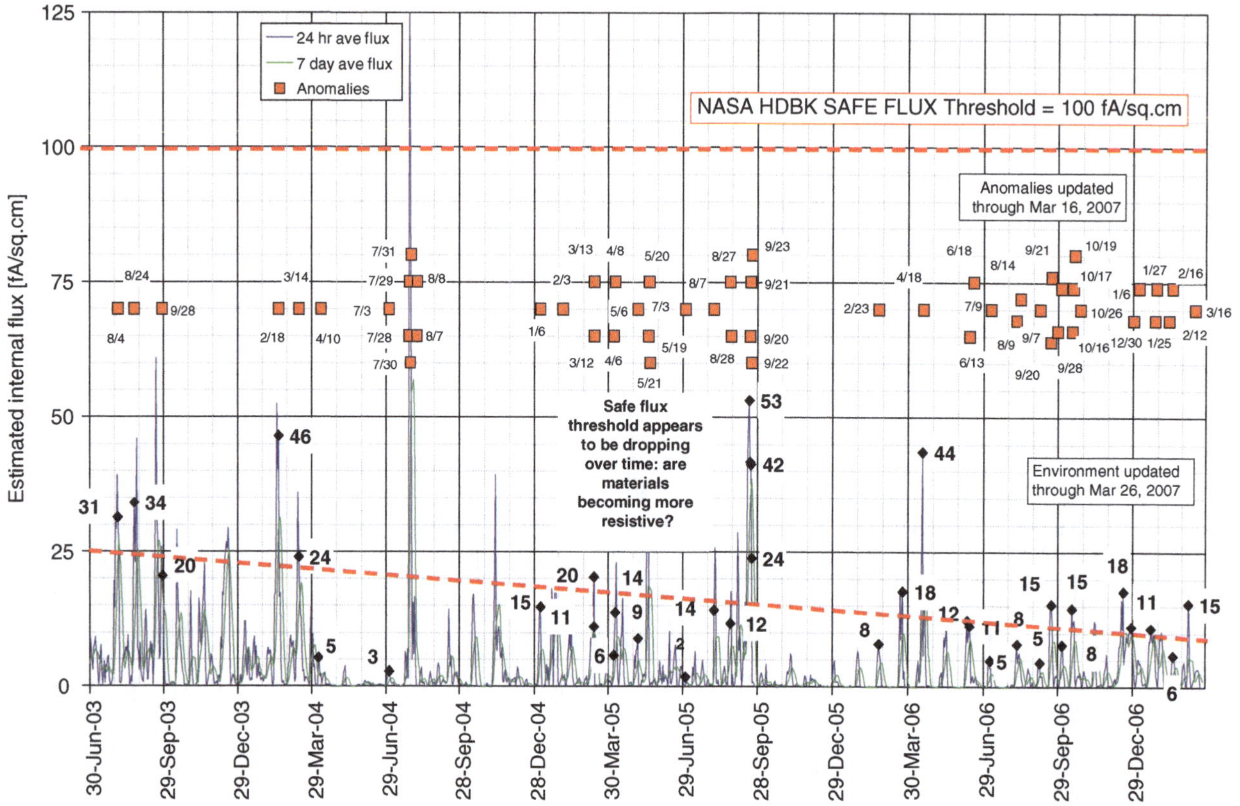

Figure 2.51 Internal energetic electron flux responsible for the anomalies was well below the NASA HDBK 4002 "Safe-Flux" Threshold. (Source: Bodeau, 2007c. Reproduced with permission of Space Weather Enterprise Forum.)

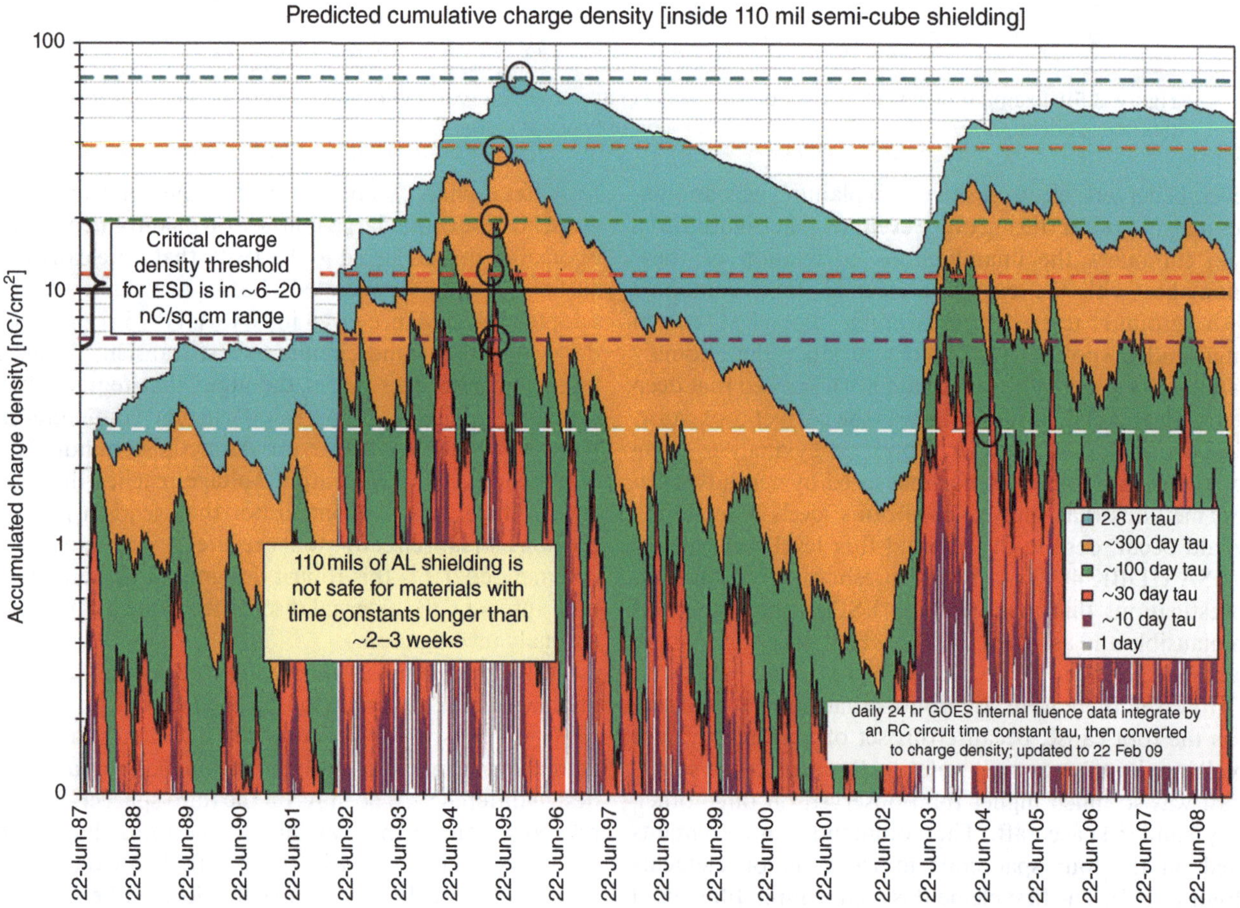

Figure 2.52 Parametric study of the buildup and decay of trapped charge inside insulators shielded by 11 0mils aluminum shielding versus the dielectric electrical time constant. (Source: Bodeau, 2010. Reproduced with permission of American Institute of Aeronautics and Astronautics, Inc.)

comparator and downstream parts (not shown in Figure 2.53), which was omitted from the new units. The higher reference voltage and small capacitor in the older design afforded those units with sufficient noise immunity that they did not respond to discharges from the ungrounded spot shields. However, the fourth spacecraft had an even newer design: the comparator reference voltage was close to an order of magnitude lower than the reference voltage in the two oldest spacecraft; and the small decoupling capacitor was also omitted. Despite being even more susceptible to ESD noise, these units had no anomalies because their spot shields were grounded, eliminating the source of the ESD. The printed circuit board dielectrics presumably can still charge, and potentially be an additional source of ESD. However, as the CRRES IDM data showed, discharges from dielectrics are smaller and slower than discharges from isolated conductors, and if circuit board dielectric discharges were occurring in these spacecrafts, they apparently were not large enough to produce phantom command anomalies.

The charge decay test methods discussed previously indicate that the bulk resistivity of insulators is far higher than indicated by ASTM methods, and the leakage of charge deposited by the space environment is far slower. However, the same radiation, in the process of penetrating and trapping inside insulators, results in ionization of the dielectric (unless the insulator is protected by a spot shield). The ionization introduces new free charge carriers that increase the conductivity of the material. This process is termed radiation-induced conductivity (RIC; Meyer et al., 1956). Researchers have performed tests using multiple energy beams to better simulate the broad energy spectrum of space radiation and have found that RIC reduces the rate and ultimate level of charging (e.g. Figure 2.54 from Payan et al., 2005a; and Payan et al., 2005b) but still produces discharges. However, these tests were performed at flux levels roughly 100 times higher than the NASA HDBK 4002 safe flux level of 0.1 pA/cm^2, suggesting that deep charging may not reach breakdown thresholds in real space level environments.

Most of these tests were performed at ambient (~25°C) temperature. It has long been known that the resistivity of insulators rises as temperature falls (e.g. Adamec & Calderwood, 1978; ASTM, 1991; Meyer et al., 1956), and the variation typically follows an Arrhenius-type exponential variation with absolute temperature.

$$\rho(T) = \rho(T_0) * \exp\left[\frac{Ea}{k_B}\left(\frac{1}{T} - \frac{1}{T_0}\right)\right]$$

Note. ρ is resistivity, T is absolute temperature (in Kelvin), k_B is Boltzmann's constant, and Ea is the activation energy in eV. The dramatic effect of temperature on resistivity is shown in Figure 2.55. (Frederickson et al., 1993, showed the operating temperature of the IDM experiment ranged between 0°C and 5°C over most of its mission.)

A study of 26 solid-state power amplifiers (SSPAs) that occurred between 1996 and 2009 on two fleets of

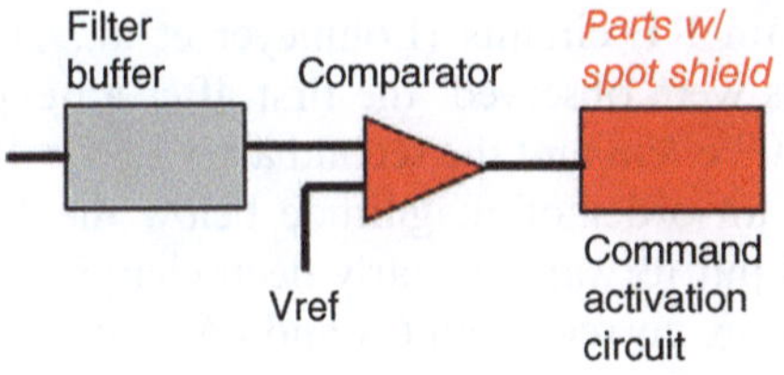

Figure 2.53 Common elements of command control circuits.

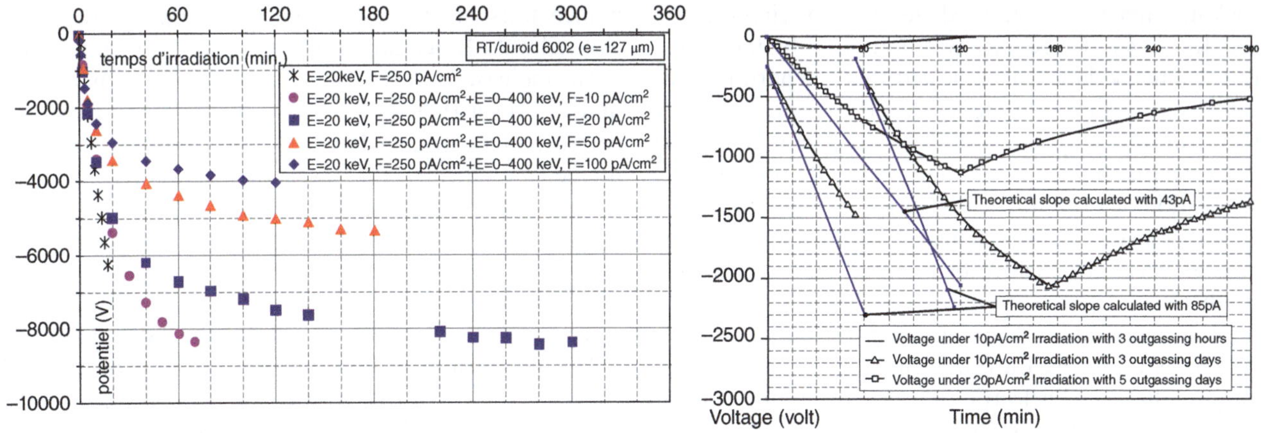

Figure 2.54 Radiation-induced conductivity from high-energy penetrating electron flux reduces the rate and severity of charging from a higher flux of lower energy nonpenetrating flux. (Sources: Left side: Payan et al., 2005a. Right side: Payan et al., 2005b. Both reproduced with permission of Japan Aerospace Exploration Agency.)

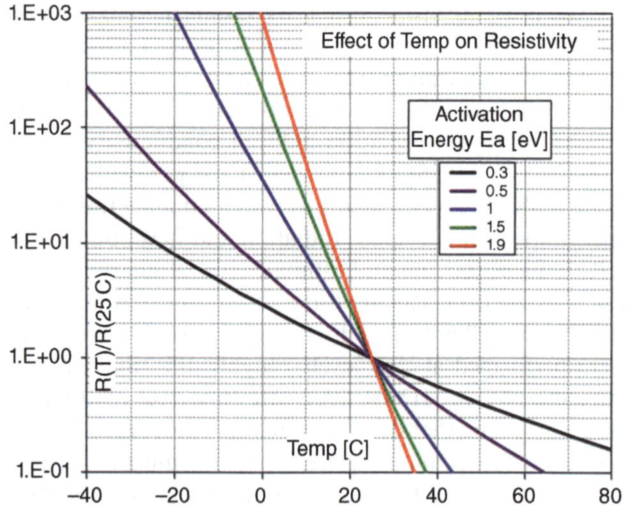

Figure 2.55 Effect of temperature on resistivity for a range of activation energies.

commercial communication satellites owned and operated by Inmarsat found a statistical connection between the anomaly occurrences and the fluence of energetic electrons over the prior 7–14 days (Lohmeyer & Cahoy, 2013). A follow-up study (Lohmeyer et al., 2015) reinforced the connection between the anomalies and energetic electrons, finding the strongest statistical correlation with the 14-day and 21-day fluences of >2 MeV electrons (from GOES satellites in GEO orbit). This suggested that deep charging was occurring in materials with a long electrical charging time constant, which in turn requires a very high electrical resistivity. A survey of the RF payload design suggested that the RF coaxes connected to the SSPA inputs and outputs were the most likely source of ESD, because of their very high resistance dielectrics (PTFE Teflon) and high electron flux exposure. The coax center conductors were assumed to be DC isolated by DC blocking capacitors, commonly used at RF circuit interfaces to coaxes. This choice was reinforced by the spacecraft telemetry data that indicated that the hard failures manifested as a drop in the SSPA RF current, without a change in the DC power supply or in the command and telemetry functionality, pointing to the RF circuitry as the location of the failures. Like prior anomaly investigations, this investigation assumed at least some section of the coax was only shielded by minimal thermal blankets and therefore exposed to higher charging flux than cables and units shielded inside the spacecraft structure.

Analysis using the European Space Agency's internal charging model, Dielectric Internal Charging Threat Analysis Tool (DICTAT; Rodgers, 2004), was performed. DICTAT is a 1-D internal charging code that determines radiation transport through various shielding to identify the maximum internal electric field in a component and compares it with a suitable breakdown threshold to determine whether ESD will occur. To determine this, DICTAT calculates the internal electron current using analytical approximations to quantify the electron transport through a shield and the charge deposited inside the dielectric. From this deposited charge, the maximum electric field within the dielectric is found. This field is then compared with the dielectric breakdown field. DICTAT accounts for both the effects of temperature and RIC on insulator conductivity. A literature search was performed to establish the parameters that define the bulk conductivity and its temperature and radiation dose rate dependencies. There was a wide range of bulk conductivities reported, and the operating temperature of the RF coaxes was unknown. (The activation energy for temperature effects used for Teflon was 1.4 eV, which makes the resistivity very sensitive to temperature per Figure 2.55). Consequently, the analysis was performed parametrically using combinations of five material (room temperature) bulk resistivities and seven operating temperatures. The results showed that the induced voltage in the coax would exceed the maximum coax voltage rating of 1900 V over a wide range of temperatures and bulk resistivities.

Brief electron beam tests were performed to determine what the actual breakdown voltage was (the breakdown voltage of the coax and blocking caps were unknown) and if the ESD arcs were energetic enough to damage downstream RF circuits (Lohmeyer et al., 2015). Two discharges were observed, the first after an exposure to ~1.46×10^{11} e-/cm^{2}and the second after 2.35×10^{11} e-/cm^2 (which is an order of magnitude below the 10^{12} e-/cm^2 threshold put forward by early deep charging investigators). The arc currents had 6A and 4A peaks and 0.25 μs durations. The corresponding peak voltages across a 50 ohm load would be 300 V and 200 V, and the total pulse energy would be up to ~200–450 μJ, more than sufficient to damage SSPA RF FETs. Post-irradiation RF performance tests of the test circuit were conducted, which showed a 2 dB degradation in insertion loss and a 5 pF (33%) increase in the DC blocking capacitor, which is significant given that the precision of the measurement is better than 5% in capacitance. Pre- and postarc RF measurements confirmed the operability of the RF blocking circuit and specifically that the DC blocking capacitor continued to isolate the RF input port, suggesting that the SSPAs could be subject to multiple arcs over a mission life, and that it was damage to the downstream RF components that was responsible for the SSPA failures.

The conclusion was that deep charging of the RF coaxes was a plausible root-cause of at least some of the SSPA failures, based on (1) the statistically significant elevated anomaly rate correlated with high 14-day and 21-day fluences preceding the anomalies; (2) the DICTAT analysis that predicts hazardous charging of a coax over a wide range of possible bulk resistivities and operating temperatures; (3) a demonstration by electron beam exposure that a representative coax (RG-141) isolated by a representative DC blocking circuit would discharge; and (4) the discharge peak voltage and energy were more than sufficient to damage RF FETs. The investigators pointed out that to make the case stronger would require a much more detailed analysis using the specific coax type and length connected to each SSPA, exposed to the specific flux inside the shielding provided by the specific satellite configuration, using the external electron flux environment accumulated following the unique launch dates of the eight satellites. This level of detail makes both the analysis and ground test prohibitively expensive. Because internal charging was found to be a credible explanation of at least some of the SSPA failures, the authors suggested that designing out the risk of those failures (by adopting the recommendations of Payan et al., 2005a, to incorporate a shunt resistance to prevent charging of the coax and blocking cap) was more practical than undertaking the expensive tests and analyses to conclusively determine if charging and discharging was responsible for the failures. This is a frequent result of anomaly investigations: once a root cause is determined credible, satellite manufacturers deem it more cost effective to incorporate design changes to mitigate the problem than to continue to acquire more detailed and definitive proof that it is the precise root cause. Subsequent in-orbit operating experience with the improved designs is the ultimate demonstration of the effectiveness of those changes.

2.3. PATH FORWARD

This historical review shows that both surface charging by external plasmas in the energy range of 1–100 keV and deep internal charging by energetic electrons of 100 keV to 1 MeV or more have caused anomalies on satellites. Identifying the cause of the anomalies required both appropriate space environment data as well as engineering analysis tools and ground tests to establish the level of charging involved and the potential hazard posed by the subsequent discharges. Once the hazards were understood, appropriate engineering solutions were incorporated in future designs and promulgated in industry-wide design guidelines, which included worst-case environments to support those analyses and tests.

A recurring theme that emerges from this review is that unintended vulnerabilities appeared in spacecraft designs as a byproduct of technological advances. Even modest design changes that were adopted because of their very desirable benefits (lower power, cost, weight, etc.) have uncovered previously unknown upset and failure thresholds associated with charging. This places a great burden upon engineers that have to apply space environments to spacecraft designs to more precisely define when the risk of anomalies and failures becomes credible and unacceptable. Despite significant progress in our knowledge of space environments and their interactions with spacecraft, we still are unable to fully answer the challenge posed by Dr. Rosen 40 years ago (Rosen, 1978): to accurately assign a hazard level of 1 to 10 for specific satellite designs so appropriate engineering and financial decisions can be made. The inability of the current state of the art to fully explain the diversity of the observed in-orbit anomaly and failure rates shows there are still gaps in our understanding of charging and discharging, and the impacts they have on satellites.

A comprehensive spacecraft charging road map has recently been developed, outlining the gaps in our knowledge and setting priorities for future efforts of the space physics and spacecraft engineering communities (Likar et al., 2018). The surface charging community identified needs in the following areas:

1. Design and effects tools that go beyond predictions of charging to estimation of effects, such as breakdown thresholds, ESD characteristics (rate, magnitude), and coupling to cables/harnesses and RF systems

2. Quick turnaround anomaly analysis tools, including environment severity definition in terms of statistical occurrence percentile and a hazard index based on observed historical effects and anomalies

3. In-situ observations of the low-energy plasma environment to improve environment data sets and to improve now-casting and forecasting models. In-situ observations of the EMI produced by discharges are also required for a wider range of materials and satellite configurations than the few historical examples cited herein, which date back decades.

4. Studies relating preflight effects estimates and on-orbit performance to benchmark those estimates and establish "error-bars" or margin requirements commensurate with the limitations of the predictions

5. Models that define environments in formats immediately useable for engineering analysis tools and ground tests, as opposed to the formats/units in common use by the space science community.

Similarly, the internal charging community identified the following needs:

1. Better environment models, such as real-time and long-term assimilative environment models, and products derived from them such as integral fluxes and charge accumulation. In addition to worst-case spectra, models that define percentiles based on occurrence rates and durations are needed.

2. Better design and effects tools. There is a significant gap between knowledge of the environment and the

ability to predict internal charging effects which has led to inconsistent design specifications across the satellite industry. A significant part of this gap is due to incomplete definition of the material properties that control the charging process, and the test protocols that quantify those properties and their dependencies on temperature, dose and dose rate, and other aging factors. A centralized, standardized database for the data would significantly improve charging assessments.

3. Quick-look anomaly analysis tools. Most anomaly investigations start with plots of fluxes or intensity indices but quickly transition to analyses based upon particle fluxes. Environment data in reduced engineering units is needed (e.g. omni-direction flux energy spectra, differential and integral, versus the full pitch angle distribution). Availability of the data and its formats continue to change as organizational changes occur, making it challenging to find the data when they are needed, or worse, the needed data are no longer produced.

4. In-situ observations of the environment. This includes further reprocessing of existing data sets in GEO and LEO, and expansion of data sets in MEO by future satellites with instruments and by retrospective mapping of data from other orbits to MEO. Reestablishing and then sustaining an anomaly database has also been identified as a critical need.

These lists show that the collaboration between the satellite industry and the space science community that began in the 1960s during the development of scientific satellites, and that waxed and waned through subsequent decades (as old anomalies disappeared and new ones appeared), looks to continue in the decades ahead.

Continuing this collaboration is critical to protect society's technological infrastructure. A workshop on the potential effects of severe space weather conducted by the National Research Council, Space Studies Board in 2008, examined the potentially catastrophic consequences of an extreme space weather event on terrestrial and space technical systems. As the workshop report noted, extreme space weather events have been observed in the past that have resulted in significant disruptions to the technologies of the day (e.g. telegraph networks in the 1860s, British radar in 1942, electric power stations and distribution in the northeastern US and Canada in 1989). Extrapolating from the history of satellite anomalies and failures caused by more frequent and much less severe environments, it is possible that large numbers of satellites could be partially or totally damaged, with dire consequences to society, by an extreme one in 100-year or 300-year event. We need to be able to credibly estimate the characteristics of extreme events and then accurately estimate their potential impacts on satellites in order to understand the cost-benefit trade-offs of more robust satellite designs.

REFERENCES

Adamec, V., & Calderwood, J. (1978). Electrical conduction and polarization phenomena in polymeric dielectrics at low fields. *J. Phys D: Appl. Phys*, *11*, 781–800.

Adamo, R. C., & Nanevicz, J. E. (1981). Preliminary comparison of material charging properties using single-energy and multi-energy electron beams. *Spacecraft Charging Technology—1980* (NASA CP-2182 and AFGL-TR-81-0279, pp. 129–132).

AIAA-S-111 (2005). Solar cell electrostatic discharge sensitivity (ESDS) test. In *Qualification and Quality Requirements for Space Solar Cells* (section 8.9). Reston, VA: American Institute of Aeronautics and Astronautics. [The currently active version is AIAA-S-111A-2014.]

AIAA-S-112 (2005). *Qualification and Quality Requirements for Space Solar Panels*. Reston, VA: American Institute of Aeronautics and Astronautics. [The currently active version is AIAA-S-112A-2013, which mandates ESD testing per ISO 11221.]

Akahoshi, Y., Nakamura, T., Fukushige, S., Furusawa, N., Kusunoki, S., Machida, Y., et al. (2008). Influence of space debris imp. The use of environmental data to predict and analyses spacecraft anomalies, act on solar array under power generation. *Int. J. Impact Eng.*, *35*(12), 1678–1682.

Andersson, L., Norberg, O., Eliasson, L., & Wintoft, P. (2000). The use of environmental data to predict and analyse spacecraft anomalies. *6th Spacecraft Charging Tech. Conf.* (AFRL-VS-TR-20001578, pp. 231–235).

Anselmo, J. C. (1997). Solar storm eyed as satellite killer. *Aviation Week Space Tech.*, *146*, 61–62.

ASTM (1991). Factors affecting insulation resistance or conductance measurements: Temperature. In *Standard Test Methods for DC Resistance or Conductance of Insulating Materials* (ASTM D 257, sec. X1.2, p. 11). 100 Barr Harbor Drive, West Conshohocken, PA 19428: ASTM International. [The currently active revision of this standard is ASTM D-257-14.]

Baker, D. N. (2000). The occurrence of operational anomalies in spacecraft and their relationship to space weather. *IEEE Trans. Plasma Sci.*, *28*(6), 2007–2016.

Baker, D. N., Allen, J. H., Kanekal, S. G. & Reeves, G. D. (1998). Space environment conditions during April and May 1998: An indicator for the upcoming solar maximum. *EOS*, *79*(40), 477, 482–483.

Baker, D. N., Belian, R. D., Higbie, P. R., Klebesadel, R. W., & Blake, J. B. (1987). Deep dielectric charging effects due to high-energy electrons in Earth's outer magnetosphere. *J. Electrostatics*, *20*, 3–19.

Baker, D. N., Kanekal, S., Blake, J. B., Klecker, B., & Rostoker, G. (1994). Satellite anomalies linked to electron increase in the magnetosphere. *EOS*, *75*(35), 401, 404, 405.

Balcewicz, P. T., Bodeau, J. M., Frey M. A., Leung, P. L., & Mikkelson, E. J. (2000). Environment on-orbit anomaly correlation efforts at Hughes. *6th Spacecraft Charging Technology Conference—1998* (AFRL-VS-TR-20001578).

Balmain, K. G., & Hirt, W. (1981). Dielectric surface discharges: Effects of combined low-energy and high-energy incident electrons. *Spacecraft Charging Technology—1980* (NASA CP-2182 and AFGL-TR-81-0279, pp. 115–128).

Besse, A. L., Rubin, A. G., & Hardy, D. A. (1985). Charging of DMSP/F6 spacecraft in aurora on 10 January 1983, *Spacecraft Environmental Interactions Technology—1983* (NASA CP-2359 and AFGL-TR-85-0018, pp. 125–130).

Bodeau, J. M. (1999). Application of space weather data to a recent anomaly investigation, *Presentation at 1999 NOAA SEC Space Weather Workshop*, 19–23 April, Boulder, CO.

Bodeau, M. (2007a). An extreme case of triggered ESD in GEO. *Proceedings of the 10th Spacecraft Charging Technology Conference*, 18–21 June 2007, Biarritz, France.

Bodeau, M. (2007b). Killer electrons from the angry Sun did not stop the pagers. *Space Weather*, *5*, S03006. doi:10.1029/2006SW000266

Bodeau, M. (2007c). Satellite industry requires accurate and timely space environmental data. Presentation to the 2007 Space Weather Enterprise Forum, 4–5 April, Washington, D.C.

Bodeau, M. (2010). High energy electron climatology that supports deep charging risk assessment in GEO (AIAA 2010-1608). *48th AIAA Aerospace Sciences Meeting Including the New Horizons Forum and Aerospace Exposition*, 4–7 January, Orlando, FL.

Bodeau, M. (2012). Current and voltage thresholds for sustained arcs in power systems. *IEEE Trans. Plasma Sci.*, *40*(2), 192–200. doi: 10.1109/TPS.2011.2176932

Bodeau, J. M. (2013). Test approach to assess solar array diode board susceptibility to sustained arcing. *J. Space Rockets*, *50*(6), 1277–1287. doi: 10.2514/1.A32442

Bodeau, M. (2014). Updated current and voltage thresholds for sustained arcs in power systems. *IEEE Trans. Plasma Sci.*, *42*(7), 1917–1921.

Bodeau, M. (2015a). Observation of sustained arc circuit failure on solar array backside in Low-Earth-Orbit. *IEEE Trans. Plasma Sci.*, *43*(9), 2961–2974. doi: 10.1109/TPS.2015.2407866

Bodeau, M. (2015b). Review of better space weather proxies for spacecraft surface charging. *IEEE Trans. Plasma Sci.*, *43*(9), 3075–3085. doi: 10.1109/TPS.2015.2441038

Bodeau, J. M., Elkman, W. R., Diaz, J. F., Hansen, D. P., Kirsten, C., & Stevens, N. J. (1989). The effect of spacecraft charging transient in GEO satellites. *Proc. 24th Intersociety Energy Conversion Conf IECEC-89*, *1*, 375–383.

BS ISO (2003). Immunity to electrostatic discharge. In *Space Systems – Electromagnetic Compatibility Requirements* (sections 5.3.17 and A.3.2.8, BS ISO 14302, 19 & 37–38).

Catani, J.-P. (2001). Electrostatic discharges & spacecraft anomalies. *Proceedings of the 7th Spacecraft Charging Technology Conference* (ESA SP-476).

Chaky, R. C., & Inouye, G. T. (1985), Characterization of EMI generated by negative metal/positive dielectric voltage stresses due to spacecraft charging. *Spacecraft Environmental Interactions Technology—1983* (NASA CP-2359 and AFGL-TR-85-0018, pp. 437–452).

Clark, D. M., & Hall, D. F. (1985). Flight evidence of spacecraft surface contamination rate enhancement by spacecraft charging obtained with a quartz crystal microbalance. *Spacecraft Environmental Interactions Technology—1983* (NASA CP-2359 and AFGL-TR-85-0018, pp. 493–508).

Coakley, P. G., Treadaway, M. J., & Robinson, P. A. (1985). Low flux laboratory testing of the IDM experiment intended for CRRES. *IEEE Trans. Nuc. Sci.*, *39*(6), 4066–4072.

Coakley, P., Treadaway, M., Wild, N., & Kitterer, B. (1985). Discharge characteristics of dielectric materials examined in mono-, dual-, and spectral energy electron charging environments. *Spacecraft Environmental Interactions Technology—1983* (NASA CP-2359 and AFGL-TR-85-0018, pp. 511–524).

Coates, A. J., Johnstone, A. D., Rogers, D. J., & Wrenn, G. L. (1989). Quest for the source of Meteosat anomalies. *Proc. (5th) Spacecraft Charging Technology Conference—1989* (PL-TR-93-2027 (I), 120–146).

Cooke, D. L., Gussenhoven, M. S., Hardy, D. A., Tautz, M., Katz, I., Jongeward, G., & Lilley Jr., J. R. (1993). Polar code simulation of DMSP satellite auroral charging. *Proceedings of the (5th) Spacecraft Charging Technology Conference—1989* (PL-TR-93-2027(I), pp. 194–203).

Damron, S. A., Adamo, R. C., & Nanevicz, J. E. (1981). Preliminary analysis of data from SRI Transient Pulse Monitor on board P78-2 SCATHA satellite. *Spacecraft Charging Technology—1980* (NASA CP-2182 and AFGL-TR-81-0279, pp. 470–477).

Daughtridge, S. C., Garrett, H. B., & Whittlesey, A. C. (1991). Environment-induced anomalies on the TDRS and the role of spacecraft charging. *J. Spacecraft and Rockets*, *28*(3), 324–329.

DeForest, S. E. (1972). Spacecraft charging at synchronous orbits. *JGR*, *77*, 651–659.

Dennison, J. R., Brunson, J., Swaminathan, P., Green, N. W., & Frederickson, A. R. (2006). Methods for high resistivity measurements related to spacecraft charging. *IEEE Trans. Plasma Sci.*, *35*(5), 2191–2203.

Evans, R., Garrett, H., Gabriel, S., & Whittlesey, A. (1993). A preliminary spacecraft charging map for the near-Earth environment. *Proceedings of the (5th) Spacecraft Charging Technology Conference—1989* (PL-TR-93-2027(I), pp. 615–621).

Fennel, J. F., Crowley Jr., D. R., Mizera, P. F. & Richardson, J. D. (1981). Electron angular distributions during charging events. *Spacecraft Charging Technology—1980* (NASA CP-2182 and AFGL-TR-81-0279, pp. 370–385).

Ferguson, D. C. (1993). SAMPIE – A Shuttle-based solar array arcing experiment. *Proceedings of the (5th) Spacecraft Charging Technology Conference—1989* (PL-TR-93-2027(I), pp. 346–361).

Fergusson, D. C., & Hillard, G. B. (2011). Spacecraft charging, arcing, and sustained arcs in Low-Earth-Orbit. In S.T. Lai (Ed.), *Spacecraft Charging, Progress in Astronautics and Aeronautics* (vol. 237, ch.3). Reston, VA: Am. Inst. of Astronautics & Aeronautics.

Finke, R. C., & Pike, C. P. (1978). *Proceedings of the 2nd Spacecraft Charging Technology Conference*, held at the U.S. Air Force Academy, 31 Oct. to 2 Nov. 1978.

Fredericks, R. W., & Kendall, D. E. (1973). Geomagnetic substorm charging effects on Defense Satellite Communication System—Phase II. *TRW Report 09070-7032-RU-00*.

Fredricks, R. W. & Scarf, F. L. (1973). Observations of spacecraft charging effects in energetic plasma regions. In R.J.L. Grard (Ed.), *Photon & particle interactions with surfaces in space* (pp. 277–308). Dordrecht, Holland: D. Reidel. [This book captured the proceedings of the 6th ESLAB Symposium, held 26–29 September, 1972, Noordwijk, The Netherlands.]

Frederickson, A. R. (1985). Discharge pulse phenomenology. *Spacecraft Environmental Interactions Technology—1983* (NASA CP-2359 and AFGL-TR-85-0018, pp. 483–509).

Frederickson, A. R., Benson, C. E., & Bockman, J. F. (2003). Measurement of charge storage and leakage in polyimides. *Nuclear Instrum. Methods, Phys. Res. B*. Elsevier.

Frederickson, A. R., & Dennison, J. R. (2003). Measurement of conductivity and charge storage in insulators related to spacecraft charging. *IEEE Trans. Nucl. Sci.*, *50*(6), 2284–2291.

Frederickson, A. R., Holeman, E. G., & Mullen, E. G. (1992). Characteristics of spontaneous electrical discharging of various insulators in space radiations *IEEE Trans. Nuc. Sci.*, *39*(6), 1773–1782.

Frederickson, A. R., Mullen, E. G., Brautigam, D. H., Kerns, K. J., Robinson Jr., P. A., & Holeman, E. G. (1991). Radiation-induced insulator discharge pulses in the CRRES Internal Discharge Monitor satellite experiment. *IEEE Trans. Nuc. Sci.*, *38*(6), 1614–1621.

Frederickson, A. R., Mullen, E. G., Kerns, K. J., Robinson Jr. P. A., & Holeman, E. G. (1993). The CRRES IDM spacecraft experiment for insulator discharge pulses. *IEEE Trans. Nuc. Sci.*, *40(3)*, 233–241.

Frost & Sullivan (2004. Commercial communications satellite bus reliability analysis. http://www.satelliteonthenet.co.uk/white/frost3.htm

Garrett, H. B. (1981). The charging of spacecraft surfaces. *Rev. Geophys. and Space Phys.*, *19*(4), 577–616.

Garrett, H. B., Schwank, D. C., Higbie, P. R., & Baker, D. N. (1980). Comparison between the 30 to 80 keV electron channels on ATS-6 and 1976-059A during conjunction and application to spacecraft charging prediction. *JGR*, *83*(A3), 1155–1162.

Garrett, H. B., & Whittlesey, A. C. (2012). *Guide to mitigating spacecraft charging effects*. Hoboken, NJ: John Wiley & Sons. ISBN 978-1-118-18645-9.

Hanlon, M. (1994). Sun's electrons beaned Aniks 1 and 2-Telesat team working to make crippled satellite correct its own orbit by summer. *Toronto Star, Final Edition*, 29 Jan.

Hilgers, A., Grystad, D., Andersson, L., & Wu, J.-G. (2000). Meteosat anomalies and time varying plasma conditions. *6th Spacecraft Charging Tech. Conf.* (AFRL-VS-TR-20001578, pp. 69–72).

Hiriart, T., & Saleh, J. H. (2009). Cyclicality in the space industry: Myth or reality. AIAA 2009-6478, *AIAA Space 2009 Conference and Exposition*, Pasadena, CA, Sept 2009. http://www.ascendworldwide.com/spacetrak.aspx

Hoeber, C. F., Robertson, E. A., Katz, I., Davis, V. A., & Snyder, D. B. (1998). Solar array augmented electrostatic discharge in GEO, *AIAA 98-1401. Proceedings of the 17th AIAA International Communications Satellite Systems Conference and Exhibit, International Communication Satellite Systems Conferences, ICSSC*, Yokohama, Japan. doi: 10.2514/6 1998-1401

Hoge, D. G. (1981). METEOSAT spacecraft charging investigation. *Spacecraft Charging Technology—1980* (NASA CP-2182 and AFGL-TR-81-0279, pp. 814–834).

Holman, A. B. (1981). Military standard for spacecraft charging status report. *Spacecraft Charging Technology—1980* (NASA CP-2182 and AFGL-TR-81-0279, pp. 772–788).

Hones, E. W., Thomsen, M. F., Reeves, G. D., Weiss, L. A., McComas, D. J., & Newell, P. T. (1996). Observational determination of magnetic connectivity of the geosynchronous region of the magnetosphere to the auroral oval. *J. Geophys. Res., Space Phys.*, *101*(A2), 2629–2640.

Inouye, G. T. (1997). Spacecraft charging anomalies on the DSCS II, launch 2 satellites. In C. P. Pike & R. R. Lovell (Eds.), *Proc. (1st) Spacecraft Charging Technology Conference* (AFGL-TR-77-0051, NASA-TMX-7337, pp. 829–852).

INTELLSAT (1994). INTELLSAT Satellites weather severe solar storm. *Pr Neswire (PRNW)*, 26 Jan.

ISO 11221 (2011). *Space Systems—Space Solar Panels—Spacecraft Charging Induced Electrostatic Discharge Test Methods*. Geneva, Switzerland.

Johnson, R. G. (1981). Review of hot plasma composition near geosynchronous altitude. *Spacecraft Charging Technology—1980* (NASA CP-2182 and AFGL-TR-81-0279, pp. 412–432).

Johnson, R. G., Strangeway, R., Kaye, S., Sharp, R., & Shelly, E. (1981). SCATHA observations of space plasma composition during a spacecraft charging event. *Spacecraft Charging Technology—1980* (NASA CP-2182 and AFGL-TR-81-0279, pp. 433–438).

Katz, I., Cassidy, J. J., Mandell, M. J., Schnuelle, G. W., Steen, P. G., & Roche, J. C. (1978). The capabilities of the NASA charging analyzer program. In R.C. Finke & C. P. Pike (Eds.), *Proceedings of the 2nd Spacecraft Charging Technology Conference* (pp. 101–122), held at the U.S. Air force Academy, 31 Oct. to 2 Nov. 1978.

Katz, I., Davis, V. A., & Snyder, D. B. (1998). Mechanism for spacecraft charging initiated destruction of solar arrays in GEO. *36th AIAA Aerospace Sciences Meeting and Exhibit*, Reno, NV, USA.

Kendal, B.R.F. (1985). Mass spectra of neutral particles released during electrical breakdown of thin polymer films. *Spacecraft Environmental Interactions Technology—1983* (NASA CP-2359 and AFGL-TR-85-0018, pp. 525–535).

Koons, H. C. (1981). Aspect dependence and frequency spectrum of electrical discharges on the P78-2 (SCATHA) satellite, *Spacecraft Charging Technology—1980* (NASA CP-2182 and AFGL-TR-81-0279, pp. 478–492).

Koons, H. C., Fennel, J. F., & Hall, D. F. (2000). A summary of the engineering results from the Aerospace Corporation experiments on the SCATHA spacecraft. *6th Spacecraft Charging Technology Conference* (AFRL-VS-TR-20001578).

Koons, H. C., & Gorney, D. J. (1992). The relationship between electrostatic discharges on spacecraft P78-2 and the electron environment. *Aerospace Corp. report TR-091 (6940-06)-2*, released as SMC-TR-92-35, 15 March, DTIC accession number AD-A255 179 (approved for public release). [Similar material is contained in a paper by these authors with the same title in *J. Spacecraft*, *28*(6), 683–688].

Kuck, G. (1978). Comments recorded during panel discussion at end of conference. In R. C. Finke & C. P. Pike (Eds.), *Proceedings of the 2nd Spacecraft Charging Technology*

Conference, held at the U.S. Air Force Academy, 31 Oct. to 2 Nov. 78.

Lafferty, J. M. (1980). *Vacuum arcs: Theory and applications.* New York: Wiley Interscience.

Laframboise, J. G., & Luc, J. (1993). High-voltage auroral zone charging of large dielectric spacecraft: A wake induced barrier effect mechanism. *Proceedings of the (5th) Spacecraft Charging Technology Conference—1989* (PL-TR-93-2027(I), p. 218) [cites the original paper by these authors of the same title in *J. Geophys. Res.*, *94*, 9033 (1989)].

Lai, S. T. (2011). Spacecraft Charging, Arcing, and Sustained Arcs in Low Earth Orbit, *Spacecraft Charging, AIAA, pp* 53.

Lam, H.-L., Boteler, D. H., Burlton, B., & Evans, J. (2012). Anik-E1 and E2 satellite failures of January 1994 revisited. *Space Weather, 10*, S10003. doi:10.129/2012SW000811

Lanzerotti, L. J., LaFleur, K., Maclennan, C. G., & Maurer, D. W. (1998). Geosynchronous spacecraft charging in January 1997. *Geophys. Res. Letters*, *25*(15), 2967–2970.

Lanzerotti, L. J., Roberts, C. S., & Brown, W. L. (1967). Temporal variations in the electron flux at synchronous altitudes. *JGR*, *72*(23), 5893–5902.

Leach & Alexander (1995). *Failures and anomalies attributed to spacecraft charging* (NASA RP-1375).

Leggett, P. (1993). Olympus: To the edge and back (ESA WPP-054). *Proc. Europ. Space Power Conf.*, Graz, Austria, 23–27 August.

Leung, P., Scott, J., Seki, S., & Schwartz, J. A. (2010). Arcing on space solar arrays. *Proc. 11th Spacecraft Charging Technol. Conf.*, Albuquerque, NM, USA, pp. 1–26.

Leung, P., Whittlesey, A. C., Garrett, H. B., & Robinson Jr., P. A. (1986). Environment-induced electrostatic discharges as the cause of Voyager 1 power-on resets. *J. Spacecraft*, *23*(3), 323–330.

Levy, L., Mandeville, J. C., Siguer, J. M., Rulet, R., Sarrail, D., Catani, J. P., & Gerlach, L. (1997). Simulation of in-flight ESD anomalies triggered by photoemission, micrometeoroid impact and pressure pulse. *IEEE NS*, *44*(6), 2201–2208.

Likar, J., Rodriguez, J., O'Brien, P., & Daly, E. (2018). Improved spacecraft charging roadmaps developed to support user community needs. *IEEE Trans. Plasma Sci.*, submitted to special issue on spacecraft charging technology.

Lohmeyer, W. Q., & Cahoy, K. (2013). Space weather radiation effects on geostationary satellite solid-state power amplifiers, *Space Weather*, *11*, 476–488. doi:10.1002/swe.20071

Lohmeyer, W. Q., Carlton, A., Wong, F., Bodeau, M., Kennedy, A., & Cahoy, K. (2015). Response of geostationary communications satellite solid-state power amplifiers to high-energy electron fluence. *Space Weather*, *13*, 298–315. doi:10.1002/2014SW001147

Love, D. P., Toomb, D. S., Wilkinson, D. C., & Parkinson, J. B. (2000). Penetrating electron fluctuations associated with GEO spacecraft anomalies. *IEEE Trans. Plasma Sci.*, *28*(6), 2075–2084.

Maejima, H., Kaweakita, S., Kusawake, H., Takahashi, M., Goka, T., Kurosaki, T., et al. (2004). Investigation of power system failure of a LEO satellite (AIAA 2004-5657). *Proc. 2nd Int. Energy Convers. Eng. Conf. (IECEC),* Providence, RI, USA, pp. 1–6. doi: 10.2514/6.2004-5657.

Mallon, C. E., Treadaway, M. J., Flanagan, T. M., Wenaas, E. P, & Denson, R. (1979). High-energy electron-induced discharges and SGEMP modifications in cables. *ECEMP Phenomenology Memo No. 29,* AFWL.

Martin, D. F., & Mallott, K. D. (1983). Space Station Freedom solar array panels plasma interaction test facility. *Proceedings of the (5th) Spacecraft Charging Technology Conference-1989* (PL-TR-93-2027(I), pp. 334–345).

Massaro, M. (1978). Comments recorded during panel discussion at end of conference. In R.C. Finke & C. P. Pike (Eds.), *Proceedings of the 2nd Spacecraft Charging Technology Conference*, 31 Oct. to 2 Nov. 1978.

Meyer, R. A., Bouquet, F. L., & Alger, R. S. (1956). Radiation induced conductivity in polyethylene and Teflon. *J. Appl. Phys.*, *27*(9), 1012–1018.

MIL-STD-1541A (1987). Electrostatic arc discharge susceptibility, section 6.7.2. In *Military Standard—Electromagnetic Compatibility Requirements for Space Systems* (pp. 25–26).

Mizera, P. F., & Boyd, G. M. (1981). Satellite potential survey, *Spacecraft Charging Technology—1980* (NASA CP-2182 and AFGL-TR-81-0279, pp. 461–469).

Nanevicz, J. E., Adamo, R. C., & Beers, B. L. (1978). Characterization of electromagnetic signals generated by electrical breakdown of spacecraft insulating materials. *Proceedings of the 2nd Spacecraft Charging Technology Conference*, pp. 868–875.

Nanevicz, J. E., Adamo, R. C., Beers, B. L., & Delmer, T. N. (1981). Electromagnetic fields produced by simulated spacecraft discharges. *Spacecraft Charging Technology—1980* (NASA CP-2182 and AFGL-TR-81-0279, pp. 894–911).

Napoli, J., & Seliga, J. (1977). RCA Satcom 3-axis spacecraft experience at geosynchronous altitude. *Proceedings of the (1st) Spacecraft Charging Technology Conference* (AFRL-TR-77-0051 and NASA TMX-73537, pp. 865–871).

NASA (1999). *Avoiding problems caused by spacecraft on-orbit internal charging effects* (NASA HDBK 4002).

NASA (2007). *Low Earth orbit spacecraft charging design handbook* (NASA-HDBK-4006).

Nat. Res. Council (2008), *Severe space weather events—Understanding societal and economic impacts workshop report*. National Academy of Sciences, http://www.nap.edu/catalog/12507.html

Odenwald, S. F. (2001). They call them satellite anomalies. In *The 23rd cycle: Learning to live with a stormy star*. New York: Columbia University Press, ISBN: 9780231120784.

Olsen, R. C. (Ed.) (1993). *Proceedings of the (5th) Spacecraft Charging Technology Conference—1989* (PL-TR-93-2027(I)), 31 Oct. to 3 Nov. 1989.

Payan, D., Reulet, R., & Dirassen, B. (2005a). Electrostatic behavior of dielectrics under GEO-like charging space environment simulated in laboratory. *9th Spacecraft Charging Technology Conference* (JAXA-SP-05-001E), Tsukuba, Japan.

Payan, D., Reulet, R., & Dirassen, B. (2005b), Charging of coaxial lines with floating core at geosynchronous altitudes. *9th Spacecraft Charging Technology Conference* (JAXA-SP-05-001E), Tsukuba, Japan.

Pike, C. P., & Bunn, M. H. (1976). A correlation study relating spacecraft anomalies to environmental data. In In A. Rosen

(Ed.), *Spacecraft charging by magnetospheric plasmas* (vol. 47, pp. 45–60), New York: AIAA Press.

Pike, C. P., & Lovell, R. R., editors (1977). *Proceedings of the (1st) Spacecraft Charging Technology Conference* (AFRL-TR-77-0051 and NASA TMX-73537), 27–29 Oct 1976.

Pike, C. P., Purvis, C. K., & Hudson, W. R. (1985). Spacecraft environmental interactions: A joint Air Force and NASA research and technology program. *Spacecraft Environmental Interactions Technology—1983* (NASA CP-2359 and AFGL-TR-85-0018), pp. 599–608.

Pilipenko, V., Yagoya, N., Romanova, N., & Allen, J. (2006). Statistical relationships between satellite anomalies at geosynchronous orbit and high energy particles. *Adv. Space Res.*, *37*, 1192–1205.

Purvis, C. K., Garrett, H. B., Whittlesey, A. C., & Stevens, N. J. (1985). Design guidelines for assessing and controlling spacecraft charging effects. *Spacecraft Environmental Interactions Technology—1983* (NASA CP-2359 and AFGL-TR-85-0018), p. 389.

Purvis, C. K., & Pike, C. P. (Eds.) (1985). *Spacecraft Environmental Interactions Technology—1983* (NASA CP-2359 and AFGL-TR-85-0018). Proceedings of the 4th Spacecraft Charging Technology Conference held at the U.S. Air Force Academy, 4–6 October 1983.

Qu, D., Tomoki, K., Satoshi, H., Jeongho, K., & Mengu, C. (2005). Laboratory test of sustained arc between back surface of solar array panel and power cable (IAC-05-E2.P.10). *Proc. 56th Int. Aeronautical Congr.*, pp. 6275–6280.

Raine, H. R. (1978). The Communication Technology Satellite flight performance. *Acta Astron.*, *5*(5-6), 343–368.

Reasoner, D. L., Lennartsson, W., & Chappell, C. R. (1976). Relationship between ATS-6 spacecraft-charging occurrences and warm plasma encounters. In A. Rosen (Ed.), *Spacecraft charging by magnetospheric plasmas* (vol. 47, pp. 89–101), New York: AIAA Press.

Reddy, J. (1981). Electron irradiation tests on European meteorological satellite. *Spacecraft Charging Technology—1980* (NASA CP-2182 and AFGL-TR-81-0279, pp. 835–855).

Robbins, A., & Short, C. D., (1977). Space environmental effects on the SKYNET 2B spacecraft. *Proceedings of the (1st) Spacecraft Charging Technology Conference* (AFRL-TR-77-0051 and NASA TMX-73537, pp. 852–863).

Rodgers, D. J. (2004). DICTAT software: Users' manual v3.0. Hampshire, U. K.: European Space Agency.

Rosen, A. (1975). Spacecraft charging: Environment induced anomalies (AIAA Paper 75-91). *AIAA 13th Aerospace Sciences Meeting*, Pasadena CA.

Rosen, A. (1976). *Spacecraft charging by magnetospheric plasmas* (Progress in Astronautics and Aeronautics, vol. 47), New York: AIAA Press. This volume contains papers selected from the joint AIAA/AGU Symposium on Spacecraft Charging by Magnetospheric Plasma, June 1975.

Rosen, A. (1978). Comments recorded during panel discussion at end of conference. In R.C. Finke & C. P. Pike (Eds.), *Proceedings of the 2nd Spacecraft Charging Technology Conference*, held at the U.S. Air Force Academy, 31 Oct. to 2 Nov. 1978.

Saflekos, N. A., Tautz, M. F., Rubin, A. G., Hardy, D. A., Mizera, P. F., & Feynman, J. (1981). Three-dimensional analysis of charging events on days 87 and 114, 1979, from SCATHA. *Spacecraft Charging Technology—1980* (NASA CP-2182 and AFGL-TR-81-0279, pp. 608–631).

Schnuelle, G. W., Stannard, P. R., Katz, I., & Mandell, M. J. (1981). Simulation of charging response of SCATHA (P78-2) satellite. *Spacecraft Charging Technology—1980* (NASA CP-2182 and AFGL-TR-81-0279, pp. 580–591).

Sevoz, M., Pelissou, P., Gerhard, A., & Kersting, S. (2012). EUROSTAR E3000 solar array connectors robustness to ESD-induced sustained arcing. *Proc. 12th Spacecraft Charging Technol. Conf.*, Kitakyushu, Japan.

Shaw, R. R, Nanevicz, J. E., & Adamo, R. C. (1976). Observations of electrical discharging caused by differential satellite charging. In A. Rosen (Ed.), *Spacecraft charging by magnetospheric plasmas* (vol. 47, pp. 61–76), New York: AIAA Press.

Shoga, M. P., Adams, P., Chenette, D. L., Koga, R., & Smith, E. C. (1987) Verification of single event upset rate estimation models with on-orbit observations, *IEEE NS*, *34*(6).

Soubeyran, A., Levy, L., & Coggiola, E. (1993). Numerical simulation of the wake of non-equipotential spacecrafts in the ionosphere. *Proceedings of the (5th) Spacecraft Charging Technology Conference—1989* (PL-TR-93-2027(I), pp. 204–217).

Stevens, N. J. (1981). Use of charging control guidelines for geosynchronous satellite design studies. *Spacecraft Charging Technology—1980* (NASA CP-2182 and AFGL-TR-81-0279, pp. 789–801).

Stevens, N. J., & Pike, C. P. (Eds.) (1981). *Spacecraft Charging Technology—1980* (NASA CP-2182 and AFGL-TR-81-0279), held at U.S. Air Force Academy, 12–14 Nov. 1980.

Treadaway, M., Leadon, R., Mallon, C., Flanagan, T., Denson, R., & Wenaas, E. (1981). Dielectric discharge characteristics in a two-electron simulation environment. *Spacecraft Charging Technology—1980* (NASA CP-2182 and AFGL-TR-81-0279, pp. 4–16).

TRW Space Sciences Dept. (1972). RGA analysis: Findings regarding correlation of satellite anomalies with magnetospheric substorms, and laboratory test results, final report (TRW Report 09670-7020-R0-00). [Similar information is contained in subsequent publications, especially those more readily accessible in the public domain such as Fredericks & Scarf (1973b), Rosen (1975), and Rosen (1976)].

Vampola, A. L. (1986). Thick dielectric charging on high-altitude spacecraft (Report SD-TR-86-48, DTIC accession number AD-A171 078; approved for public release, distribution unlimited). The report was republished as Vampola (1987a), Thick dielectric charging on high-altitude spacecraft. *J. Electrostatics*, *20*, 21–30.

Violet, M. D., & Frederickson, A. R. (1993). Spacecraft anomalies on the CRRES satellite correlated with the environment and insulator samples. *IEEE Trans. Nucl. Sci.*, *40*(6), 1512–1520.

Wenaas, E. P., Treadaway, M. J., Flanagan, T. M., Mallon, C. E., & Denson, R. (1979). High-energy electron-induced discharges in printed circuit boards. *IEEE Trans. Nucl. Science*, NS-26, pp. 5152–5155.

Whittlesey, A. C. (1978). Voyager electrostatic discharge protection program. *IEEE Int'l. Symp. on EMC*, Atlanta, GA, USA, 20–22 June, pp. 377–383.

Wilkinson, D. C. (1993). Trends in spacecraft anomalies. *Proceedings of the (5th) Spacecraft Charging Technology Conference—1989* (PL-TR-93-2027(I), pp. 1–17).

Wilkinson, D. C., Daughtridge, S. C., Stone, J. L., Sauer, H. H., & Darling, P. (1991). TDRS-1 single event upsets and the effect of the space environment. *IEEE Trans. Nuc. Science*, *38*(6).

Williamson, P. R., Banks, P. M., Storey, L.R.O., & Raitt, W. J. (1985). Vehicle charging on STS-3 mission. *Spacecraft Environmental Interactions Technology—1983* (NASA CP-2359 and AFGL-TR-85-0018, pp. 19–32).

Wintle, H. J. (1983). Conduction processes in polymers. In R. Bartinikas & R. M. Eichorn (Eds.), *Engineering dielectrics vol IIA, Electrical properties of solid insulating materials: Molecular structure and electrical behavior* (ASTM STP 783, ch. 3).

Wrenn, G. L. (1995). Conclusive evidence for internal dielectric charging anomalies on geosynchronous communications spacecraft. *J. Space. and Rockets*, *32*(3), 514–520.

Wrenn, G. L., Rodgers, D. J., & Ryden, K. A. (2002). A solar cycle of spacecraft anomalies due to internal charging. *Annales Geophysicae*, *20*, 953–956.

Wrenn, G. L., & Smith, R.J.K. (1996). Probability factors governing ESD effects in geosynchronous orbit. *IEEE Trans. Nuc. Science*, *43*(6), 2783–2789.

3

Effects of Space Radiation on Humans in Space Flight

Lawrence W. Townsend

ABSTRACT

Space radiation effects on human spaceflight include risks to crews from particulate radiation present in the space environment. Sources are all naturally occurring elements present in the galactic cosmic ray environment, charged particles emitted during solar eruptions, and particles trapped in the Earth's magnetic field. Historically, protection afforded by Earth's magnetic field results in low doses to crews in low Earth orbit. However, extended exposure to galactic cosmic rays and solar particle events during potential deep space missions could pose significant risks of developing a fatal cancer. Additionally, infrequent large solar energetic particle events could give radiation doses large enough to manifest acute radiation syndromes in crew members, thereby resulting in radiation sickness and possibly death. This chapter looks at analyses of solar energetic particle events, often referred to as space weather. Most solar events pose little or no health hazards to space crews. But a few during each solar cycle could expose lightly shielded crews to doses sufficient to pose some health risks. Rare extremely large solar energetic events could expose crews to radiation levels that may be mission threatening or life threatening. This chapter presents potential biological effects of solar and galactic cosmic radiation on humans in deep space.

3.1. INTRODUCTION

Within Earth's magnetic field, the radiation belts are composed of protons and electrons with maximum energies lower than those found outside the magnetosphere. Hence, radiation exposures within the belts for missions in low Earth orbit (LEO) are mainly a concern for increased lifetime cancer risks. LEO missions also encounter some exposures to galactic cosmic radiation, mainly at higher orbital latitudes, such as the orbit of the International Space Station (ISS). Recently, measured doses ranged from 44 to 105 mGy, depending upon location within the ISS (https://www.nasa.gov/mission_pages/station/research/experiments/1043.html). Thus, they will not be further considered. For missions beyond Earth's magnetosphere, such as the moon landings during the Apollo era, the crews minimized their exposure by rapidly traversing the radiation belts during their flights to the moon.

Beyond the Earth's magnetosphere, the space radiation environment is a mixture of galactic cosmic radiation (GCR), originating from sources outside the solar system, and solar energetic particles (SEPs), emitted by the Sun into the heliosphere. The former consist of 2% electrons and positrons, and all elemental species (87% protons, 12% helium, and ~1% heavier ions) with kinetic energies ranging up to and exceeding 10s of GeV/nucleon. Of these, for human exposures, particles heavier than nickel (Z = 28) are of little concern since their fluxes are very low. Figure 3.1 displays relative abundances of these ion species at energies of 1 GeV/nucleon. Elemental species with atomic numbers greater than nickel (Z = 28) are not displayed since their abundances are several orders of magnitude smaller than iron (Z =26).

Department of Nuclear Engineering, The University of Tennessee, Knoxville, Tennessee, USA

Space Physics and Aeronomy Collection Volume 5: Space Weather Effects and Applications, Geophysical Monograph 262, First Edition.
Edited by Anthea J. Coster, Philip J. Erickson, and Louis J. Lanzerotti.

DOI: 10.1002/9781119815600.ch3

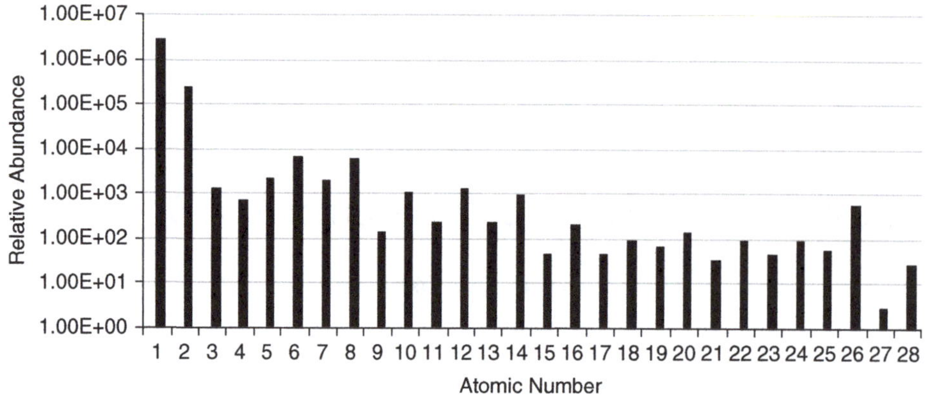

Figure 3.1 Relative abundances of galactic cosmic ray element species at 1 GeV/nucleon, plotted from Table 3.1 of NCRP Report 153 (2006) (Silicon = 1000).

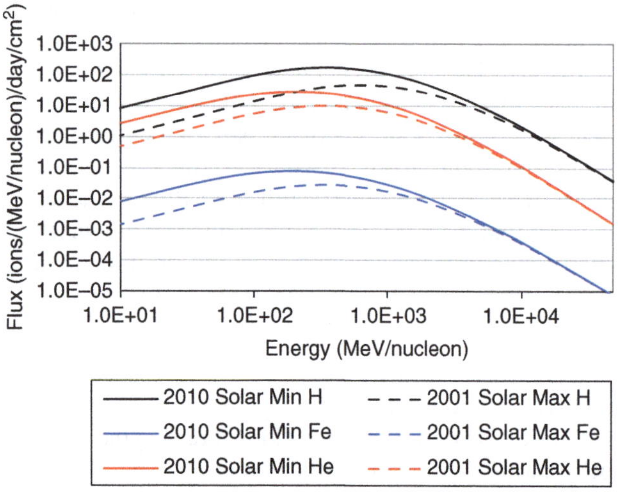

Figure 3.2 Energy spectra for representative GCR elements displaying typical variations between solar maximum and solar minimum periods. (Source: OLTARIS. Reproduced with permission of NASA.)

The intensity of these radiations varies with the solar cycle. During periods of minimum solar activity, the GCR intensity is a maximum. However, during periods of maximum solar activity, the increased modulation created by the increased solar wind intensity reduces the fluxes of the GCR particles at lower energies incident on the heliosphere from deep space. This variation in intensity is depicted in Figure 3.2, obtained using the OLTARIS website (https://oltaris.nasa.gov/) at NASA Langley Research Center (On-Line Tool for the Assessment of Radiation In Space). It shows the fluxes, at solar minimum and solar maximum of several representative ions in the GCR spectrum, as a function of their kinetic energies. Note the flux differences between solar minimum and solar maximum conditions at energies below ~1000 MeV/nucleon due to the modulation provided by the solar wind. At energies above ~1000 MeV/nucleon, the effects of solar modulation on the particle fluxes are much less.

Chronic exposures of crew members to GCR particles, and the secondary particles they produce in nuclear collisions, pose a long-term cancer risk, as well as possible risks to the eye lens and to the central nervous and cardiovascular systems. However, the relatively low fluxes of these particles do not pose any risk to crew members of developing acute radiation syndrome symptoms. As defined by the Centers for Disease Control and Prevention, "Acute Radiation Syndrome (ARS) (sometimes known as radiation toxicity or radiation sickness) is an acute illness caused by irradiation of the entire body (or most of the body) by a high dose of penetrating radiation in a very short period of time" (https://emergency.cdc.gov/radiation/pdf/ars.pdf).

Energetic particles emitted by the Sun consist of electrons, protons, and some heavier ion species. Of these particle types, energetic protons are typically the most abundant, generally comprising over 95% of the incident fluence. Alpha particles are next in abundance, with heavier ions contributing ~1% or less to the event fluence (Cucinotta et al., 1994; Townsend et al., 1994). The health risk to human crew members in deep space is mainly from the protons, except that alpha particles can contribute significantly if the event has a large particle fluence and a hard energy spectrum (here, "hard" refers to a particle spectrum which decreases slowly with increasing particle kinetic energies). Electrons in an SEP event are not a significant source of exposure to crew members.

The following section describes the space weather environment as it affects humans in space. It includes a definition of space weather and an overview of SEP event spectra and their frequency. Relevant radiation units and quantities are discussed in section 3.3. The effects of radiation on humans are presented in some detail in section 3.4. section 3.5 presents radiation limits for

NASA and other international space agencies. Actual and potential SEP radiation exposures in deep space are presented in section 3.6. Finally, a summary and conclusions are presented.

3.2. THE SPACE WEATHER ENVIRONMENT

The major influence on the radiation environment within the heliosphere is the Sun. The solar wind emitted by the Sun, consisting of a continuous plasma stream of charged particles, modulates the fluxes of GCR particles entering the heliosphere from interstellar space. The Sun also emits sudden large bursts of electromagnetic energy in the form of solar flares, which span the entire electromagnetic spectrum from radio waves to gamma rays. From a crew radiation exposure perspective, however, solar flares are not a concern. Sometimes a solar event involves the emission of charged particles, such as protons, electrons, and heavier ions, and is referred to as a solar energetic particle (SEP) event. These sporadic events can be damaging to spacecraft components and may present a significant health risk for their human crews. Space weather has been described in the following way:

> The term **space weather** generally refers to conditions on the sun, in the solar wind, and within Earth's magnetosphere, ionosphere and thermosphere that can influence the performance and reliability of space-borne and ground-based technological systems and can endanger human life or health. (https://www.nasa.gov/mission_pages/rbsp/science/rbsp-spaceweather.html)

Thus, SEP events represent the most significant space weather concern for human missions beyond Earth's magnetosphere. The National Oceanic and Atmospheric Administration Space Weather Prediction Center (NOAA SWPC) defines the start of an event to be the first of three consecutive data points with fluxes greater than or equal to 10 proton flux units (pfu, where proton fluxes > 10 MeV are provided in pfu; 1 pfu = 1 particle cm^{-2} s^{-1} sr^{-1}) (ftp://ftp.swpc.noaa.gov/pub/indices/SPE.txt). The end of an event is the last time the flux was greater than or equal to 10 pfu. The proton fluxes measured by the instruments onboard the GOES (Geostationary Operational Environmental Satellite) platforms are integral values averaged over 5-minute intervals, located in the outer radiation belt, which is mainly composed of electrons. Therefore, the non-SEP proton and heavier ion fluxes measured there are mainly GCR background.

As thoroughly discussed in *Solar Energetic Particles* (Reames, 2017), there are two classes of SEP events: impulsive and gradual. In *impulsive events*, particles are accelerated along open magnetic field lines by magnetic-reconnection events. *Gradual events* involve the acceleration of particles by shock waves expelled from the Sun by coronal mass ejections. These two classes of events typically differ in their duration, particle abundances, isotopic compositions, and electron/proton ratios. As will become evident later in this work, the main radiation protection concerns are large and prolonged particle fluxes from gradual events.

Figure 3.3 displays a typical impulsive event proton spectrum, in this case, for an event beginning 30

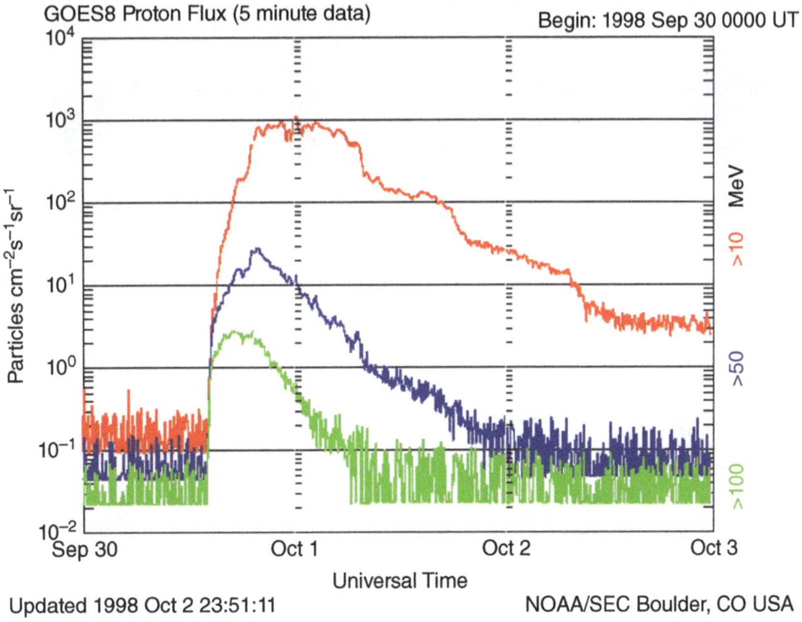

Figure 3.3 Energetic proton spectra for the 30 September 1998 impulsive solar energetic particle event. (Reproduced with permission of NOAA SWPC.)

September 1998. Note the rapid rise as the event commences. This is followed by a rapid decrease in proton flux levels over time. The overall duration of the event is short, especially for the protons with energies >100 MeV, and the integral flux levels are low. As such, these events pose little or no health risk to shielded crews in space.

A typical gradual event proton spectrum is displayed in Figure 3.4, in this case for the event beginning 26 September 2001. Note that the initial rise in the fluxes is not a rapid as the impulsive event, but the proton fluxes remain elevated for a much longer time, and the decline in the fluxes over time is much slower. Note also that the proton flux levels are larger than for the impulsive event. There is also a shock enhancement at approximately 34 hours after event onset. Such large fluxes, especially at higher energies and longer duration for these events, indicate that these events could pose a serious health risk to crews in space.

Another major concern for crew health is the occurrence of a series of large, energetic events spaced closely in time. Such a series of events occurred in October 1989, when three separate, overlapping events happened within a 5-day period. Elevated flux levels persisted for approximately 10 days. Figure 3.5, taken from Neal et al. (2008), displays the >10 MeV, >50 MeV, and >100 MeV proton fluxes for this series of events.

In a 2008 report, the National Research Council recommended that the King spectrum (King 1974) for the August 1972 SEP event be replaced as the NASA design standard for assessing SEP risk since it was not representative of a worst-case event (NAS/NRC, 2008). The council noted that exposures from other events that have occurred in the era of human spaceflight could exceed those from the August 1972 event. The council further opined that the October 1989 event could be a candidate

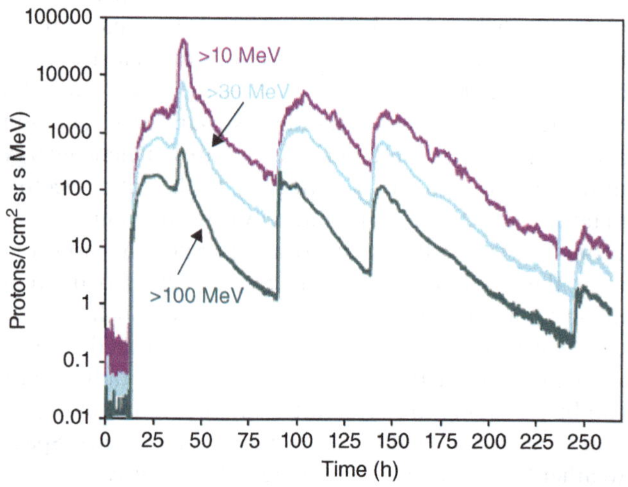

Figure 3.5 Proton energy spectra for the 19 October 1989 series of SEP events, including the shock enhancement that occurred at approximately 25 hours into the event (37 hours on the plot). (Source: Neal et al., 2008. Reproduced with permission of IEEE.)

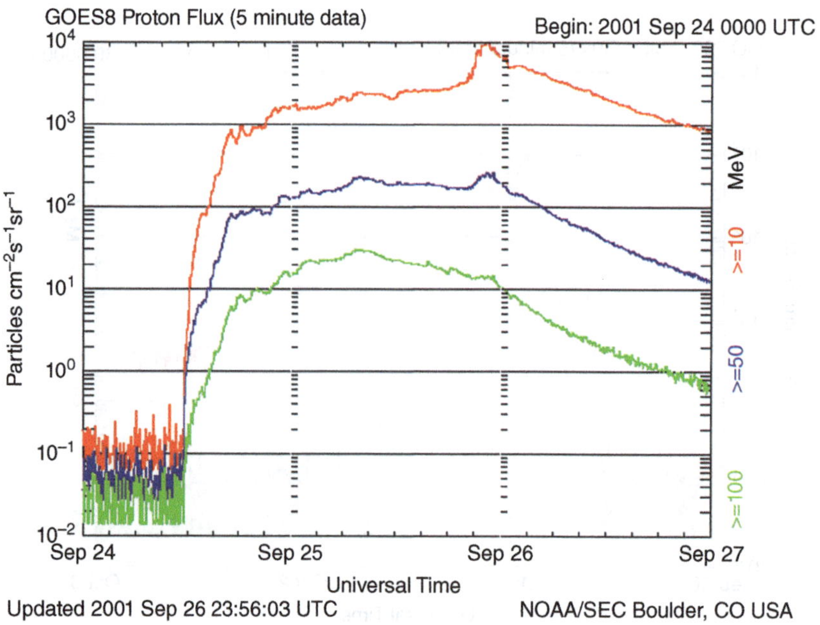

Figure 3.4 Energetic proton spectra for the 24 September 2001 gradual solar energetic particle event. The bump in the spectra 34 hours into the event indicates a shock enhancement has occurred. (Reproduced with permission of NOAA SWPC.)

for a new design standard. In January 2017, a Technical Interchange Meeting was held at NASA Langley Research Center, where a group of SEP event experts from government and academia assessed the current knowledge of SEP events and recommended that NASA adopt the summed spectra for the 1989 October 19, 22, and 24 events as the new design spectrum for evaluating crew risk and shielding requirements (Townsend, Adams, et al., 2018). It was also recommended that the Tylka model (Tylka et al., 2010) for the event spectra be used since it better represented the high-energy components of the events, approximately representing a 90th-percentile event for a 1-year mission, and a 99th-percentile event for a 30-day mission. Analyses of possible crew exposures from such an event are presented in section 3.6.

SEP events can occur at any time during the solar sunspot cycle. However, most occur near solar maximum and on both the ascending and descending phases of the cycle. Figure 3.6 displays sunspot numbers since January 2000. The maxima (~2001 and ~2014) and minimum (~2010) are clearly displayed. Between 2000 and 2010, 79 SEP events occurred. Only a few (e.g. July 2000, November 2001, and October 2003) posed any risk to crews in deep space. Calculated doses from these three events will be presented in section 3.6. For the current solar cycle (since 2010), 36 SEP events have occurred. None have posed any significant health risk to crews in space.

3.3. RADIATION QUANTITIES, UNITS, AND SYMBOLS

Radiation effects in humans depend upon the energy deposited in the tissue by the radiation, and the response of the body to that energy deposition. Radiation units, symbols, and quantities related to these energy deposits and their effects are defined by the International Commission on Radiation Units and Measurements (2011; Allisy et al., 1993). Relevant ones for space radiation are the following:

- **Absorbed Dose (D)** is the mean energy (Joule) imparted (deposited) per unit mass (units: one Gray = Gy = one J kg^{-1}).
- **Energy Imparted** is the sum of all energy deposited in a volume of mass m by the incident particle and its reaction products, plus any change in the rest energies of those particles.
- **Linear Energy Transfer (LET)** is mean energy lost by a charged particle per unit distance traveled, due to collisions with atomic electrons (unit: J m^{-1}; commonly use keV/μm)
- **Quality Factor (Q)** is a factor to weight the **absorbed** dose for the biological effectiveness of the charged particles producing the absorbed dose. It is a function of the LET of the radiation.

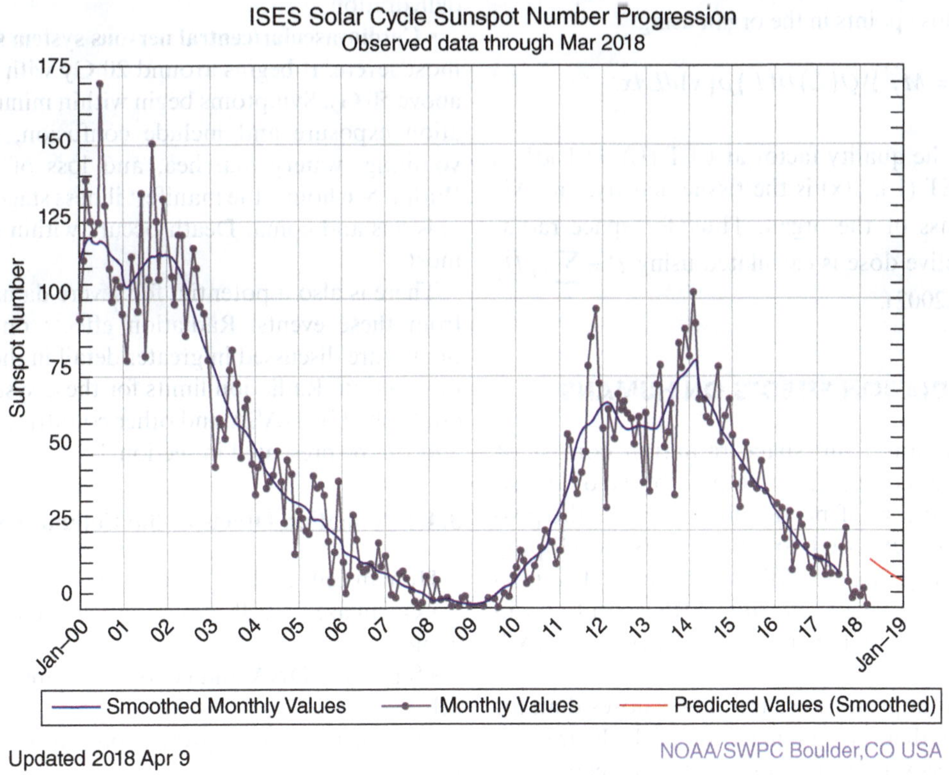

Figure 3.6 Monthly sunspot numbers and smoothed monthly sunspot numbers versus time for the 2000 to 2018 time period. Note the predicted future values beyond the measured monthly values denoted by the red line in the lower right corner of the figure.

• **Dose Equivalent (H)** is defined as H ≡ Q×D. H is defined for routine radiation protection practices (prevention of stochastic effects such as cancer) and is not applicable for acute (high-level) exposures, such as occur in radiation accidents (unit: Sievert, Sv).

• **Equivalent Dose (H_T)** is given by $H_T = \sum_R w_R D_T$, where D_T is the organ-averaged dose for the tissue (organ) denoted by T and w_R is a unitless weighting factor to account for differences in biological effectiveness of different types of radiation, and is specific to the type of incident radiation (e.g. proton, neutron, etc.). For protons, $w_R = 2$ (NCRP, 1993) (unit: Sievert, Sv).

• **Effective Dose (E)** is given by $E = \sum_T w_T H_T$, where w_T is a tissue weighting factor that accounts for the proportionate detriment (stochastic) of tissue T when the whole body is irradiated uniformly. Tissue weighting factors range from 0.01 for skin to 0.12 for the colon and bone marrow, to 0.20 for the gonads (NCRP, 1993). However, since space radiation fields consist of an environment of mixed components of low- and high-LET particles, and are much more complex than other radiation fields to which individuals are typically exposed (e.g. radiotherapy), NCRP recommended that NASA use organ dose equivalent, rather than equivalent dose, for assessing effective dose. Organ dose equivalent ($\bar{H}_T$) is then obtained by averaging over all dose points in the organ using

$$\bar{H}_T = M_T^{-1} \iint Q(L) D(L) \rho(x) dL dx,$$

where Q(L) is the quality factor at LET (L), D(L)dL is the dose at LET (L), ρ(x) is the tissue density, and M_T is the total mass of the organ. Thus, for space radiations, the effective dose is calculated using $E = \sum_T w_T \bar{H}_T$ (NCRP, 2000, 2002).

3.4. RADIATION EFFECTS ON HUMANS

Risks to humans from space radiation consist of potential tissue damage and the possibility of developing a fatal cancer. Of particular concern is the possibility of crews exposed to SEP events developing acute radiation syndrome (ARS) symptoms. The possibility of an SEP event causing ARS symptoms in space crews was recognized over 50 years ago (NAS/NRC, 1967).

Symptoms from protons occur at lower doses than for the reference radiation, gamma doses, which are discussed below. Sample results for SEP event exposures, presented in section 3.6, will be corrected for the increase efficacy of proton by use of a multiplicative relative biological effectiveness factor of 1.5 (NCRP, 2000).

Tissues at risk for ARS are the bone marrow, skin, gastrointestinal, cardiovascular/central nervous systems (https://emergency.cdc.gov/radiation/pdf/ars.pdf).

• **Bone marrow syndrome** (hematopoietic syndrome) is the least severe. Mild symptoms can occur at doses as low ~0.3 Gy and the full syndrome begins at approximately 0.7 Gy. The severity of symptoms increases with increasing dose. Prodromal symptoms include nausea, vomiting, and diarrhea. Considering potential exposures from large SEP events that have previously occurred in the space era, exposures to space crews from SEP events are most likely to result in prodromal effects. In the latent stage, bone marrow stem cells will begin dying if the dose is above ~1 Gy. In the manifest illness stage, bone marrow damage increases with higher doses, and death within weeks is possible in the final recovery/death stage.

• **Gastrointestinal syndrome** begins around 6 Gy, with full expression above 10 Gy. In the prodromal stage, symptoms of severe anorexia, vomiting, cramping, and diarrhea begin within two hours. This lasts about two days before progressing to the latent stage when bone marrow stem cells and cells lining the gut are dying. Progressing to the manifest illness stage occurs within a week. Within 2 weeks of exposure, symptoms of anorexia, malaise, severe diarrhea, dehydration, and electrolyte imbalance result in death due to infection and dehydration.

• **Cardiovascular/central nervous system syndrome** is the most severe. It begins around 20 Gy with full expression above 50 Gy. Symptoms begin within minutes of the radiation exposure and include confusion, severe nausea, vomiting, watery diarrhea, and loss of consciousness. Within 5–6 hours, the manifest illness stage results in convulsions and coma. Death occurs within several days at most.

There is also a potential for severe damage to the skin from these events. Radiation effects on these tissues/organs are discussed in greater detail in the remainder of this section. Radiation limits for these systems have been established by NASA and other countries' space agencies and will be presented in section 3.5.

3.4.1. Radiation Effects at the Cellular Level

Human cells undergo a four-stage cycle (Alpen, 1998):

• G_1 phase – cell grows larger and organelles are copied.

• S phase – DNA and centrosome are replicated in the cell.

• G_2 phase – cell grows, produces organelles and proteins.

• M phase – cell divides into two cells.

Cells are most radiosensitive when getting ready to divide (G_2 and M phases) and least sensitive in late S phase.

Rapidly dividing cells, such as the bone marrow, skin, gastrointestinal (GI) tract, and the eye lens are the most radiosensitive (Alpen, 1998).

Ionizing radiation (charged particles, neutrons, and photons with sufficient energy to ionize atoms and molecules) causes damage to cells by two mechanisms: (1) direct action by ionizing the cell constituents directly and (2) indirect action by ionizing water within the cell, creating free radicals which can attack DNA in the cell (Anderson, 1984). Direct damage causing cell death can occur in the DNA, cell membrane, and nucleus. Severely damaged cells often undergo apoptosis (programmed cell death). DNA damage may involve a break in a single strand of the DNA, which can often be repaired, thereby restoring normal cell function. It may also involve breaks in both strands (double strand break), which is much more difficult to properly repair and may result in cell death. Mutations and chemical changes in the cells are not repairable and may result in cancer. Chromosome damage may also occur in cells, which can lead to mutations. Significant damage to the blood-forming organs, GI tract, and cardiovascular/central nervous system (CNS) can result in premature death.

3.4.2. Radiation Effects on Blood-Forming Organs

As summarized in *Evidence Report: Risk of Acute Radiation Syndromes Due to Solar Particle Events* (https://humanresearchroadmap.nasa.gov/Evidence/reports/Acute.pdf), blood cell counts decrease as radiation doses increase, due to destruction of human stem cells, which are highly radiosensitive. The dose (D_0) at which the surviving fraction is 37% is 0.6–1.6 Gy. The hematopoiesis system is unable to recover if 99% of the hematopoietic stem cells are killed, unless a transfusion of compatible stem cells is carried out. However, a complete recovery is possible at lower doses, if a sufficient number of these cells survive (Fliedner et al., 2002).

Different cells respond differently to radiation based upon variation in transit times, the numbers of cell in peripheral blood, and their lifetimes. Normal erythrocytes, which are radioresistant, have a ~120-day lifetime. Hence, even if developing cells are destroyed, the loss of these cells is low enough that anemia is unlikely to occur, even for moderate irradiation (Fliedner et al., 2002). Granulocytes are also radioresistant. They are important for immune protection in the body. However, their concentration in blood has a 6.6-hour half-life. Hence, the survival of these depends upon the dose received. Five to six days after a lethal dose (~10 Gy), granulocytes completely disappear from blood. Recovery from moderate doses (<2 Gy) is likely (Fliedner et al., 2002). Thrombocytes (platelets) have a maximum lifespan of ~12 days. Lethal doses do irreversible damage to the thrombopoeitic system. Moderate doses are recoverable. A reduction in platelet counts results in increased bleeding, requiring medical support for survival. Of all cells in peripheral blood, lymphocytes are the most radiosensitive. They have long lifetimes (~ years) but take longer to recover to normal levels (Hu et al., 2012) and are important for the immune system.

3.4.3. Radiation Effects on the Gastrointestinal System

The epithelial cell lining of the entire GI system is radiosensitive due to its rapid cell turnover. This cell renewal depends upon a stem cell population residing in the crypts of Lieburkuhn. Doses greater than 4 Gy destroy these crypts and the mucosal lining of the gut (Waselenko et al., 2004).

Early GI symptoms include nausea, vomiting, loss of appetite, and diarrhea, and may begin within hours of exposure. Nausea and vomiting may be indicative of CNS involvement. Early nausea and vomiting indicate a severe exposure, and early diarrhea is indicative of severe damage. It often results in death (Conklin & Walker, 1987). For moderate exposures, late symptoms of effects on the GI tract, which appear a week or more after exposure, include cramping and diarrhea and indicate a major loss of stem cells in the gut crypts (Fliedner et al., 2001). Bloody diarrhea may occur, which results in an increased loss of fluids and electrolytes and can be life threatening.

3.4.4. Radiation Effects on the Cardiovascular/Central Nervous Systems

Radiation-induced cardiovascular disease (CVD) has been identified as a side effect of radiotherapy involving high doses to the chest to treat cancer (Boerma et al., 2016; Little et al., 2012). Deterministic or tissue effects observed include fibrosis of the pericardium and myocardium, pericardial adhesions, and damage to microvascular and heart valves. Most of these effects occur a decade or more after exposure, except for pericarditis, which occurs within months (Little et al., 2008). Studies at much lower doses (0.1–0.2 Gy) have found evidence of CVD in Chernobyl emergency workers (Ivanov et al., 2006) and Japanese atomic bomb survivors (Ozasa et al., 2012). Ischemic heart disease, stroke, and hypertension were the main CV effects (Boerma et al., 2016). There is some evidence suggestive of CVD for exposures at occupational levels above 0.5 Gy (AGIR, 2010).

Based upon high-dose radiotherapy exposures with low-LET radiation, the brain has been considered to be highly resistant to radiation, since early effects usually did not develop at daily doses of 2 Gy up to a total of 50 Gy (NCRP, 2000). However, existing laboratory data, which largely use rodents as test subjects, are uncertain as to the existence of a threshold dose for clinically significant effects.

Estimates of a threshold dose range between 0.1 and 0.5 Gy (Cucinotta et al., 2014). Acute effects are possible at higher doses, and death is usually the result of doses above 50 Gy (https://emergency.cdc.gov/radiation/pdf/ars.pdf).

3.4.5. Radiation Effects on the Skin

Damage to the skin can occur from exposure to SEP events behind thin shielding, as will be apparent in section 3.6. If exposures are high enough, skin damage can be severe. The outer layer of the skin, the epidermis, functions as a barrier protecting the underlying cell layers, and protecting against infections. It contains a basal cell layer, which is the reservoir for stem cells. Stem cells in the epidermis are especially sensitive to radiation and easily depleted. The next layer is the dermis, then the hypodermis. The dermis has numerous blood vessels and is penetrated by the sebaceous glands and hair follicles (Alpen, 1998). For SEP events, the likely skin exposures will be substantial only in the case of large events and minimal shielding, such as in a space suit. Likely symptoms of concern from these exposures will include erythema, moist desquamation, and epilation. Moist desquamation of the epidermis results in loss of the epidermis layer and oozing of fluid from the dermis layer. The underlying capillary bed is damaged (telangiecasia), as are hair follicles (causing epilation) and sweat glands. Late effects of radiation on the skin are manifestations of damage to the microvascular structure in the layer beneath the dermis (Hopewell, 1986). Table 3.1, using data taken from Brown & Rzucidlo (2011), displays threshold doses for a variety of symptoms of skin injuries from radiation. The values in the table are for exposures delivered in minutes to several hours. Since most SEP event exposures are delivered over longer time periods of hours to days, the threshold doses for these effects may be higher (https://humanresearchroadmap.nasa.gov/Evidence/reports/Acute.pdf).

Table 3.1 Skin dose thresholds. (Source: Brown & Rzucidlo, 2011. Reproduced with permission of Elsevier.)

Effect	Dose (Gy)	Postexposure Time of Onset
Early transient erythema	2	hours
Main erythema	6	10 d
Permanent epilation	7	3 w
Dry desquamation	14	4 w
Moist desquamation	18	4 w
Ulceration	24	6 w
Dermal atrophy	10	>12 w
Invasive fibrosis	10	>1 y

3.4.6. Radiation Effects on the Eye Lens

The major concern with space radiation exposure to the eye is damage to the eye lens, mainly through cataract formation. It has been well known that the severity of cataract formation increases with dose. Epidemiological studies of Japanese atomic bomb survivors indicated that low to moderate gamma ray doses would produce cataracts (Preston et. al., 2003). Early estimates of the threshold for cataract formation were 0.6–1.5 Gy (Junk et al., 1998). More recently, the threshold dose appears to be <1 Gy (Worgul et al., 2007). At higher doses, damage to the retina (radiation retinopathy) is possible. The threshold dose for this disease is ~15–30 Gy, with manifestation around 6 months to 3 years after exposure (Gupta et al., 2007).

3.4.7. Radiation Carcinogenesis

A major concern with deep space missions is risk of astronauts developing a fatal cancer, primarily from chronic exposure to the galactic cosmic radiation (GCR) environment. Cancer is unlike organ/tissue exposures, which have thresholds, and whose severity increases with dose. Instead, cancer is stochastic: its probability of occurrence, not severity, increases with dose. There are large uncertainties associated with the biological risks from the GCR environment, especially effects from the high-energy heavy ion component, the so-called HZE particles. Hence, the established career limit is a 3% risk of exposure-induced death (REID) at the 95% confidence limit. The latter is imposed to account for the uncertainties. Actual effective dose limits are then determined on a case-by-case basis for the expected radiation environment, mission duration, and sex and ages of the crew members. Although the major concern is exposure to the ever-present GCR environment, large SEP events also contribute to cancer risk. A detailed overview of the risk of cancer from space radiation is found in NASA's *Evidence Report: Risk of Radiation Carcinogenesis* (https://humanresearchroadmap.nasa.gov/Evidence/reports/Cancer.pdf).

3.5. RADIATION LIMITS

An excellent overview of radiation limits for various spacefaring countries is presented in McKenna-Lawlor et al. (2014). Space agencies represented include NASA (USA), ESA (EU), RSA (Russia), CSA (Canada), and JAXA (Japan). Limits for these space agencies are presented here. Note the similarities and differences between the various agency limits.

3.5.1. National Aeronautics and Space Administration Radiation Limits (USA)

Radiation limits adopted for USA astronauts by NASA are the result of expert guidance provided by the National Council on Radiation Protection and Measurements (NCRP, 1989, 2000, 2001, 2002, 2006, 2010, 2014, 2016) and the National Academy of Science/National Research Council (NAS/NRC, 2000, 2006, 2008, 2012). As discussed in section 3.4, the major health concerns are tissue damage resulting in the impairment or failure of various body organs to properly function in the near term, and chronic exposures to the crew, resulting in a long-term REID from cancer. Radiation limits for crew members adopted by NASA and other spacefaring nations are intended to minimize the risk of developing a fatal cancer and to prevent short-term tissue effects that could immediately impact a mission (e.g. radiation sickness), as well as prevent long-term noncancer and degenerative disease effects, especially for the cardiovascular and central nervous systems.

Table 3.2 displays the current NASA Space-Permissible Exposure Limits (SPELs) for short-term and noncancer career exposures of astronauts to the space radiation environment (NASA, 2015). The rationale behind these limits is discussed in section F.8 of NASA STD-3001A, Change 1, *NASA Space Flight Human System Standard, Volume 1: Crew Health*.

> SPEL for radiation have the primary functions of preventing in-flight risks that jeopardize mission success and limiting chronic risks to acceptable levels based on legal, ethical or moral, and financial considerations. Both short-term and career exposure limits are applied using assessments of the uncertainties in projection models and a reasonable "worst-case" space environment to be encountered on specific missions. Uncertainties are due to gaps in knowledge of biological effects of GCR, heavy ions, and the nature of SPEs. Although specific exposure limits are identified based on mortality risk, in all cases decisions concerning vehicle, habitat, and mission design are made such that resulting crew radiation exposures are ALARA (As Low As Reasonably Achievable). As an operating practice, ALARA is a recognized NASA requirement.

The short-term dose limits in Table 3.2 are used to prevent clinically significant noncancer health effects (e.g. performance degradation, sickness, or death in-flight). Lifetime limits for cataracts, heart disease, and damage to the CNS are imposed to limit or prevent risks of degenerative tissue diseases (e.g., stroke, coronary heart disease, etc.). Career limits for the heart limit the risk of death from heart disease below ~3% to 5% (NASA, 2015).

For the eye lens, skin, and blood-forming organs (BFO), the units are in milliGray-Equivalent (mGy-Eq), which is calculated by multiplying the absorbed dose, D, by a unitless relative biological effectiveness (RBE) factor

Table 3.2 Dose limits for short-term or career noncancer effects (in mGy-Eq or mGy).

Organ	30 days (mGy-Eq)	1 year (mGy-Eq)	Career (mGy-Eq)
Lens*	1000	2000	4000
Skin	1500	3000	6000
BFO	250	500	NA
Heart**	250	500	1000
CNS***	500 mGy	1000 mGy	1500 mGy
CNS*** ($Z \geq 10$)	-	100 mGy	250 mGy

*Lens limits are intended to prevent early (<5 yr) severe cataracts (e.g. from a solar particle event).
**Heart doses calculated as average over heart muscle and adjacent arteries.
***CNS limits should be calculated at the hippocampus.

Table 3.3 RBE values for noncancer effects of the eye lens, skin, blood forming organs, and cardiovascular systems.

Radiation Type	Recommended RBE Values	Range of RBE Values
1 to 5 MeV neutrons	6	4–8
5 to 50 MeV neutrons	3.5	2–5
Heavy ions ($Z > 1$)	2.5	1–4
Protons >2 MeV	1.5	—

to account for differences in biological damage effectiveness for incident heavy charged particles from the space radiation environment and secondary neutrons produced in nuclear collisions over damage from that produced by gammas and x-rays (reference radiations). Table 3.3 presents RBE values for the lens, skin, BFO, and cardiovascular (heart) systems. RBE values for the central nervous system (brain) are unknown. Therefore, the limits for the brain are expressed directly in absorbed dose (mGy).

For operations in cis-Lunar space, the experts participating in the Technical Interchange Meeting at NASA Langley Research Center also recommended, "The habitat shall provide protection to ensure that gray equivalent to astronaut blood forming organs (BFO) does not exceed 250 mGy-Eq for the design SPE" (Townsend, Adams, et al., 2018). The meeting report also stated,

> This requirement is imposed to prevent clinically significant non-cancer tissue effects, such as performance degradation, sickness, or death during the mission, resulting from SPEs. . . .
>
> The requirement that gray equivalent to BFO does not exceed 250 mGy-Eq during any 30 day period is expected to drive shielding design; a shield design which ensures that this exposure limit has not been exceeded will also ensure that the other PELs for non-cancer tissue effects have not been exceeded.

NASA career radiation limits, relative to cancer mortality, are specified in terms of effective dose, which is related to REID. These limits are not a fixed quantity (e.g. 1 Sv) but are specific to the expected radiation environment to be encountered during the mission, mission duration, age, and sex of the astronauts. As stipulated in NASA STD-3001A (2015), "Planned career exposure to ionizing radiation shall not exceed 3 percent Risk of Exposure-Induced Death (REID) for cancer mortality at a 95 percent confidence level to limit the cumulative effective dose (in units of Sievert) received by an astronaut throughout his or her career." The effect of applying the 95% confidence limit (95% CL) to the 3% risk dose limit is to reduce that limit by an approximate factor of 3. For purposes of illustrating how these are used, sample 3% REID at the 95% CL for 40-year-old females will be used to compare with calculated career effective doses in section 3.6. The limits are 19.9 cSv for females and 26.4 cSv for males (W. de Wet, private communication).

3.5.2. European Space Agency

ESA space radiation limits are shown in Table 3.4. These are based on guidance provided in NCRP and ICRP reports (NCRP, 2000, 2002; ICRP, 1990). The career limit is independent of gender and age. The equivalent dose limits for the organs are consensus for the International Space Station.

3.5.3. Russian Space Agency

RSA radiation limits are displayed in Table 3.5. The career limit of 1 Sv is age and gender independent. The limit is based upon 10% excess risk and includes noncancer effects, as well as cancer mortality (McKenna-Lawlor et al., 2014). The BFO acute equivalent dose limit is for one-time acute exposure.

3.5.4. Canadian Space Agency

CSA has adopted NASA's short-term limits (displayed in Table 3.2) for Canadian astronauts. For lifetime limits, CSA follows the ICRP 60 recommendation (ICRP, 1990) of 1 Sv. Lifetime limits for specific organs are 4 Sv for the eye and 6 Sv for the skin.

Table 3.4 ESA radiation exposure limits.

Period	BFO	Eye	Skin	Career
30 d	0.25 Sv	0.5 Sv	1.5 Sv	—
Annual	0.5 Sv	1.0 Sv	4.0 Sv	—
Career	—	—	—	1 Sv

Table 3.5 RSA radiation exposure limits.

Period	BFO	Eye	Skin
30 d	0.25 Sv	0.5 Sv	1.5 Sv
Annual	0.5 Sv	1.0 Sv	4.0 Sv
Acute	0.15 Sv	—	—
Career	—	2 Sv	6 Sv

Table 3.6 JAXA organ radiation exposure limits.

Interval	BFO	Eye	Skin	Testis
Single/brief	NA	0.5 Sv	2.0 Sv	NA
Annual	0.5 Sv	2.0 Sv	7.0 Sv	1.0 Sv

Table 3.7 JAXA career effective dose limits (Sv).

	Age (y)			
Gender	27–29	30–34	35–39	≥40
Male	0.6/3.2%	0.9/3.1%	1.0/3.1%	1.2/3.0%
Female	0.6/2.9%	0.8/3.1%	0.9/3.1%	1.1/3.1%

3.5.5. Japan Aerospace Exploration Agency

Radiation exposure limits for specific organs in tissue dose equivalent, as specified by JAXA, are displayed in Table 3.6. Career-effective dose limits for a one-year mission, presented in Table 3.7, follow the NASA criterion that the radiation exposure shall not exceed a 3% REID at the 95% confidence limit.

3.6. REPRESENTATIVE SEP EVENT RADIATION EXPOSURES

At present, the Apollo missions to Earth's moon, between December 1968 and December 1972, are the only human spaceflights to occur beyond the protection afforded by Earth's magnetic field. Of these missions, Apollo 8 and 10 orbited the moon but did not land on the surface. Apollo 11, 12, and 14–17 did land on the lunar surface. Apollo 13, due to an accident in-flight, did a fly-by of the moon and returned to Earth. Radiation exposures on these missions, as measured by on-board dosimetry, yielded skin doses ranging from 0.16 rads (0.16 cGy) on Apollo 8 to 1.14 rads (1.14 cGy) on Apollo 14 (English et al., 1973). The total duration for all of these missions was less than 3 months, spread out over a 4-year time period. No major SEP events occurred during these missions. The only measurable exposure from an SEP event occurred during Apollo 12, when a slight increase in radiation levels was detected by sensors outside the spacecraft, but dosimetry within the spacecraft indicated no increase in crew doses (English et al., 1973). There was one large SEP event that occurred during this time, the event of August 1972. However, it occurred between the Apollo 16 and Apollo 17 missions and thus had no effect on any of the lunar missions.

In this section, analyses of potential radiation exposures from representative SEP events that have occurred during the past 6 decades of the humans-in-space era will be presented, and their possible effects on crews examined. The first two events, 24 August 1998 and 17 May 2012, represent typical, average SEP events. Both are ground-level events, which means that neutrons produced by interactions of the incident protons in Earth's atmosphere were detected by neutron monitors at the surface of the Earth. The 1998 event was an impulsive event and the 2012 event was a gradual one. Next, analyses of three of the largest events that have occurred since 2000 are made. These events are the event of 14 July 2000, the so-called "Bastille Day" event, the event of 4 November 2001, and the sum of the series of four events that occurred between 28 October and 2 November 2003, often referred to as the "Halloween" 2003 event. Then, a comparison is made of the August 1972 event, which is the current NASA design basis event (NASA, 2015), with the summed event from the 19–24 October 1989 series of events, which has been proposed as a new design basis event to replace the August 1972 event (Townsend, Adams, et al., 2018). Finally, estimated exposures from a plausible "worst case" event, based upon historical records from 774–775 AD, are presented (Townsend et al., 2016).

3.6.1. SEP Events of 24 August 1998 and 17 May 2012

Figure 3.7 displays the proton spectra for the 24 August 1998 and 17 May 2012 SEP events. Note the sharp increase in proton flux for both events, which remains elevated for the gradual event of August 1998 and decreases quickly for the impulsive event of May 2012. Table 3.8 presents calculated doses for these two events obtained using the OLTARIS website. Doses for female crew members are displayed. Organ doses for male crew members are not displayed but are slightly smaller due to the additional self-shielding provided by the larger body mass of the males.

None of the organ doses displayed in Table 3.8, for either event, present a hazard to the health of crews in deep space. All of the organ doses are well below the short-term and noncancer career limits displayed in Table 3.2. In addition, the career effective doses displayed are far below 3% REID limits for any mission duration. For the period from 1997 through 2009, this 1998 event had the 21st largest peak proton flux of the 93 SEP events that occurred. For the period since then, the May 2012 event was the 7th largest of over 35 events. Hence, the doses for these two events also demonstrate that most SEP events are of limited threat for crews protected by even nominal shielding of 5 g cm^{-2} aluminum.

3.6.2. SEP Events of 14 July 2000, 4 November 2001, and 28 October–2 November 2003

SEP event protection concern for space crews is focused on the potential for a very large event to occur, such as happened in mid-July 2000, or with the series of events that began in late October 2003. In this situation, one or more organ dose limits may be exceeded. The career effective dose limit may also be exceeded for crew members, especially younger ones. Also, depending on the shielding available, acute radiation syndrome symptoms could occur if the crew is lightly shielded (e.g. in a space suit). Table 3.9 presents dose estimates for these events obtained from the OLTARIS website.

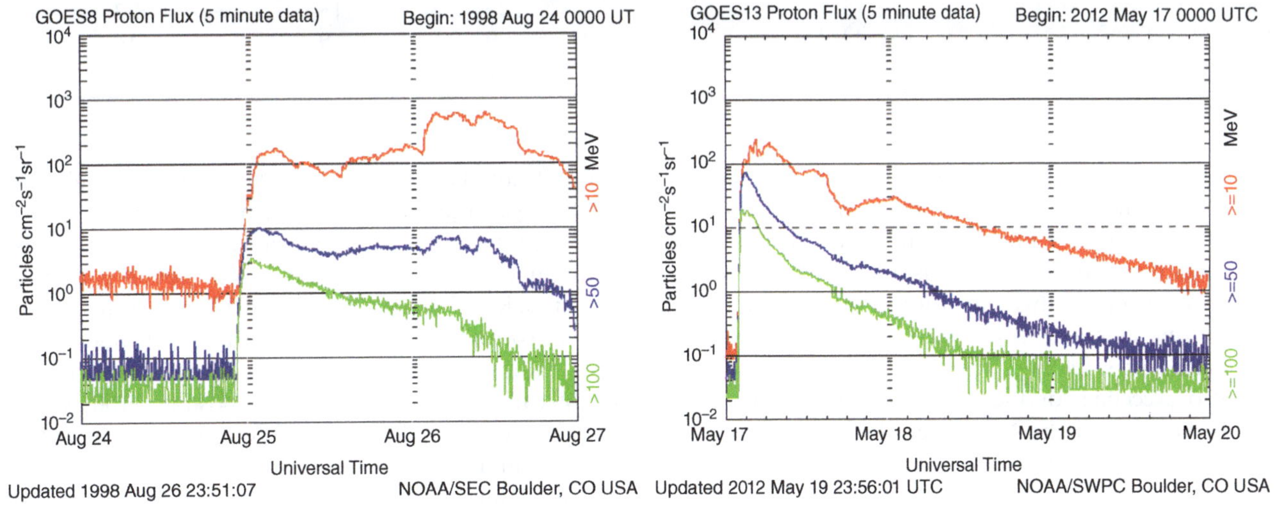

Figure 3.7 Proton energy spectra for the SEP events of 24–27 August 1998 (left panel) and 17–20 May 2012 (right panel).

Table 3.8 Organ dose and effective dose in cis-Lunar space for female crew members for the 24 August 1998 and the 17 May 2012 events, shielded by aluminum.

Al Shield ($g\ cm^{-2}$)	Skin (cGy-Eq.)	Eye (cGy-Eq.)	BFO (cGy-Eq.)	CNS (cGy)	Heart (cGy-Eq)	Effective Dose (cSv)
24 August 1998 SEP Event						
0.4	95	12	0.31	0.12	0.15	2.4
5	0.61	0.38	0.09	0.06	0.08	0.18
10	0.19	0.15	0.06	0.03	0.06	0.13
20	0.08	0.07	0.04	0.02	0.04	0.10
17 May 2012 SEP Event						
0.4	21	5	1	0.6	0.5	1.1
5	2	1	0.5	0.3	0.4	0.4
10	0.7	0.6	0.3	0.2	0.3	0.3
20	0.3	0.3	0.2	0.1	0.2	0.2

Table 3.9 Organ dose and effective dose in cis-Lunar space for female crew members for the 14 July 2000, 4 November 2001, and the summed 28 October–2 November 2003 events, shielded by aluminum. Values exceeding 30-day short-term organ limits and career effective dose limits for 40-year-old females are in bold type.

Al Shield ($g\ cm^{-2}$)	Skin (cGy-Eq)	Eye (cGy-Eq)	BFO (cGy-Eq)	CNS (cGy)	Heart (cGy-Eq)	Effective Dose (cSv)
14 July 2000 SEP Event						
0.4	**7484**	**826**	**96**	45	**36**	**209**
5	**155**	**106**	**32**	21	22	**34**
10	52	43	18	13	14	**22**
20	18	16	9	6	8	14
4 November 2001 SEP Event						
0.4	**650**	**515**	**53**	21	16	**120**
5	85	55	14	10	9	16
10	25	20	8	5	6	10
20	7	7	4	2	3	6
Summed 28 October to 2 November 2003 SEP Event						
0.4	**4557**	**570**	**67**	31	**25**	**131**
5	111	75	22	15	15	**23**
10	53	43	12	9	9	14
20	12	11	6	4	5	9

All three events had peak proton fluxes in excess of 30,000 pfu and are the three largest SEP events since 2000. Note that the thin shielding provided by a space suit (~0.4 g cm^{-2} of aluminum) is inadequate for protection from events of this magnitude. For all three events, the skin, eye lens, and BFO 30-day limits are exceeded for the space suit shielding. Thirty-day limits for the heart are also exceeded for the July 2000 and October 2003 events, but not for the November 2001 event. No limits are exceeded for the CNS from any of these three events. Career effective dose limits are exceeded for females and males of all ages and are likely to result in a substantially increased REID. The organ at greatest risk from these events is the skin. Skin exposures exceeding 45 Gy for the July 2000 and October 2003 events would cause substantial damage to the skin and body limbs. From Table 3.1, symptoms include permanent epilation, dry and moist desquamation, tissue necrosis, and possible fibrosis (Brown & Rzucidlo, 2011; ICRP, 2012). Since space radiation is isotropically incident, these are whole body exposures. Exposures to the eye lens will result in increased risk of cataract formation. BFO exposures ~100 cGy or less will result in mild prodromal effects (nausea, vomiting, etc.), and bone marrow stem cells may begin dying.

Behind aluminum shield thicknesses of 5–10 g cm^{-2}, all organ doses for the three events are below 30-day limits, except for the skin and eye lens doses from the July 2000 event, which are slightly higher than the limits. All organ exposures are below applicable limits, for all three events, within a 20 g cm^{-2} SEP storm shelter.

3.6.3. SEP Events of August 1972 and 18–24 October 1989

As previously discussed in section 3.2, the NASA design standard for designing and evaluating radiation shielding efficacy was to evaluate the shielding using the King spectrum representation for the August 1972 SEP event (King, 1974; NASA, 2015). It was also mentioned that the National Research Council recommended using a different event whose exposures could exceed those from the 1972 event (NAS/NRC, 2008). In 2018, a recommendation was made to replace the August 1972 event proton spectrum with the summed spectrum from the October 1989 series of events, the latter representing a 90% event for a 1-year mission and a 99% event for a 30-day mission (Townsend, Adams, et al., 2018).

Table 3.10 displays organ doses and effective doses for the August 1972 event and the summed event of October 19–24, 1989, obtained from the OLTARIS website. Figure 3.8 displays the integral proton fluence spectra for the two events. Note that the summed October 1989 event has more particles and a harder energy spectrum (slower decrease in fluence with increasing energy) than the 1972 event. For both events, organ doses behind space suit shielding (0.4 g cm^{-2} aluminum) are well above limits for all organs. In particular, extensive skin damage including moist desquamation, permanent epilation, ulceration, fibrosis, and tissue necrosis are likely. The doses to the BFO are likely to cause damage to the bone marrow and begin destroying stem cells. Prodromal effects, including nausea, vomiting, and diarrhea, are likely. Nominal shielding of 5 g cm^{-2} aluminum reduces the organ doses significantly. However, except for CNS dose, the remaining organ doses are at or above their respective limits.

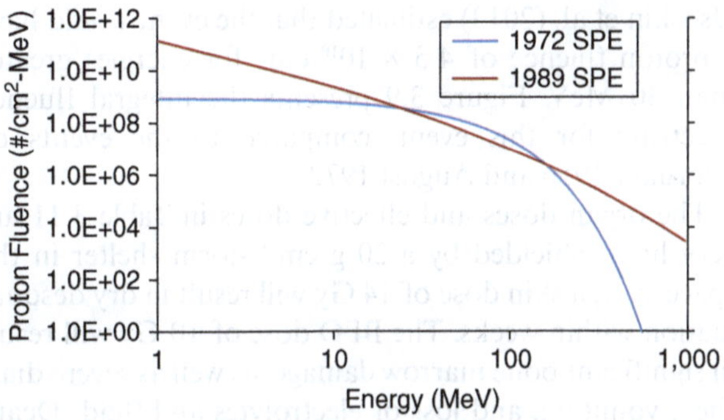

Figure 3.8 Proton integral fluence spectra for the August 1972 and 19–24 October 1989 SEP Events. (Source: Townsend, deWet, et al., 2018. Reproduced with permission of the University of Tennessee.)

Mild skin erythema and prodromal effects are possible. Note that organ doses behind an SEP "storm shelter" of 20 g cm^{-2} aluminum are all below limits. Here, the term *storm shelter* refers to a heavily shielded space used to protect the crew during a large SEP event in which exposures could exceed radiation limits. However, the female effective dose limit is slightly exceeded.

3.6.4. Extreme SEP Event of 775 AD

The possibility of SEP events larger than the events considered herein warrants consideration. Evidence of excess ^{14}C production found in tree rings for a variety of locations worldwide, as well as records of aurora sightings, and ^{10}Be production in ice cores indicate that an extreme event, likely of solar origin, occurred in 774 or 775 AD (Usoskin et al, 2013). This event, much larger in proton fluence than any events that have been observed in the years since humans began space travel, had a very hard energy spectrum, comparable to that of the February 1956 SEP event, the hardest SEP spectrum on record.

Table 3.10 Organ doses and effective dose for female crew members for the August 1972 and summed October 1989 event spectra shielded by aluminum. Values exceeding 30-day short-term organ limits and career effective dose limits for 40-year-old females are in bold type. (Source: Townsend, de Wet, et al., 2018. Reproduced with permission of the University of Tennessee.)

Al Shield (g cm^{-2})	Skin Dose (cGy-Eq.)	Eye Dose (cGy-Eq.)	BFO Dose (cGy-Eq.)	CNS Dose (cGy)	Heart Dose (cGy-Eq)	Effective Dose (cSv)
August 1972 SEP Event						
0.4	**3696**	**1606**	**260**	**111**	**71**	**338**
5	**376**	**279**	**74**	41	**30**	**81**
10	103	84	**28**	16	15	**35**
20	18	17	8	4	6	15
Summed October 1989 SEP Event						
0.4	**3917**	**874**	**136**	**68**	**59**	**229**
5	**176**	**142**	**57**	38	**38**	**59**
10	71	66	**34**	24	**26**	**37**
20	28	28	18	22	16	**23**

Usoskin et al. (2013) estimated that the event would have a proton fluence of 4.5×10^{10} cm^{-2} for energies greater than 30 MeV. Figure 3.9 presents the integral fluence spectrum for this event, compared to the events of February 1956 and August 1972.

The organ doses and effective doses in Table 3.11 are very high. Shielded by a 20 g cm^{-2} storm shelter in the spacecraft, a skin dose of 14 Gy will result in dry desquamation within weeks. The BFO dose of 10 Gy will result in significant bone marrow damage, as well as severe diarrhea, vomiting, and loss of electrolytes and fluid. Death is probable. Shielded by a 40 g cm^{-2} storm shelter, significant bone marrow damage will occur, and there will be severe diarrhea and vomiting. Death is possible. Clearly, an event of this magnitude would be catastrophic. Fortunately, it is highly unlikely, since there is no evidence of another SEP event of this magnitude for the past 1200 years. An analysis of the impact of this event, to astronauts on the surface of Mars, has been published (Townsend et al., 2016).

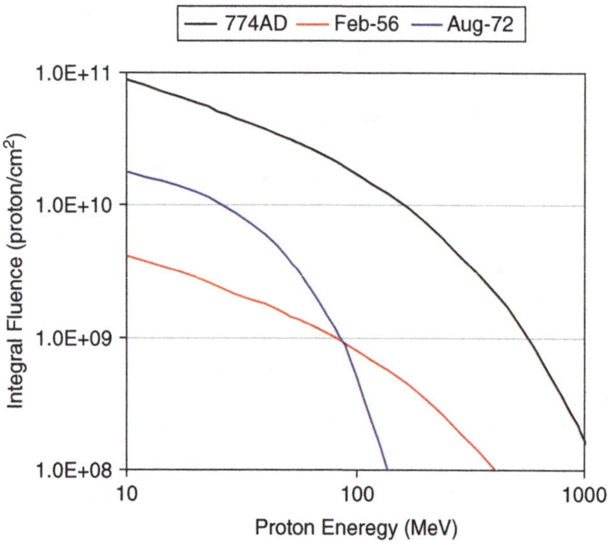

Figure 3.9 August 1972, February 1956, and AD775 SEP event proton spectra.

Table 3.11 Organ dose and effective dose in cis-Lunar space for female crew members for the 775 AD event.

	Aluminum Shield Thickness	
Organ	20 g cm^{-2}	40 g cm^{-2}
Skin (Gy-Eq)	14.1	7.8
Eye (Gy-Eq)	13.7	7.7
BFO (Gy-Eq)	10.1	6.3
CNS (Gy)	7.2	4.1
Heart (Gy-Eq)	9.6	6.1
Effective Dose (Sv)	10.9	7.7

3.7. SUMMARY AND CONCLUSIONS

The space radiation environment presents a wide variety of hazards to human crews in deep space. The ever-present GCR environment (space climate), consisting of electrons and all naturally occurring elements with energies up to and beyond 10s of GeV/ nucleon, poses a chronic exposure risk to crews of developing a fatal cancer. Since their kinetic energies are so high, they have long ranges in shielding materials, and through nuclear interactions, produce secondary particles that may possess even longer ranges. Hence, the GCR environment is extremely difficult to shield from for crews on missions in deep space. The current career limits are based upon limiting the REID from cancer to 3% at the 95% confidence limit and are converted to actual effective dose limits for crew members based upon their age and sex, the expected radiation environment for their mission, and the duration of the mission. The 95% confidence limit is applied because of the large uncertainties in biological risk associated with the GCR environment, especially the effects of the high-energy, heavy ion component, for which there is a paucity of human exposure data. Dose levels from the GCR environment are less than 20 cGy per year and are below the applicable organ dose limits.

SEP events, because their spectra are less energetic than the GCR spectrum, are more easily shielded. The main radiation concern from these events is the incident proton spectrum, although at times, the helium component is large enough to contribute measurably to the dose. As demonstrated by the analyses of the August 1998 and May 2012 events, most events are easily averted by spacecraft shielding and pose little risk to crews. However, large events typically occur one to four times during each solar cycle of approximately 11 years' duration. Some of these events can pose significant risk to crews that are lightly shielded, for example, by only a space suit when performing extravehicular activities. Analyses of potential exposures from the three largest events since 2000 indicate that there is the potential for acute radiation syndrome effects if shielded only by a space suit. But within a storm shelter (20 g cm^{-2}), organ doses can be reduced to less than the short-term organ dose limits.

Unlike the GCR environment, particles from all but extreme SEP events can be attenuated by a storm shelter, and thus pose little risk to crews in deep space. The occurrence of an extreme event, such as the one that may have occurred in 775 AD, would likely be catastrophic and could result in failure of the mission and loss of the crew. Fortunately, such events appear to be extremely rare.

REFERENCES

AGIR (2010). Circulatory disease risk. Report of the independent Advisory Group on Ionising Radiation. London: Health Protection Agency. RCE-16.

Allisy, A., Jennings, W. A., Kellerer, A. M., & Müller, J. W. (1993). Report 51. *Journal of the International Commission on Radiation Units and Measurements. 26*(2).

Alpen, E. L. (1998). *Radiation biophysics*, 2nd ed. San Diego, CA: Academic Press.

Anderson, D. W. (1984). *Absorption of ionizing radiation*. University Park Press.

Boerma, M., Sridharan, V., Mao, X. W., Nelson, G. A., Cheema, A. K., Koturbash, I., et al. (2016). Effects of ionizing radiation on the heart. *Mutat Res.*, *770*(Pt B), 319–327.

Brown, K. R., & Rzucidlo, E. (2011). Acute and chronic radiation injury. *J Vasc Surg.*, *53*(1 Suppl):15S–21S.

Conklin, J. J., Walker, R. I. (1987). Diagnosis, triage and treatment of casualties. In R. A. Conard & R. I. Walker (Eds.). *Military radiobiology* (pp. 231–240). Orlando, FL: Academic Press.

Cucinotta, F. A., Alp, M., Sulzman, F. M., & Wang, M. (2014). Space radiation risks to the central nervous system. *Life Sciences in Space Research*, *2*, 54–69.

Cucinotta, F. A., Townsend, L. W., Wilson, J. W., Golightly, M. J., & Weyland, M. (1994). Analysis of radiation risk from alpha particle component of solar particle events. *Advances in Space Research*, *14*(10), 661–670.

English, R. A., Benson, R. E., Bailey, J. V., & Barnes, C. M. (1973). *Apollo experience report: Protection against radiation*. Washington, DC: National Aeronautics and Space Administration. pp. i–15.

Fliedner, T. M., Friesecke, I., & Beyrer, K. (2001). *Medical management of radiation accidents: Manual on the acute radiation syndrome*. London: British Institute of Radiology.

Fliedner, T. M., Graessle, D., Paulsen, C., & Reimers, K. (2002). Structure and function of bone marrow hemopoiesis: Mechanisms of response to ionizing radiation exposure. *Cancer Biother Radiopharm*, *17*(4), 405–426.

Gupta, A., Dhawahir-Scala, F., Smith, A., Young, L., & Charles, S. (2007). Radiation retinopathy: Case report and review. *BMC Ophthalmol*, *7*, 6.

Hopewell, J. W. (1986). Mechanisms of the action of radiation on skin and underlying tissues. *Br J Radiol Suppl*, *19*, 39–47.

Hu, S., Smirnova, O. A., & Cucinotta, F. A. (2012). A biomathematical model of lymphopoiesis following severe radiation accidents: Potential use for dose assessment. *Health Physics*, *102*(4), 425–436.

ICRP. (1990). *Recommendations of the international commission on radiological protection*. International Commission on Radiological Protection. ICRP Publication 60.

ICRP. (2012). ICRP Statement on tissue reactions/early and late effects of radiation in normal tissues and organs – Threshold doses for tissue reactions in a radiation protection context. *Annals of the ICRP*, *41*(1–2).

International Commission on Radiation U, Measurements. (2011). Report 85: Fundamental quantities and units for ionizing radiation. *J ICRU*, *11*(1), 1–31.

Ivanov, V. K., Maksioutov, M. A., Chekin, S. Y., Petrov, A. V., Biryukov, A. P., Kruglova, Z. G., et al. (2006). The risk of radiation-induced cerebrovascular disease in Chernobyl emergency workers. *Health Phys*, *90*(3), 199–207.

Junk, A., Kundiev, Y., Vitte, P., & Worgul, B. (1998). Ocular radiation risk assessment in populations exposed to environmental radiation contamination. *Proceedings of the Advanced Research Workshop, Kiev, Ukraine*. Netherlands: Kluwer Academic Publishers.

King, J. H. (1974). Solar proton fluences for 1977–1983 space missions. *Journal of Spacecraft and Rockets*, *11*(6), 401–408.

Little, M. P., Azizova, T. V., Bazyka, D., Bouffler, S. D., Cardis, E., Chekin, S., et al. (2012). Systematic review and meta-analysis of circulatory disease from exposure to low-level ionizing radiation and estimates of potential population mortality risks. *Environ Health Perspect*, *120*(11), 503–1511.

Little, M. P, Tawn, E. J., Tzoulaki, I., Wakeford, R., Hildebrandt, G., Paris, F., et al. (2008). A systematic review of epidemiological associations between low and moderate doses of ionizing radiation and late cardiovascular effects, and their possible mechanisms. *Radiat Res*, *169*(1), 99–109.

McKenna-Lawlor, S., Bhardwaj, A., Ferrari, F., Kuznetsov, N., Lal, A. K., Li, Y., et al. (2014). Feasibility study of astronaut standardized career dose limits in LEO and the outlook for BLEO. *Acta Astronautica*, *104*(2), 565–573.

NAS/NRC. (1967). *Radiobiological factors*. Washington, DC: National Academy of Science/National Research Council.

NAS/NRC. (2000). *Radiation and the International Space Station: Recommendations to reduce risk*. Washington, DC: National Academy of Science/National Research Council.

NAS/NRC. (2006). *Space radiation hazards and the vision for space exploration: Report of a workshop*. Washington, DC: National Academy of Science/National Research Council.

NAS/NRC. (2008). *Managing space radiation risk in the new era of space exploration*. Washington, DC: National Academy of Science/National Research Council.

NAS/NRC. (2012). *Technical evaluation of the NASA model for cancer risk to astronauts due to space radiation*. Washington, DC: National Academy of Science/National Research Council.

NASA. (2015). *NASA space flight human-system standard volume 1, revision A w/change 1: Crew health*. Washington DC: National Aeronautics and Space Administration.

NCRP. (1989). *Guidance on radiation received in space activities* (Report No. 98). Bethesda, MD: National Council on Radiation Protection and Measurements.

NCRP. (1993). *Limitation of exposure to ionizing radiation* (Report No. 116). Bethesda, MD: National Council on Radiation Protection and Measurements.

NCRP. (2000). *Radiation protection guidance for activities in low-Earth orbit* (Report No. 132). Bethesda, MD: National Council on Radiation Protection and Measurements.

NCRP. (2001). *Fluence-based and microdosimetric event-based methods for radiation protection in space* (Report No. 137). Bethesda, MD: National Council on Radiation Protection and Measurements.

NCRP. (2002). *Operational radiation safety program for astronauts in low-Earth orbit: A basic framework* (Report No. 142). Bethesda, MD: National Council on Radiation Protection and Measurements.

NCRP. (2006). *Information needed to make radiation protection recommendations for space missions beyond low-Earth orbit* (Report No. 153). Bethesda, MD: National Council on Radiation Protection and Measurements.

NCRP. (2010). *Potential impact of individual genetic susceptibility and previous radiation exposure on radiation risk for astronauts* (Report No. 167). Bethesda, MD: National Council on Radiation Protection and Measurements.

NCRP. (2014). *Radiation protection for space activities: Supplement to previous recommendations* (Commentary No. 23). Bethesda, MD: National Council on Radiation Protection and Measurements.

NCRP. (2016). *Potential for central nervous system effects from radiation exposure during space activities Phase I: Overview* (Commentary No. 25). Bethesda, MD: National Council on Radiation Protection and Measurements.

Neal, J. S, Nichols, T. F., & Townsend, L. W. (2008). Importance of predicting the dose temporal profile for large solar energetic particle events. *Space Weather*, *6*(9), S09004.

Ozasa, K., Shimizu, Y., Suyama, A., Kasagi, F., Soda, M., Grant, E. J., et al. (2012). Studies of the mortality of atomic bomb survivors. Report 14, 1950–2003: An overview of cancer and noncancer diseases. *Radiat Res*, *177*(3), 229–243.

Preston, D. L., Shimizu, Y., Pierce, D. A., Suyama, A., & Mabuchi, K. (2003). Studies of mortality of atomic bomb survivors. Report 13: Solid cancer and noncancer disease mortality: 1950–1997. *Radiation Research*, *160*(4), 381–407.

Reames, D. V. (2017). *Solar energetic particles* (Lecture Notes in Physics, 932). Springer International.

Townsend, L. W., Adams, J. H., Blattnig, S. R., Clowdsley, M. S., Fry, D. J., Jun, I., et al. (2018). Solar particle event storm shelter requirements for missions beyond low Earth orbit. *Life Sci Space Res (Amst)*, *17*, 32–39.

Townsend, L. W., Cucinotta, F. A, Wilson, J. W., & Bagga, R. (1994). Estimates of HZE particle contributions to SPE radiation exposures on interplanetary missions. *Advances in Space Research*, *14*(10), 671–674.

Townsend, L. W., de Wet, W. C., & Zaman, F. (2018). Estimates of radiation exposures to crews on missions in cis-Lunar space from the October 1989 series of solar particle events. *48th International Conference on Environmental Systems*, 8–12 July 2018, Albuquerque, NM.

Townsend, L. W., Porter, J. A., de Wet, W. C., Smith, W. J., McGirl, N. A., Heilbronn, L. H., & Moussa, H. M. (2016). Extreme solar event of AD775: Potential radiation exposure to crews in deep space. *Acta Astronautica*, *123*, 116–120.

Tylka, A. J., Dietrich, W. F., & Atwell, W. A. (2010). Band function representations of solar proton spectra in ground-level events. *38th Scientific Assembly of the Committee on Space Research (COSPAR)*, Bremen, Germany, July 18–25.

Usoskin, I. G., Kromer, B., Ludlow, F., Beer, J., Friedrich, M., Kovaltsov, G. A., et al. (2013). The AD775 cosmic event revisited: The Sun is to blame. *Astronomy & Astrophysics*, *552*.

Waselenko, J. K., MacVittie, T. J., Blakely, W. F., Pesik, N., Wiley, A. L., Dickerson, W. E., et al. (2004). Medical management of the acute radiation syndrome: Recommendations of the Strategic National Stockpile Radiation Working Group. *Ann Intern Med*, *140*(12), 1037–1051.

Worgul, B. V., Kundiyev, Y. I., Sergiyenko, N. M., Chumak, V. V., Vitte, P. M., Medvedovsky, C., et al. (2007). Cataracts among Chernobyl clean-up workers: Implications regarding permissible eye exposures. *Radiation Research*, *167*(2), 233–243.

4

Space Weather Radiation Effects on High-Altitude/-Latitude Aircraft

Christopher J. Mertens[1] and W. Kent Tobiska[2]

ABSTRACT

The effects from ionizing radiation exposure are a concern faced by aircrew, high-altitude pilots, frequent flyers, and commercial space travelers. This chapter describes the physics of atmospheric radiation's primary sources, including galactic cosmic rays (GCRs) and solar energetic particles (SEPs), as well as the possibility of a newly identified third source of secondary radiation caused by precipitating radiation belt charged particles. The primary radiation processes of transport through the heliosphere, modulation by the solar wind, and magnetic shielding by the Earth's geomagnetic field are described. The SEP primary radiation differential energy spectrum is described, and radiation belt precipitated charged particles are discussed with information from a large database of atmospheric radiation measurements. The transport of charged particles from all sources through the magnetic shielding from the Earth's geomagnetic field is described and the concept of geomagnetic cutoff rigidity is thoroughly explained. The transport of radiation through the atmosphere from precipitating charged particles below the mesopause is presented from a theoretical perspective, including the interactions of neutrons and charged particles with target species in the atmosphere. Radiation dosimetry is described, including dose quantities and quality factor. Additional discussions are presented with an introduction on radiation effects on avionics, the models capable of reporting the aviation radiation environment, the measurements that have been made of the atmospheric radiation by specific instruments, and introductory remarks on the regulatory activity internationally and in the US related to the effects of radiation faced by aircrew, high-altitude pilots, frequent flyers, and commercial space travelers.

4.1. INTRODUCTION

Air safety has improved significantly over the past decades, with the exception of effects from space weather, which includes ionizing radiation. Radiation exposure is a natural hazard faced by aircrew, high-altitude pilots, frequent flyers, and commercial space travelers.

Multiple sources of ionizing radiation contribute dose exposure in the aerospace environment that reaches from Earth's surface into space. Galactic cosmic rays (GCRs) and solar energetic particles (SEPs) are the dominant ionizing radiation sources (Figure 4.1). This is especially true for travel above commercial aviation altitudes starting at 8 km (26,000 ft.; Friedberg & Copeland, 2003, 2011; Tobiska et al., 2016). GCRs are produced outside the

[1] *Space Radiation Group, NASA Langley Research Center, Hampton, Virginia, USA*

[2] *Space Environment Technologies, Pacific Palisades, California, USA*

Space Physics and Aeronomy Collection Volume 5: Space Weather Effects and Applications, Geophysical Monograph 262, First Edition.
Edited by Anthea J. Coster, Philip J. Erickson, and Louis J. Lanzerotti.

DOI: 10.1002/9781119815600.ch4

solar system in high-energy explosive events and mostly consist of energetic protons that penetrate the heliosphere. They are slowly modulated by the strength of the Sun's interplanetary magnetic field (Simpson, 1983). SEPs come from solar activity such as coronal mass ejections related to flaring events or from interplanetary magnetic field shocks (Gopalswamy, 2004; Reames, 2013). In the latter case, fast coronal mass ejections plow through both ambient background and high-speed stream solar wind fields to create a shock front that produces accelerated energetic protons. A third radiation source has been recently hypothesized, based on measurements, and likely originates from the wave-particle interaction in the Van Allen radiation belts (RB) leading to precipitated charged particles (PCPs; Tobiska et al., 2018); substantial study is underway to understand the source of this third radiation source of RBPCPs.

The presence of primary ionizing dose in the aerospace environment, which can be altered by the composition of the terrestrial atmosphere and the dynamic variability of the Earth's magnetosphere resulting in secondary and tertiary ionizing radiation environments, causes concern for human health and vehicle avionics. It is this motivation that directs the study of radiation environment effects on high-altitude and high-latitude aircraft.

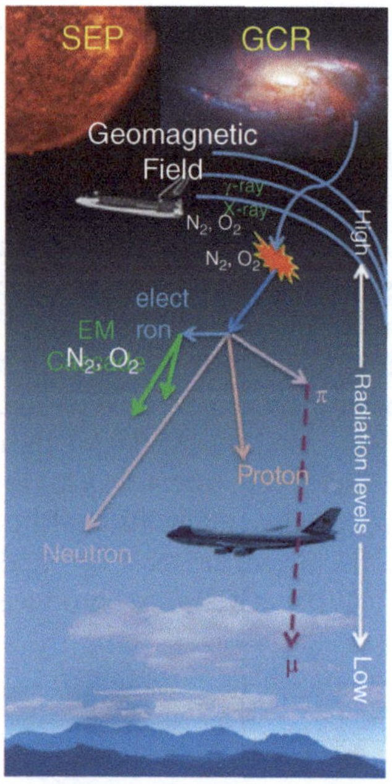

Figure 4.1 The complex radiation conditions at and above commercial aviation altitudes from galactic cosmic rays (GCRs) and solar energetic particles (SEPs).

4.2. THE PHYSICS OF ATMOSPHERIC RADIATION

4.2.1. Galactic Cosmic Rays

GCRs consist of approximately 90% protons and 8% helium nuclei, with the remainder being heavier nuclei and electrons (Gaisser, 1990). After these particles penetrate the magnetic fields of the solar system and the Earth to reach the Earth's atmosphere, their collisions with air molecules create cascades of secondary and tertiary radiations (Reitz et al., 1993). The collisions are primarily due to Coulomb interactions of the GCR particle with orbital electrons of the atmospheric molecules, delivering small amounts of energy to the orbital electrons. This energy transfer process leaves behind electron-ion pairs (Wilson et al., 1991). The ejected electrons often have sufficient energy to extend ionizing events. The GCR ions lose a small fraction of their energy and endure many atomic collisions before slowing down. Occasionally the GCR ion will collide with an air molecule nucleus where large energies are exchanged, resulting in dramatic changes to the ion and nucleus. The remnant nucleus emits further air nuclear constituents by decaying through the usual radioactivity channels (Wilson et al., 1991). An important secondary particle created in the GCR-air interactions is the neutron. Because of charge neutrality, the neutron is able to penetrate deep into the atmosphere, causing further ionization events and contributing over half the atmospheric radiation exposure at typical commercial airline altitudes (Wilson et al., 2003). Furthermore, neutron exposures create a relatively high health risk since neutron interactions produce abundant ions in the cell that is impacted and repair is less efficient for these events (Mertens et al., 2012; Wilson, 2000).

The intensity of the atmospheric radiations, their energy distribution, and their effects on aircraft crew and passengers vary with altitude, location in the geomagnetic field, and the Sun's magnetic activity (solar) cycle (Heinrich et al., 1999; Reitz et al., 1993; Wilson, 2000). The atmosphere provides material shielding depending on the atmospheric depth overhead. The geomagnetic field provides shielding by deflection of low-momentum charged particles back into space. Because of the orientation of the dipolar geomagnetic field, the polar and high-latitude regions experience increased particle penetration. At each geographic location, the minimum momentum per unit charge, i.e. magnetic rigidity, that a vertically incident particle can have to reach a given magnetic latitude is called the vertical geomagnetic cutoff rigidity. The local incident charged particle flux varies widely with geomagnetic location and solar modulation. When solar activity is high, GCR flux arriving from outside the solar system is low and vice versa. The balance between outward

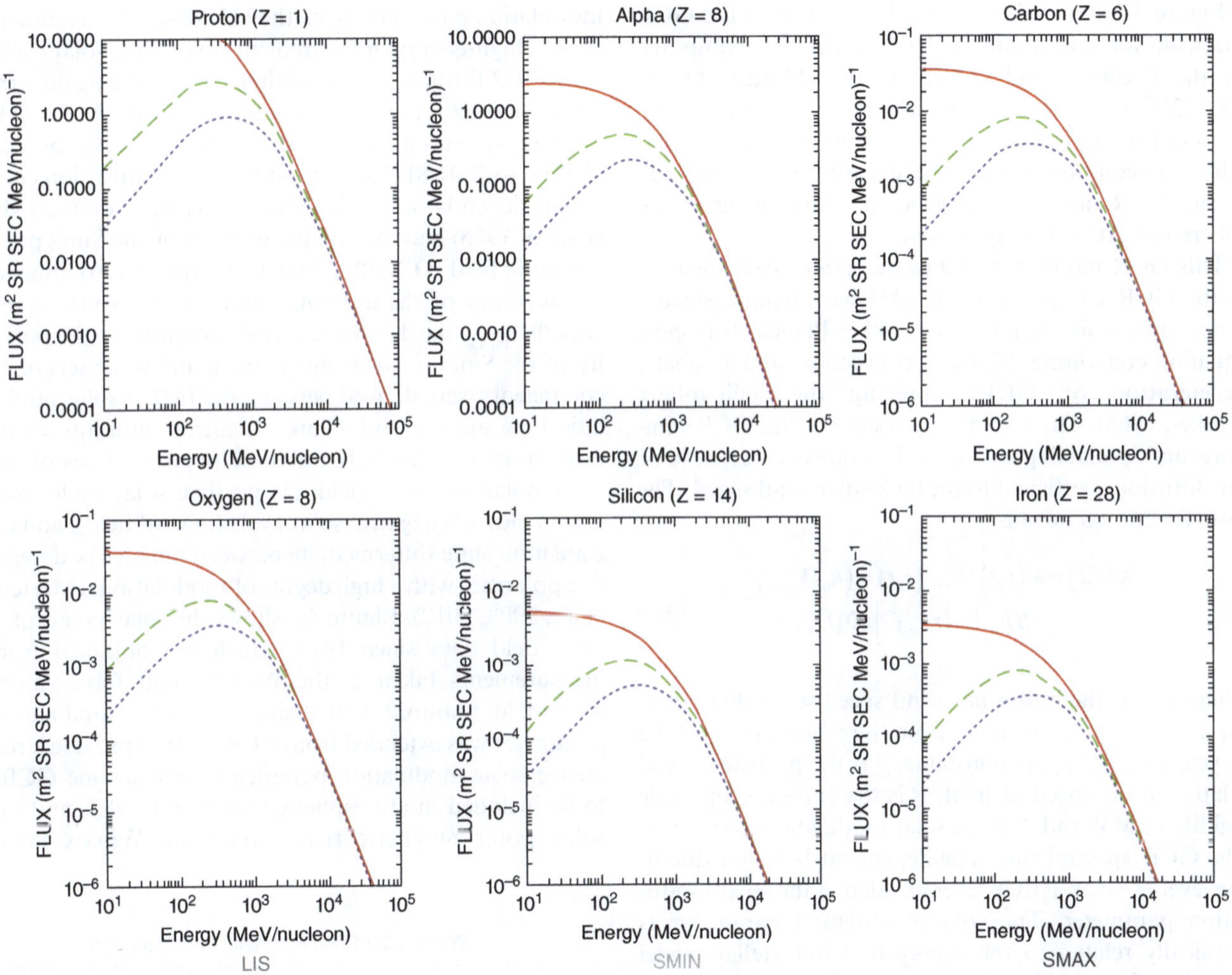

Figure 4.2 GCR spectral flux for various nuclei predicted by the Badhwar and O'Neill model for solar cycle 23. The local interstellar spectrum (LIS) is denoted by the red lines. Solar minimum spectra are represented by June 1996 conditions and are denoted by green lines. Solar maximum spectra are represented by June 2000 conditions and are denoted by blue lines.

convective flux of solar wind and the inward diffusive flux of GCRs is responsible for the dynamic anticorrelation between incident GCRs and solar cycle activity (Clem et al., 1996; Parker, 1965).

4.2.2. Heliospheric GCR Transport

GCRs are transported through the heliosphere to the near-Earth space environment and can be modeled using an expanded version of the Badhwar and O'Neill model (Badhwar & O'Neill, 1991, 1992, 1993, 1994, 1996). This model, referenced here as the GCR model, was updated (O'Neill, 2006) to use ground-based neutron monitor count rates from the Climax neutron monitor site. This provided a measurement constraint on the simulated solar cycle modulation of the GCR spectrum at 1 astronomical unit (AU). Comparisons between the GCR model and NASA Advanced Composition Explorer (ACE) satellite CRIS instrument measurements of the GCR spectra have shown that this step has enabled accurate predictions of GCR spectra in the geospace environment on monthly to seasonal time scales (O'Neill, 2006). O'Neill (2006) was extended by incorporating four high-latitude neutron monitor count rate measurements (Thule, Oulu, Izmiran [or Moscow], and Lomnicky) into the GCR model predictions at 1 AU (Mertens et al., 2009, 2012). These high-latitude neutron monitor locations are sensitive to the GCR spectral region most influenced by solar cycle variability, and the data from these stations are available in real time or near real time.

Figure 4.2 shows the solar cycle variation in the GCR spectrum for several nuclei. The spectra were computed by the Badhwar and O'Neill model (Mertens et al., 2007, 2008). The figure shows the local interstellar spectrum (LIS) and the spectra at solar maximum and solar minimum conditions. The solar cycle modulation of LIS as the GCR nuclei are transported through the heliosphere to 1 AU is clearly evident.

This GCR model propagates the LIS of each element in the GCR composition to 1 AU by solving a steady-state, spherically symmetric Fokker-Planck transport equation containing diffusion, convection, and adiabatic deceleration of GCRs entering the heliosphere (Parker, 1965). The transport physics enables GCR temporal and spatial dependence to be expressed as a ratio of the diffusion coefficient to the bulk solar wind speed. The ratio functional form is

$$\tilde{k}(r,t) \equiv k(r,t)/V_{SW}(r,t) = (k_0/V_{SW}) \times \beta R\left[1+(r/r_0)^2\right]/\Phi(t), \quad (4.1)$$

where V_{SW} is the bulk solar wind speed set to 400 km s^{-1} for all time; r is the distance from the Sun in AU, t is time in years, k_0 and r_0 are constants, β is the particle's speed relative to the speed of light, R is the particle's magnetic rigidity in MV, and Φ is the solar modulation parameter. The GCR spectral flux time-dependent behavior due to the level of solar activity is embedded in the solar modulation parameter. The solar modulation parameter is physically related to the energy that interstellar nuclei must have in order to overcome the heliospheric potential field. The latter is established by the large-scale structure of the interplanetary magnetic field (IMF). The solar modulation parameter is determined by fitting the solution of the Fokker-Planck equation for a specified GCR nuclei to the corresponding spectral flux measurements throughout the solar cycle as described by O'Neill (2006).

Once the solar modulation parameter was derived based on satellite oxygen spectra, the LIS for the remaining elements (i.e. lithium [Z = 3] through nickel [Z = 28]) were similarly determined by fitting the solutions of the Fokker-Planck equation to the ACE CRIS instrument spectral flux measurements. A simple power law form of the differential LIS can be assumed:

$$j_{\text{LIS}}(E) = j_0 \beta^{\delta} (E + E_0)^{-\gamma}, \quad (4.2)$$

where E is the particle kinetic energy per nucleon and E_0 is the rest mass energy per nucleon (938 MeV/n). The free parameters (γ, δ, and j_0) were determined from the fit of the GCR model to satellite measurements.

The GCR model was extended beyond the time period of satellite measurements by first deriving the solar modulation parameter from the IMP-8 satellite channel 7 (Z > 8, high-energy) measurements over three solar cycles from 1973 through 2001 and then calibrating against the solar modulation parameter derived from ACE/CRIS satellite measurements for the period of data overlap (1997.6 to 2001.8). The heliospheric GCR diffusion coefficient depends on the large-scale structure of the IMF (Parker, 1965) and thus on the polarity of the Sun's polar magnetic field (O'Neill, 2006). Consequently, to improve the accuracy of the neutron count rate fits to the reference Φ_{CLIMAX}, the data were sorted according to the polarity of the Sun's polar magnetic field, and three sets of fit coefficients were derived between the IMP-8 solar modulation parameter and Climax neutron monitor count rates from 1973 to 2001 for three polarity states of the solar polar magnetic field: (1) positive solar cycle (outward field), (2) negative solar cycle (inward field), and (3) transition state (intermediate between positive and negative polarities with a high degree of modulation) (Mertens et al., 2009, 2012). Figure 4.3 shows the solar polar magnetic field data since 1978, which are obtained from measurements taken at the Wilcox Solar Observatory located at Stanford University. The solar modulation parameter was extended from 1958-2009 to provide a reference solar modulation parameter for real-time GCRs to be included in atmospheric radiation models and the solar polar magnetic field data from Wilcox Solar

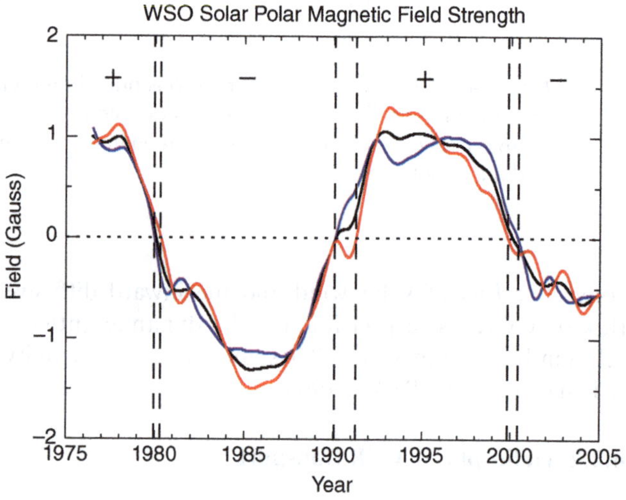

Figure 4.3 Wilcox Solar Observatory solar polar magnetic field measurements. The blue line is the field strength of the northern solar hemisphere. The red line is the negative of the southern hemisphere field strength. The black line is the average solar field strength of the northern and southern hemispheres (A = (N-S)/2). The vertical dashed lines separate different solar polar magnetic polarity states. The plus and minus signs in the legend denote the positive and negative polarity states, respectively.

Observatory must be added to the atmospheric radiation model input data streams.

High-latitude neutron monitor sites are particularly valuable for real-time GCR models. By cross-correlating the Climax-based solar modulation parameter (Φ_{CLIMAX}) with neutron count rates measured at high-latitude sites (Mertens et al., 2009, 2012), they maximize the solar cycle information content contained in the GCR spectrum via the ground-based neutron count rates. Neutrons detected on the ground are secondary particles produced by nuclear fragmentation reactions between the incoming GCR particles and atmospheric constituents (Wilson et al., 1991). At high latitudes, the geomagnetic shielding of the incoming GCR particles is low. Thus, the ground-level neutron count information is high for the low and medium energy region of the GCR spectrum. This is desirable because this GCR spectrum energy range is most modulated by the solar wind, and IMF and is closely related to solar activity cycle (Mertens et al., 2008).

4.2.3. Solar Energetic Particles

Solar energetic particles (SEPs) originate from coronal mass ejections (CME) from active regions on the solar surface (Kahler, 2001; Wilson, Mertens, et al., 2005a). The CME propagates through interplanetary space carrying the local surface magnetic field frozen into the ejected mass along with it. There is a transition or shock region often associated between the normal background and sectored magnetic structure of interplanetary space and the fields frozen into the ejected mass. The plasma is accelerated at this shock front and forms SEPs. As an accelerated region passes an observation location, the flux intensity is observed to dramatically increase. The SEP energy spectrum obtained in the acceleration process is related to the plasma density and CME velocity. During a CME event, the number flux distribution that is incident upon Earth's atmosphere is a combination of the GCR, SEP, and RBPCP distributions. The SEP-air interaction mechanisms are the same as GCR-air interactions described above, where the secondary and tertiary radiations caused vary with altitude and geomagnetic field cutoff rigidity.

Because the particles from SEP events are transported from the Sun to the near-Earth environment through the inner heliosphere, their energy spectrum results from injected particle seed populations that are stochastically accelerated in a turbulent magnetic field. That field is associated with a CME-driven interplanetary shock (Tylka & Lee, 2006). Ellison and Ramaty (1985) provided an analytical expression representing the differential energy spectrum for shock acceleration with the form

$$d^2J/dEd\Omega = C_a E^{-\gamma_a} \exp(-E/E_0). \qquad (4.3)$$

The differential energy spectrum on the left-hand side of equation (4.3) has units of $(\text{cm}^2\text{-sr-hr-MeV/n})^{-1}$, and the energy (E) has units of MeV/n (i.e. MeV/nucleon). The constant value, C_a, relates to the injected seed population upstream of the shock. The power-law energy dependence of the spectrum occurs from shock acceleration of the seed population by random first-order Fermi acceleration (scattering) events in a turbulent magnetic field. The power index (γ_a) parameterizes the shock compression ratio. The exponential turnover in equation (4.3) represents high-energy limits to the acceleration mechanism, with particle escape from the shock region being a prime example. Using the above analytical form, the three parameters (C_a, γ_a, and E_0) can be determined by fitting equation (4.3) to ion flux measurements.

Mewaldt et al. (2005) noted that the Ellison-Ramaty spectral form did not properly fit the GOES ion flux measurements in the highest energy channels during the Halloween 2003 SEP events. As a result, Mewaldt et al. proposed using a double power-law spectrum. The low-energy spectrum follows the Ellison-Ramaty form, while the high-energy spectrum has a power-law energy dependence using a different power index, such that

$$d^2J/dEd\Omega = C_b E^{-\gamma_b}. \qquad (4.4)$$

The power-law expressions in equations (4.3) and (4.4) can be merged into one continuous spectrum by requiring that the differential energy spectra in equations (4.3) and (4.4) and their first derivatives are continuous at the merge energy. Mertens et al. (2012) show this combined form with the expression

$$d^2J/dEd\Omega = CE^{-\gamma_a} \exp(-E/E_0), E \le (\gamma_b - \gamma_a) E_0 \qquad (4.5)$$

$$= CE^{-\gamma_b} \left\{ \left[(\gamma_b - \gamma_a) E_0 \right]^{(\gamma_b - \gamma_a)} \exp(\gamma_b - \gamma_a) \right\},$$
$$E > (\gamma_b - \gamma_a) E_0. \qquad (4.6)$$

Physically, the double power-law spectrum in equations (4.5) and (4.6) represents SEP sources from two different injected seed populations. The low-energy spectrum, with γ_a power index and the e-folding energy E_0, is possibly associated with solar corona (solar wind) seed populations while the high-energy spectrum, with γ_b power index, is likely associated with flare suprathermal seed populations (Tylka et al., 2005).

Another widely used analytical representation of a SEP energy spectrum is a Weibull distribution (Townsend et al., 2003, 2006), which has been successful at fitting satellite ion flux measurements. The differential energy spectrum is given by

$$d^2J/dEd\Omega = C k\alpha E^{\alpha-1} \exp(-kE^{\alpha}). \qquad (4.7)$$

The Weibull distribution has an analytical form similar to the Ellison-Ramaty distribution in equation (4.3), and the former's exponential energy dependence could be due to dissipation of high-energy SEP ions through scattering from self-generated waves. However, some researchers recommend further investigation (Xapsos et al., 2000).

Models such as the NASA Langley Research Center's Nowcast of Atmospheric Ionizing RAdiation System (NAIRAS) physics-based radiation model (Mertens et al., 2013) using a continuous, real-time data stream fits four analytical SEP spectral fluence rate functions to satellite ion flux measurements. The analytical forms that are fit to measurements are (1) single power-law in equation (4.4), (2) Ellison-Ramaty in equation (4.3), (3) double power-law in equations (4.5) and (4.6), and (4) Weibull in equation (4.7). The free parameters for each analytical differential energy distribution are derived by a nonlinear least-square fit to differential-directional ion flux measurements using a Marquardt-Levenberg iteration technique (Brandt, 1999). NAIRAS, for example, utilizes available real-time measurements of proton and alpha differential-directional particle flux (cm^2-sr-sec-MeV/n)$^{-1}$ for the SEP spectral fitting described above. SEP spectral fluence rates (cm^2-hr-MeV/n) incident on Earth's magnetosphere are obtained by time-averaging the particle flux measurements in 1-hour time bins and projecting the incident flux onto the vertical direction assuming an isotropic angular distribution for the solar ions.

As the SEPs arrive at the top of the Earth's atmosphere, their characteristics have been predetermined by several items. First, the geomagnetic field has a significant influence on SEP atmospheric ionizing radiation exposure. For example, the magnetospheric magnetic field can weaken the overall geomagnetic field which results in an increase in radiation levels. The cutoff rigidity values will be zero where open geomagnetic field lines exist. Thus, effective dose rates based on the solar on the Earth main field, e.g. the International Geomagnetic Reference Field (IGRF), are underestimated even for magnetically quiet times. During strong geomagnetic storms, the region of open field lines is further expanded and this produces larger exposure rates at much lower latitudes. Second,

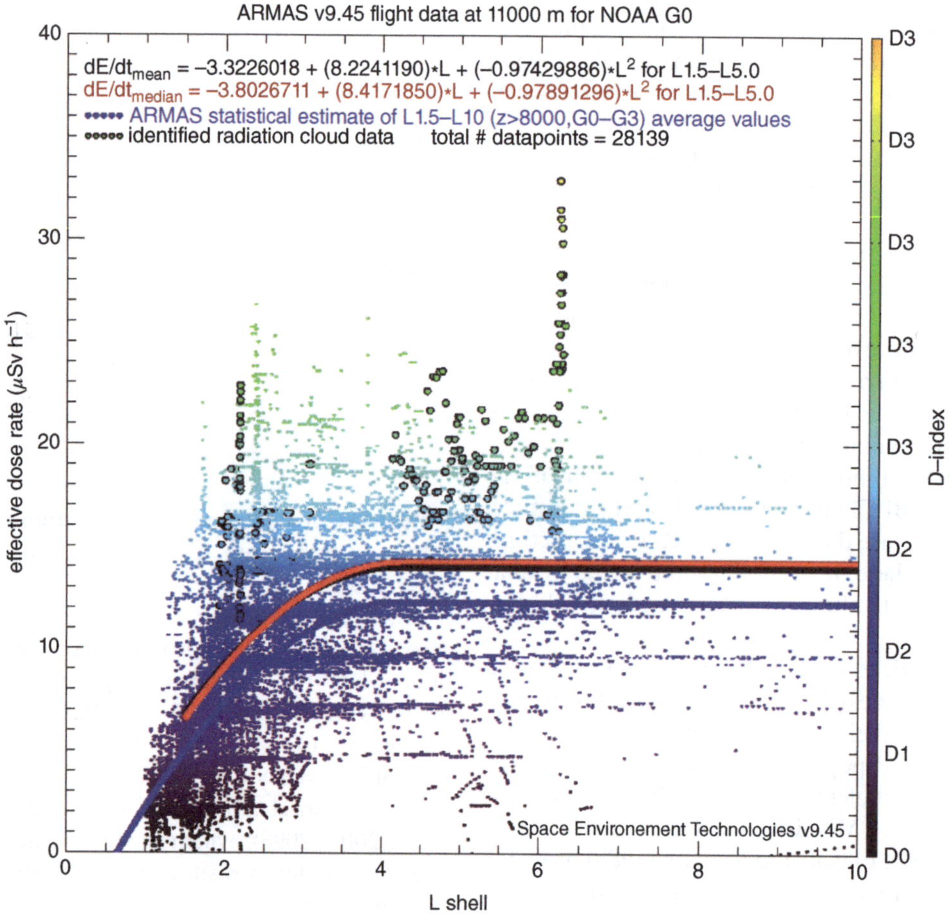

Figure 4.4 ARMAS effective dose rates substantially increase between $4 \leq L \leq 5$ above typical 11 km GCR values for quiet geomagnetic conditions during solar cycle 24 decline.

there is a strong altitude dependence upon atmospheric shielding as discussed in detail later.

4.2.4. Radiation Belt Precipitated Charged Particles

Recent measurements (Dachev, 2017; Dachev et al., 2017a; Tobiska et al., 2016, 2018) have suggested the possibility of another radiation source besides GCRs and SEPs. Liulin instrument International Space Station (ISS) measurements (Dachev, 2017; Dachev et al., 2017b) have seen radiation effects from relativistic electrons originating in the outer radiation belt. Automated Radiation Measurements for Aerospace Safety (ARMAS) instrument aircraft and balloon measurements have collected a large number of atmospheric observations at higher magnetic latitudes with L-shells between 1.5 and 5. These data are gathered during quiet and minor geomagnetic disturbances, including during high-speed streams. For National Oceanic and Atmospheric Administration (NOAA) G-scale G0 conditions at 11 km altitude near L = 4, the tissue-relevant effective dose rates are statistically higher than the GCR background by 15% (Figure 4.4) for many observations (Tobiska et al., 2018). We propose that this is a new, third radiation source likely originating from charged particles that are associated with the Van Allen radiation belts. Highly variable, dynamic mesoscale radiation events at aviation altitudes are attributed to these particles as a source, where their manifestation is analogous to an aircraft flying through radiation "clouds," although physically the aircraft appears to be flying through a γ-ray beam.

A common characteristic of these enhanced radiation event periods at higher magnetic latitudes ($2 \le L \le 6$) is that the radiation levels during flight begin at the GCR background level, rise to almost a factor of 2 higher, and then return to background GCR levels within less than an hour. This is consistent with flight through a secondary, Bremstrahlung γ-ray beam produced by precipitating relativistic electrons that are stopped in the mesosphere, for example. Throughout the ARMAS database, there have been over 100 cases of these types of enhanced radiation events that are candidates for study related to effects from radiation belt particles (Tobiska et al., 2018).

A recent and extensive study comparing ARMAS measurements (Tobiska et al., 2018) with the NAIRAS model (Mertens et al., 2013) extracted GCRs and suggested relativistic electron precipitation (REP) statistics. The Tobiska et al. (2018) study described a subset of observations of enhanced radiation short-term events lasting tens of minutes to under two hours of dose rate enhancements occurring between 9 and 14 km altitude in a well-defined magnetic latitude band. The dose rate enhancements were not attributed to GCRs or SEPs. They explored, for example, how energetic electrons from the outer Van Allen radiation belt can be precipitated by electromagnetic ion cyclotron (EMIC) waves (Tsurutani et al., 2016) associated with particle injections from geomagnetic substorms. Using modeling and observations, EMIC waves were shown as an example of how to supply a population of relativistic electrons that can subsequently become a source of radiation enhancements above the GCR background. The primaries are absorbed between 50 and 60 km, and a fraction of their secondaries are bremsstrahlung γ-rays that can penetrate down to 10–14 km at commercial aviation altitudes.

Lines of evidence for an energetic electron radiation source in one example were considered from many events that all have a common feature, i.e. the data in each example started near statistical and climatological GCR baseline values, and from there the dose rates were observed to rise significantly above and later fall back to that baseline over tens of minutes to under 2 hours (Figure 4.5, top panel). Data in this figure were taken on aircraft from the National Science Foundation National Center for Atmospheric Research Gulfstream 5 (G-5), while other ARMAS measurements were made on the NOAA G-4, the NASA Armstrong Flight Research Center DC-8 and G-3, the Federal Aviation Administration (FAA) William J. Hughes Technical Center Bombardier Global 5000 (BG5), and commercial B-737, B-747, B-767, B-757, A-320, A-380, and E-175 aircraft. The criteria for selecting potential events included altitudes above 9 km, level flight during the event, magnetic latitudes corresponding to inner and outer belt L-shells (magnetic latitudes between 43° and 67° for both hemispheres, i.e. L = 1.9 to 6.8), and a significant event rise (at least 40%) above then return to GCR background values as represented by modeled results for a flight.

Figure 4.6 shows the ratio of ARMAS (measurement) to NAIRAS (model with only GCR contribution at this time) effective dose rates throughout each minute of many ~50$^+$ magnetic latitude flights. These ratios compare ARMAS to NAIRAS for each altitude level above 8 km and for geomagnetic conditions of G0 and G1 on the NOAA G-scale. Most ratios are above 1.0 (blue) and the mean of 1537 1-minute ratio time records for all events is 1.91. In other words, there is nearly a factor of 2 increase in dose rate (green values) while flying through a radiation event compared with flying at the same altitude and under the same geomagnetic conditions under only GCR background conditions. Measurements show a defined magnetic latitude band where these events occur. The boundaries of this band are L-shells 2–7, where L-shell is the distance from the equator to the crossing point of the magnetic field line (McIlwain, 1961; Tascione, 1994). These boundaries are labeled "L2" and "L7" in Figure 4.6, and the band is marked in yellow in both hemispheres. L-shells from ~3–7 are the regions where outer radiation belt charged particles

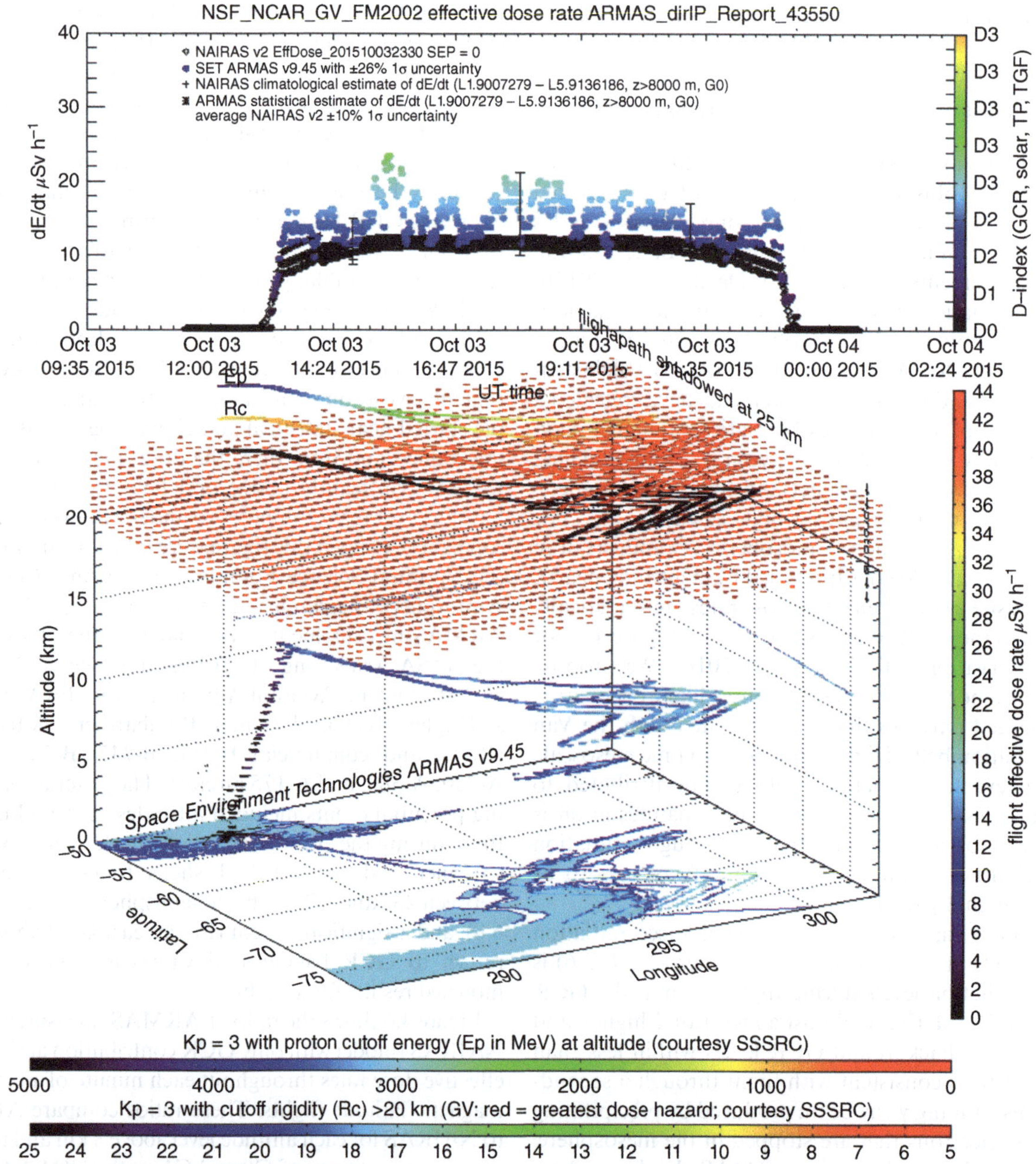

Figure 4.5 The ARMAS 3 October 2015 G5 flight at 11.5 km and magnetic latitudes near –63° with doubled effective dose rate for ~1/2 hour; effective dose rate vs. time compared with NAIRAS (top panel black diamonds) and the flight geographical context with information related to proton cutoff energies and cutoff rigidities (bottom panel).

map into the upper atmosphere, while L-shells from ~1.3–2 are the regions where inner radiation belt charged particles map into the upper atmosphere. This L-shell band is consistent with Katsiyannis et al. (2018), who observed low Earth orbit relativistic electrons from the PROBA-V satellite EPT and LYRA instruments, and Dachev et al. (2017a, 2017b), as well as Dachev (2017), who have called these precipitation bands when observed by the Liulin instrument on the ISS.

One typical flight (Figure 4.5, top panel) on 3 October 2015 at 15–16 UT shows an event with an effective dose rate that rose and declined by a factor of 2 above GCR background levels within 33 minutes and centered on 15:29 UT. The entire flight dataset was collected for 13.1 h between 3 October at 11:30 UT and 4 October at 00:34 UT. The event occurred for a fraction of the total flight. Uncertainties in the ARMAS data are ±24% and are described by Tobiska et al. (2016).

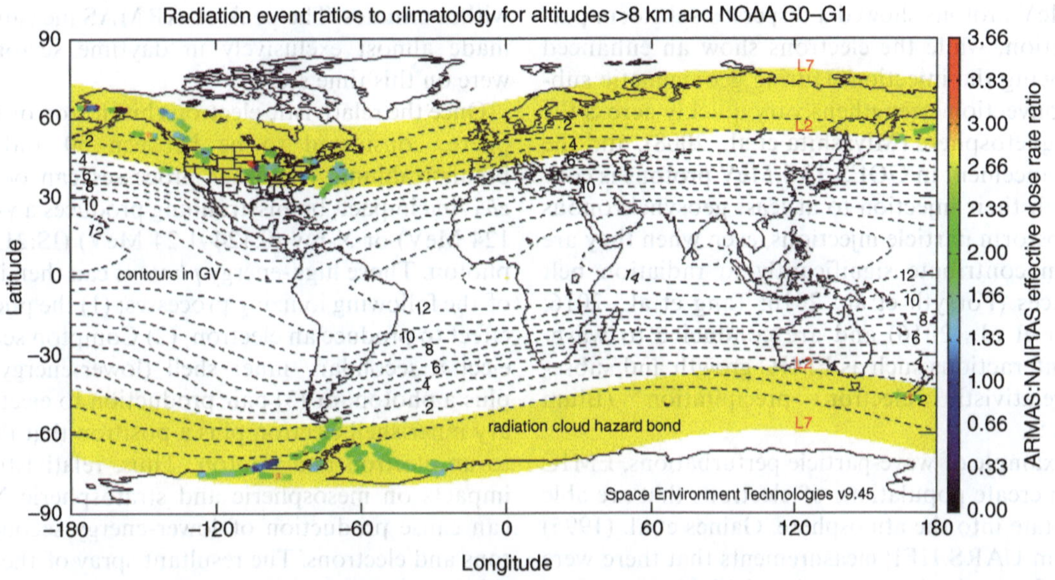

Figure 4.6 One-minute ratios of ARMAS (measurements) to NAIRAS (model; GCR at this time) for all radiation events in all altitudes above 8 km and for geomagnetic conditions of G0 and G1 on the NOAA G-scale. Cutoff rigidities are in units of GV (dashed lines). (Shea & Smart, 2012).

For the range of cutoff rigidities from Chile to Antarctica (Rc = 3.89 to 0.24, respectively; Figure 4.5), the mean radiation quality factor (Q, relevant to the dose rate in tissue) was 2.03 with a Rc dependency (Tobiska et al., 2016). The total flight effective dose was 157 µSv for the 3 October 2015 flight, i.e. the exposure of approximately one and a half chest X-rays. For this event, starting at 15:07 UT (–60.1° magnetic latitude) the peak dose rate (top panel colored dots in Figure 4.5) occurred at 15:29 UT (11:38 local clock time) when the aircraft was flying at an L-shell of 4.8 corresponding to –62.7° magnetic latitude. Background dose rates were once again reached at 15:45 UT (–64.6° magnetic latitude). The background levels were very close to the values of NAIRAS shown in the Figure 4.5 top panel (black "+" NAIRAS model, black "◊" NAIRAS values for date/time/location, black "∗" ARMAS statistical mean values for locations and geomagnetic conditions). These three background estimates lie nearly on top of one another and represent a consensus for the GCR contribution to the radiation environment. From the plane's perspective, which was flying at a constant altitude in this mesoscale region, the increased dose rate for 33 minutes was analogous to flying through a radiation cloud.

In previous work (Tobiska et al., 2016, 2018) it was observed that the dose rate enhancements in these types of measurements were not SEP related since no solar event had occurred within the previous few hours to days. In addition, changes in GCRs do not normally produce these magnitudes of short-term and localized variability along a flight path. There were no large tropospheric storms or hurricanes in the regions that might produce terrestrial γ-ray flashes or sprites (Chilingarian et al., 2015) as a radiation source. Mertens, Kress, et al. (2010) show that high geomagnetic activity can affect the aviation radiation environment. However, Tobiska et al. (2016) identified only minor Dst disturbances during the events. Because there was essentially no change in background GCRs during each of the ARMAS events during their short durations and there were no SEPs at the time of each event, and because the events demonstrate dynamic variability above the background GCR level over short timescales and across mesoscale regions, these events must be connected to magnetospheric processes.

Substorm injections can provide a source of MeV electrons in the outer radiation belt and are associated with high-speed streams of particles from the Sun (Hajra, Tsurutani, Echer, Gonzalez, Brum, et al., 2015; Hajra, Tsurutani, Echer, Gonzalez, & Santolik, 2015; Tsurutani et al., 2006). Direct injection of MeV particles by substorms has also been reported (Ingraham et al., 2001). Exploring the 3 October 2015 magnetospheric response, as reflected by geomagnetic activity on this day peaking at Kp = 3+, the Dst index was –24 nT at the time of the event. No solar energetic proton events were reported during this period and no enhancements were detected in either the NOAA GOES >10 MeV proton flux measurements or the Global Positioning System (GPS) proton data. The GOES-13 0.6 MeV electrons were rising and the >1 MeV proton flux was elevated following a rapid enhancement during the 3 October 2015 observations.

The >1 MeV protons showed a dispersed substorm particle injection, while the electrons show an enhanced flux following the injection. During geomagnetic substorms, convection strengthens vary quickly across the entire magnetosphere (Miyashita et al., 2008), and the rapid enhancement in GOES-13 1 MeV protons is consistent with their injection to at least geosynchronous orbit. Substorm particle injections, even when they are small, can contribute significantly to radiation belt disturbances (Forsyth et al., 2016; Tang et al., 2016; Tsurutani et al., 2016) and are conducive to wave-particle interactions such as EMIC growth and subsequent relativistic electron precipitation (Blum et al., 2015).

As an example of wave-particle perturbations, EMIC waves can create populations of electrons that are able to precipitate into the atmosphere. Gaines et al. (1995) found from UARS HEP measurements that there were 2 orders of magnitude increase in >1 MeV electrons at $L = 4$ during geomagnetic storm periods, followed by significant increases in relativistic electrons in the inner radiation belt ($2 \le L \le 3$) a day later. High fluxes of relativistic electrons existed at least a week after a major geomagnetic storm, even while Dst was –30 to –20 nT, and Gaines et al. concluded that ion pair production below 60 km can easily be an order of magnitude more than during quiet geomagnetic conditions. Clilverd et al. (2007) were further able to demonstrate dayside REP from EMIC waves, originating in the $4 \le L \le 5$ outer radiation belt that were associated with a CME-magnetosphere coupled event. Xu et al. (2018) studied monoenergetic beams of 0.1–10 MeV precipitating electrons into the atmosphere to understand the process of bremsstrahlung radiation and its resultant ionization production and atmospheric effects, including pitch angle dependence in the ionization rate profile. EMIC waves can be ducted through the ionosphere before reaching the ground (Kim et al., 2011), thus allowing for their observations by ground-based magnetometers far from the field lines where they originate (Rodger et al., 2007).

An important feature contributing to electron precipitation losses is the spread in their pitch angle distribution. Electrons with a wide range of equatorial pitch angles can bounce (mirror) from one conjugate hemisphere to the other as they drift between magnetic field lines. However, EMIC waves will modify the pitch angles such that more electrons find their pitch angles pushed towards the loss cone, leading to atmospheric loss (precipitation) of the particle. While EMIC waves may be generated in the midnight sector of the magnetosphere during substorm particle injections, it is typical for wave growth to occur that results in transport of relativistic electrons to the dayside magnetosphere within minutes. The peak of ARMAS measured events, made almost exclusively in daytime sector L-shells, were on this time scale.

Once the relativistic electrons hit the top of the atmosphere, considered to be between 50 and 100 km, bremsstrahlung and other processes can occur. From *electron* impact, bremsstrahlung produces a γ-ray (1.24–124 MeV) or X-ray (0.124–1.24 MeV) (IS:21348, 2007) photon. Those high-energy *photons* can then induce one of the following ionizing processes: (1) the photoelectric effect to produce an electron, (2) Compton scattering to eject a secondary inner shell (lower-energy) electron plus a photon, or (3) pair production to eject a secondary inner shell electron plus a positron that then decays to an electron and photon. Thus, relativistic electron impacts on mesospheric and stratospheric N_2 and O_2 can cause production of lower-energy secondary photons and electrons. The resultant spray of these secondaries and tertiaries can then impact other atmospheric target atoms, causing additional lower-energy photons and particles. Artamonov et al. (2016), using the CRAC:EPII model, have even demonstrated ion pair production in the lower atmosphere between 155 and 205 g cm^{-2} (10–12 km) due to >10 MeV primary electrons producing secondary radiation.

4.2.5. Magnetospheric Transport

Lower energy cosmic rays are effectively attenuated by the geomagnetic field (internal field plus magnetospheric contributions) as these charged particles are transported through the magnetosphere and into the neutral atmosphere. The geomagnetic field provides a form of momentum shielding, or attenuation, by deflecting the lower-energy charged particles back out to space via the Lorentz force. This spectral filtering effect is quantified by a canonical variable, in the mathematical sense, called the geomagnetic cutoff rigidity, R_c. Once the cutoff rigidity is known, the minimum access energy to the neutral atmosphere is determined for each incident charged particle through the relativistic energy equation (Mertens, Kress, et al., 2010).

The geomagnetic cutoff rigidity quantity is determined by considering the motion of a charged particle in a magnetic field. The particle motion is solved with Newton's equation of motion for a charged particle subject to the Lorentz force. For a positively charged particle, the equation of motion is

$$dp/dt = (Ze/c)\mathbf{v} \times \mathbf{B} \tag{4.8}$$

in cgs units. The bold-faced quantities are vectors, and × designates the vector cross-product. The charged particle momentum and velocity are $\boldsymbol{p}$ and $\boldsymbol{v}$, respectively, and **B** is the magnetic field strength. The magnitude of the charge of an electron is denoted e, and Z is the number of electron charge units. The equation of motion in equation (4.8) can be written equivalently as

$$(R/B)\,d\hat{v}/dt = \hat{\mathbf{v}}'\hat{\mathbf{B}}, \tag{4.9}$$

where the ^ symbol denotes unit vectors and

$$\mathrm{R} \equiv pc/Ze \tag{4.10}$$

is defined as the rigidity. The canonical aspect of the rigidity is evident in the above equation. For a given magnetic field strength, **B**, charged particles with the same rigidity follow identical trajectories.

Motions of charged particles in a pure magnetic dipole field were examined by Störmer (1965). Because of the azimuthal symmetry in a pure dipole field, the azimuthal angular momentum is a conserved quantity. A main feature of Störmer theory is that regions of bounded and unbounded motion can be derived analytically from the integral of motion found from the conservation of azimuthal angular momentum (Störmer, 1965; Van Allen, 1968). It can be shown that the minimum rigidity that a vertically arriving particle must have in order to reach an altitude z above the Earth's surface is

$$R_{vc} = \frac{\mathcal{M}}{(R_e+z)^2}\cos^4\lambda_m \approx 15\cos^4\lambda_m\,(\mathrm{GV}). \tag{4.11}$$

In the above equation, R_{vc} designates the vertical geomagnetic cutoff rigidity, M is the Earth's magnetic dipole moment, R_e is the average radius of the Earth, and λ_m denotes magnetic latitude. Therefore, vertically arriving charged particles with energies (E) less than the cutoff energy (E_{vc}) will be deflected by the Lorentz force and not reach altitude z. The cutoff energy for each charged particle of charge Z and mass number A is determined from the canonical cutoff rigidity through the relativistic energy equation, such that

$$E_{vc} = \left[\sqrt{R_{vc}^2\left(Z/A\cdot \mathrm{amu}\cdot c^2\right)^2+1}-1\right]\cdot \mathrm{amu}\cdot c^2, \tag{4.12}$$

where E is kinetic energy per nucleon (MeV/n), R_{vc} is vertical geomagnetic cutoff rigidity (MV), c is the speed of light in vacuum, and amu = 931.5 MeV/c^2 (atomic mass unit). Thus, the geomagnetic field has the effect of filtering out lower-energy charged particles as they are transported through the magnetosphere and into the neutral atmosphere.

The vertical cutoff rigidity at the Earth's surface derived from Störmer theory is plotted versus geographic latitude in Figure 4.7. The maximum cutoff rigidity is at the equator since a vertically arriving charged particle is perpendicular to the dipole magnetic field lines at the equator. The effect of the vector cross product in the Lorentz force in equation (4.8) is that charged particle motions perpendicular to magnetic field lines will experience the maximum deflection, while particle motions parallel to the magnetic field will experience no deflecting force whatsoever. Figure 4.7 illustrates that a vertically arriving proton at the equator must have a kinetic energy of ~15 GeV to arrive near the surface of the Earth. In the polar regions, vertically arriving charged particles travel parallel to the magnetic field lines. Therefore, the cutoff rigidity is zero and particles of all energies can arrive on the Earth's surface at the magnetic poles.

The Earth's magnetic field is not a pure dipole field. On the contrary, the internal field is comprised of dipolar and nondipolar contributions (Langlais & Mandea, 2000). The dipole moment is off-center and tilted with respect to the rotational axis. Furthermore, the magnetic field is distorted at large radial distances ($r \geq 4R_e$) by its interaction with the solar wind. A balance between the solar wind dynamic pressure and the magnetic field pressure, from the internal magnetic field, is established by inducing five magnetospheric magnetic field current systems (Tsyganenko, 1989, 2002). These current systems generate their own magnetic fields, which add vectorally to the internal magnetic field. The complexities of the actual internal magnetic field, with dipolar and nondipolar contributions, and the magnetospheric magnetic field contributions prohibit an analytical solution for the vertical geomagnetic cutoff rigidity and numerical methods must be employed.

The internal magnetic field can be specified by the IGRF model (Langlais & Mandea, 2000). In the IGRF, the internal field is represented by a magnetic potential function ($\Phi_M(r, \theta, \varphi, t)$). Outside of the internal source region, i.e. for $r \geq R_e$, the magnetic potential function must be a solution of the Laplace equation. In spherical coordinates, this solution is expressed in the following form:

$$\Phi_M(r,\theta,\phi,t) = R_e\sum_{n=1}^{N}\sum_{m=0}^{n}\left(\frac{R_e}{r}\right)^{n+1} \times\left[g_n^m(t)\cos m\phi + h_n^m(t)\sin m\phi\right]P_n^m(\cos\theta). \tag{4.13}$$

In the above equation, r denotes the radial distance from the center of the Earth, θ and φ denote the geocentric colatitude and longitude at a given location. Schmidt-normalized associated Legendre functions of degree n and order m are denoted by $P_n^m(\cos\theta)$, and g_n^m and h_n^m denote the Gauss coefficients. The magnetic field components are given by the gradient of the potential function, such that

$$B_r = \frac{\partial\Phi_M}{\partial r}, B_\theta = \frac{1}{r}\frac{\partial\Phi_M}{\partial\theta}, B_\phi = \frac{1}{r\sin\theta}\frac{\partial\Phi_M}{\partial\phi}. \tag{4.14}$$

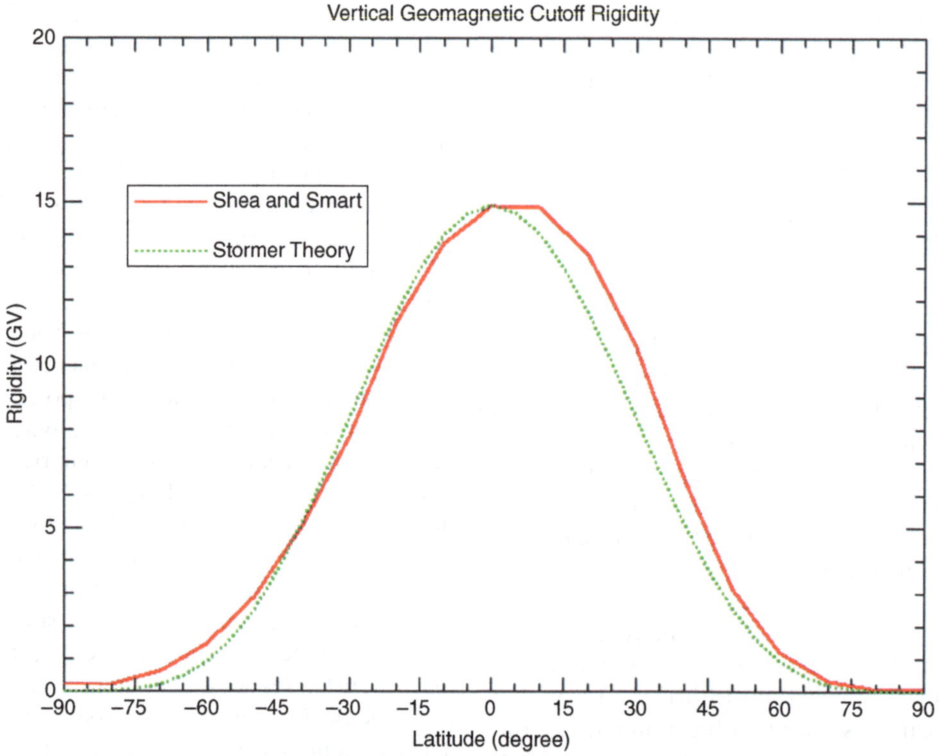

Figure 4.7 Zonal-averaged vertical geomagnetic cutoff rigidity. The solid red line corresponds to the quiescent cutoff rigidities computed from particle trajectories and the IGRF model. The green dotted line is the analytical solution for the vertical cutoff rigidity using Störmer Theory.

The Gauss coefficients are derived from a global set of magnetic field measurements, using the method of least-squares, and are updated every 5 years by IAGA Division 5 (Langlais & Mandea, 2000; ISO 16695 (2013)). Secular variations in the Gauss coefficients are also derived from magnetic field measurements so that derivatives of the Gauss coefficients can be computed. In this way, the temporal dependence of the internal geomagnetic field is represented by

$$g_n^m(t) = g_n^m(T_0) + \dot{g}_n^m(t - T_0) \quad (4.15)$$

$$h_n^m(t) = h_n^m(T_0) + \dot{h}_n^m(t - T_0) \quad (4.16)$$

where the second terms in equations (4.15) and (4.16) include first-order derivatives of the Gauss coefficients. The epoch of the IGRF model is denoted by T_0 and t is such that $T_0 \le t \le T_0 + 5$, where time is expressed in decimal years.

The vertical cutoff rigidities in a realistic magnetic field are determined by numerical solutions of charged particle trajectories in the field using the techniques advanced by Smart and Shea (1994, 2005). Figure 4.8 shows the vertical cutoff rigidities at 20 km in the internal IGRF field. The longitudinal variations in the cutoff rigidity are due to a combination of geocentric offset and relative tilt of the magnetic dipole, with respect to the rotational axis, and the nondipolar contributions to the internal magnetic field. A zonal average of the IGRF cutoff rigidities

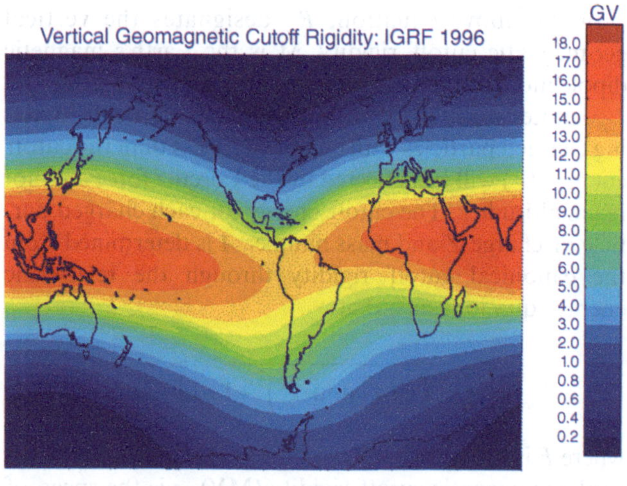

Figure 4.8 Global grid of quiescent vertical geomagnetic cutoff rigidities (GV) calculated from charged particle trajectory simulations using the IGRF model for the 1996 epoch (solar cycle 23 minimum).

in Figure 4.8 are compared to the analytical Störmer theory in Figure 4.7. The simple Störmer theory represents the latitudinal behavior of the vertical cutoff rigidity quite well. The displacement of the Störmer theory cutoffs in Figure 4.7 relative to the numerical solutions of the cutoffs in the IGRF field is due to the fact that the true dipole contribution to the internal magnetic field is off-centered and tilted with respect to the rotational axis.

Models such as NAIRAS using real-time magnetic cutoff rigidities compute them from numerical solutions of charged particle trajectories in a dynamically varying magnetic field that includes both the internal and the magnetospheric magnetic field (combined, they are called the geomagnetic field) contributions (Kress et al., 2010; Mertens et al., 2009; Mertens, Kress, et al., 2010). The cutoff rigidity code was developed by the Center for Integrated Space Weather Modeling (CISM) at Dartmouth College. The CISM-Dartmouth geomagnetic cutoff model can be run using several different empirical, semi-physics-based, and physics-based models (Kress et al., 2004). In particular, the specification of the geomagnetic field due to Earth's internal field source is provided by the IGRF model (Langlais & Mandea, 2000). The real-time dynamical response of the magnetospheric magnetic field to solar wind conditions and IMF can be provided by the semi-physics-based TS05 model (Tsyganenko & Sitnov, 2005), or by the Lyon-Fedder-Mobarry (LFM) global MHD (magnetohydrodynamic) simulation code (Fedder & Mobarry, 2004). Routines are often developed and tested to couple the geomagnetic cutoff model with the different magnetic field models.

As an example, the LFM MHD code may be run as a stand-alone model or coupled with other geospace models. The LFM magnetospheric magnetic fields may be coupled with the Thermosphere-Ionosphere Nested Grid model (Wang et al., 2004) and/or with the Rice Convection Model (Toffoletto et al., 2004), which models the ring current. The semi-physics-based TS05 model provides more accurate cutoff rigidities than the stand-alone LFM MHD model, as determined by comparisons with satellite observations during a Halloween 2003 geomagnetic storm.

Figure 4.9 shows the vertical cutoff rigidity over the northern hemisphere for three different models of the geomagnetic field during the Halloween 2003 storm period. The left column is cutoff rigidity computed using the IGRF field. Since total flight-path exposure at aviation altitudes do not change significantly (<1%) for cutoffs less than 0.05 GV, the cutoffs are set to zero at geographic locations poleward of the 0.05 GV contour (see the bold-white 0.05 GV color contour in Figure 4.9). The middle column in Figure 4.9 shows the cutoff rigidities computed using the TS05 field under geomagnetically quiet conditions on 28 October (0200 UT), prior to the onset of the Halloween 2003 SEP event. One can see that even during magnetically quiet conditions, the cutoff rigidities predicted from the TS05 field are lower than predicted from the IGRF field, and the polar cap region (i.e. inside the bold-white 0.05 GV contour in Figure 4) is expanded to lower latitudes. A weaker field predicted by the TS05 model, compared to IGRF, is due in part to the diamagnetic effect of the magnetospheric ring current included in the TS05 model. Lower cutoff rigidities correspond to less momentum shielding and higher radiation exposure levels. The right column in Figure 4.9 shows the cutoff rigidities during peak geomagnetic storm conditions, 29 October (2100 UT), during an SEP event. The cutoffs are lower at all latitudes compared with the two previous simulations, and the polar cap region has expanded to much lower latitudes than during the magnetically quiet period.

An important aspect of these model studies is the change in cutoff rigidity due to the magnetospheric field effect on atmospheric radiation exposure and the identification of accurate and computationally efficient geomagneticcutoff rigiditymodelswithsolarwind–magnetospheric dynamical responses. The ~1 GV suppression in cutoff at midlatitudes during a geomagnetic storm means that high-level SEP radiation exposure normally confined to the polar cap region will be extended to midlatitudes.

4.2.6. Atmosphere Transport

The atmosphere absorbs most of the precipitating charged particles, particularly below the mesopause near 85 km. The particles' interaction is predominantly with neutral N_2 and O_2. Collisions with these target molecules create lower energy secondary and tertiary particles as well as photons. If the primary charged particle has enough energy, there will be many generations of secondary particles, called a shower (Figure 4.1). The secondaries include neutrons (n), p^+, e^-, e^+, α (and other nuclear fragments), pions (π), muons (μ), γ-rays, and X-rays. Under typical GCR-dominated conditions when there is no SEP contribution to atmospheric ionizing radiation, the primary particles lose energy, the secondary population increases, and the total ionization increases until a maximum ionization rate occurs by 20 km (65,000 ft) called the Regener-Pfotzer maximum (Regener & Pfotzer, 1935). When there are increased SEPs, the relatively high flux and low average energy of the SEP particles compared to GCRs can move the Regener-Pfotzer maximum to higher altitudes even though RBPCPs can penetrate down to 10 km. Below the Regener-Pfotzer maximum down to the Earth's surface, the ionization rate continues to decrease because particles and photons are absorbed in an increasingly thick atmosphere. All of these particles are able to collide with an aircraft hull and

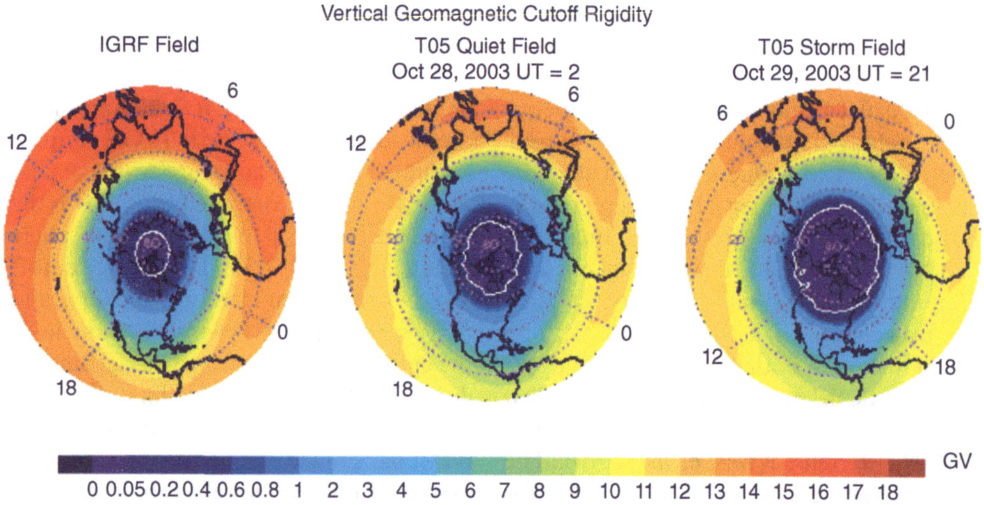

Figure 4.9 Simulated vertical geomagnetic cutoff rigidity shown over the northern hemisphere in October 2003. The cutoff rigidities in the left column were calculated using the IGRF model. The cutoffs in the middle column were calculated using the TS05 model during a geomagnetically quiet period. Cutoff rigidities in the right column were calculated using the T05 model during the largest geomagnetically disturbed period of SEP event. Also shown are the magnetic latitude circles and the meridians at 0, 6, 12, and 18 magnetic local time.

its interior components, people, or fuel to further alter the radiation spectrum that affects tissue and avionics (IARC, 2000; UNSCEAR, 2000).

There is a strong altitude dependence upon atmospheric shielding. The exposure rate altitude dependence can be used as a risk management tool by the aviation community, since radiation exposure can be significantly reduced by descending to lower altitudes. The exposure rate increases on average by 160% per km between 5 and 11 km. Between 11 and 15 km, the exposure rate increases on average by approximately 75% per km. Another canonical way of looking at atmospheric shielding is that there is a doubling of dose rates every 2 km (6500 ft) higher or a halving of dose rates every 2 km lower.

The transport of charged particles through the neutral atmosphere is described by a coupled system of linear, steady-state Boltzmann transport equations, which can be derived on the basis of conservation principles (Wilson et al., 1991). The transport equation for the directional fluence $\varphi_j(x, \Omega, E)$ of particle type j is given by (Mertens et al., 2007, 2008):

$$\Omega \bullet \nabla \Phi_j(x,\Omega,E) = \sum_k \iint \sigma_{jk}(\Omega,\Omega',E,E')\Phi_k(x,\Omega',E')\, d\Omega' dE' - \sigma_j(E)\Phi_j(x,\Omega,E), \quad (4.17)$$

where $\sigma_j(E)$ and $\sigma_{jk}(\Omega,\Omega',E,E')$ are the projectile-target macroscopic interaction cross-sections. The $\sigma_{jk}(\Omega,\Omega',E,E')$ are double-differential particle production cross-sections that represent all processes by which type k particles moving in direction Ω' with energy E' produce a particle of type j moving in direction Ω with energy E, including radioactive decay processes. The total interaction cross-section $\sigma_j(E)$ for each incident particle type j is

$$\sigma_j(E) = \sigma_{j,at}(E) + \sigma_{j,el}(E) + \sigma_{j,r}(E), \quad (4.18)$$

where the first term refers to projectile collisions with atomic electrons of the target medium, the second term refers to elastic ion-nucleus scattering, and the third term contains all relevant nuclear reactions. The corresponding differential cross-sections are similarly ordered.

Consider the transport of GCR and SEP ions through the atmosphere. In this case, the second term in equation (4.18) represents elastic ion-nucleus Coulomb scattering between the incident ions and the atoms that comprise the neutral atmosphere. Figure 4.10 shows the characteristic elastic scattering length versus kinetic energy of various ions colliding with the neutral atmosphere (Mertens et al., 2008). Ion-nucleus scattering becomes important in the atmosphere only at low energies. For example, the length of the Earth's atmosphere in units of areal density is ~ 1000 g/cm^2. Thus, Figure 4.10 shows that cosmic ray ions will not elastically scatter off an atmospheric nucleus before reaching the surface unless the ion kinetic energy is well below 1 MeV/amu. However, ions with kinetic energy less than 1 MeV/amu are stopped via ionization and/or atomic excitation energy loss processes at high altitudes before a scattering event can take place (Figure 4.11). Multiple Coulomb scattering and coupling with ionization energy loss become important factors in the transport of ions within living tissue

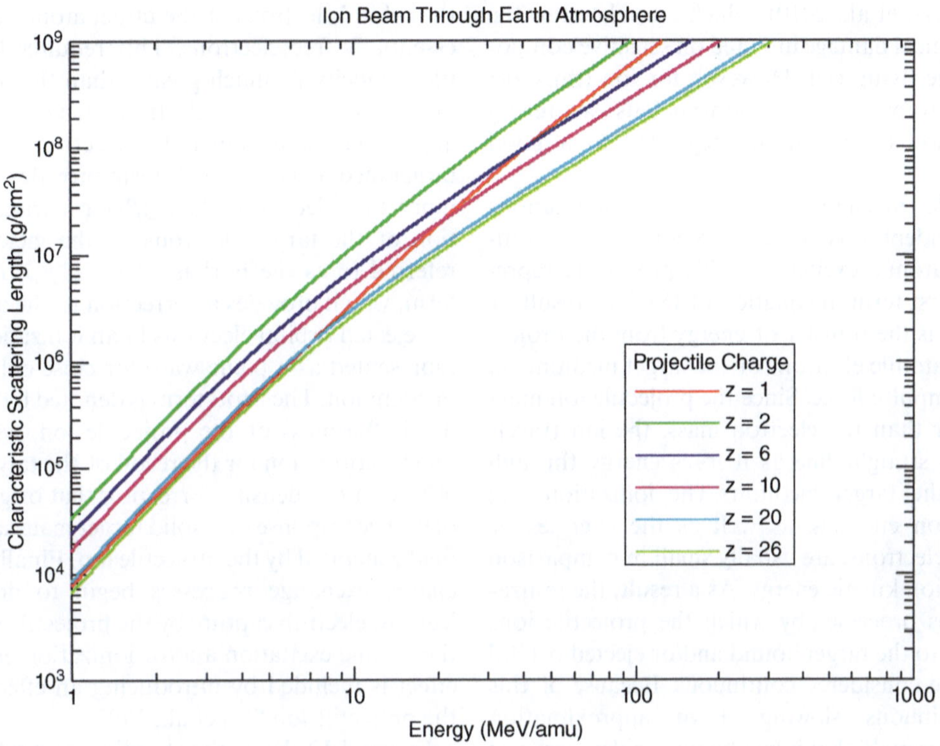

Figure 4.10 Characteristic scattering length for ion beam transport through the Earth's atmosphere. The charge number (z) of the ion is specified in the legend.

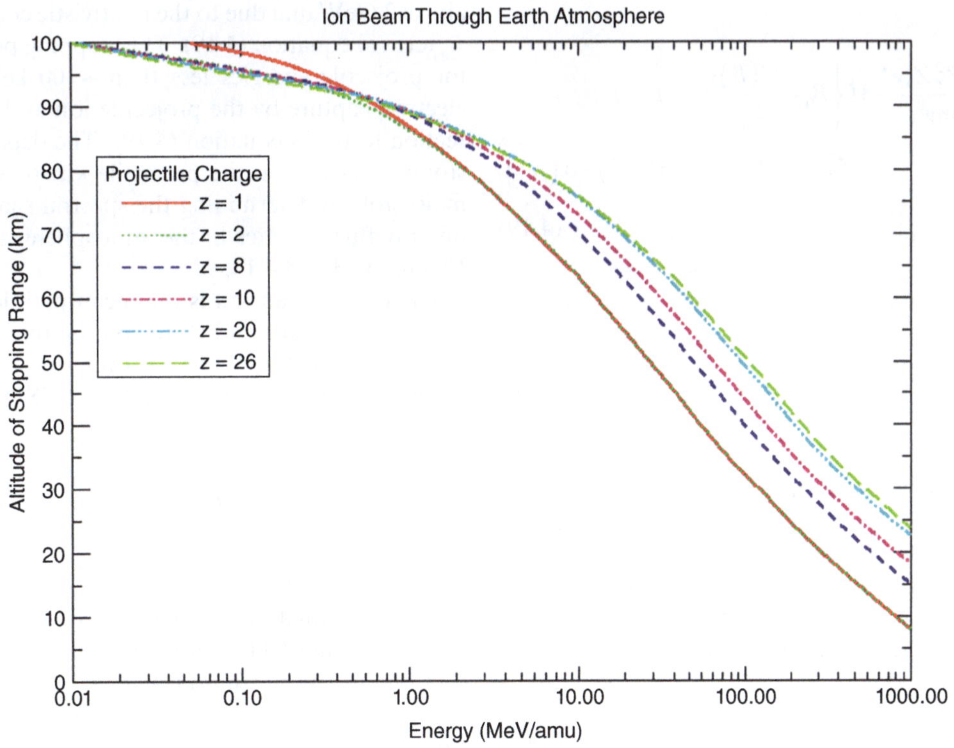

Figure 4.11 Average range versus kinetic energy of an ion beam incident at the top of Earth's atmosphere. The charge number (z) of the ion is specified in the legend.

(Mertens, Moyers, et al., 2010), which are related to the degree of biological damage inflicted on sensitive components within the living cell. However, for ion transport through the atmosphere, the ion-nucleus scattering term in equation (4.18) can be neglected to a good approximation.

The principle mechanism for atomic interactions between the incident ions and the target medium is ionization and/or atomic excitation. This process is represented by the first term in equation (4.18). The result of this interaction is the transfer of energy from the projectile ions to the atomic electrons of the target medium via the Coulomb impulse force. Since the projectile ion mass is much greater than the electron mass, the ion travels essentially in a straight line as it loses energy through ionization of the target medium. The ionization and atomic excitation energies, as well as the energies of ejected orbital electrons, are usually small in comparison to the incident ion kinetic energy. As a result, the ionization energy loss processes by which the projectile ions transfer energy to the target bound and/or ejected orbital electrons can be considered continuous. Because of this so-called continuous slowing down approximation (CSDA), the energy dE that is lost by the incident ion and transferred to the orbital electrons of the target medium by ionization and/or atomic excitation within an element of path dx is given by the stopping power S (Tai et al., 1997), i.e.

$$S = -\frac{dE}{dx} = \frac{4\pi Z_P^2 Z_T e^4}{mv^2} M \left\{ B_0 - \frac{C(\beta)}{Z_T} + Z_P L_P(\beta) + Z_P^2 L_T(\beta) + \frac{1}{2}\left[G(M_P,\beta) - \delta(\beta) \right] \right\}, \tag{4.19}$$

where

$$B_0 = \ln\left(\frac{2mc^2\beta^2}{I\left(1-\beta^2\right)} \right) - \beta^2. \tag{4.20}$$

In the above equations, Z_P and Z_T are the projectile ion charge and the number of electrons per target atom, respectively, v is the projectile velocity, c is the speed of light, $\beta = v/c$, N is the density of atoms in the target medium, and I is the mean ionization potential of the target medium. The electron charge and mass are denoted, respectively, by e and m.

The various terms in equation (4.19) have the following interpretation (Tai et al., 1997; Wilson et al., 1991; Mertens et al., 2012). The B_0 term is the high-energy asymptotic limit of the stopping power, assuming that the orbital electrons of the target atoms can be treated as essentially free electrons. This requires that the projectile's velocity be much greater than the orbital velocities of the bound atomic electrons, which is an inadequate approximation for inner-shell electrons of heavy element target media. The $C(\beta)/Z_T$ term provides a correction for inner-shell electrons. The $L_P(\beta)$ term arises from polarization of the target electrons by the incident ion and is referred to as the Barkas effect. The $L_T(\beta)$ is the Bloch term, which provides a correction to the assumption that the ejected orbital electrons in an ionization event can be represented as a plane wave for close collisions with the incident ion. The Mott term is denoted by $G(M_P,\beta)$, where M_P is the mass of the projectile ion, which includes a kinetic correction for the recoil of the target nucleus. The $\delta(\beta)$ term is a density correction that originates from the dielectric response of a solid target material to the electric field generated by the projectile ion. Finally, at low-energy charge exchange processes begin to dominate, which leads to electron capture by the projectile ion and reduces the atomic excitation and/or ionization energy loss. This effect is included by introducing an effective charge for the projectile ion (Tai et al., 1997).

Figure 4.12 shows the stopping power for various cosmic ray ions incident on Earth's atmosphere. The stopping power decreases inversely with projectile energy between ~100 keV/amu and 2 GeV/amu. The stopping power begins to increase with increasing projectile energy above 2 GeV/amu due to the relativistic corrections in the B_0 term in equation (4.19). The stopping power decreases for projectile energies less than ~100 keV/amu due to electron capture by the projectile ion and the other correction terms in equation (4.19). The dependence of the stopping power on the projectile kinetic energy plays a major role in determining the spectral shape of the cosmic ray fluence rates in the atmosphere, as indicated by Figures 4.13 and 4.14.

The range of an ion is the mean path length traveled in the target medium before coming to rest after losing its initial kinetic energy through ionization and/or atomic excitation energy loss. In the CSDA, the range is defined by

$$R_j(E) = A_j \int_0^E \frac{dE'}{S_j(E')}, \tag{4.21}$$

where A_j is the atomic mass number of ion particle type j. The above equation is referred to as the range-energy relation. Figure 4.11 shows the range of various ions incident on the neutral atmosphere as a function of ion kinetic energy. At 1 GeV/amu, only protons and alpha particles can reach the typical cruising altitudes of ~10–12 km, or ~200 g/cm^2, for commercial aircraft before coming to rest due to ionization energy loss. Below 1 GeV/amu, all

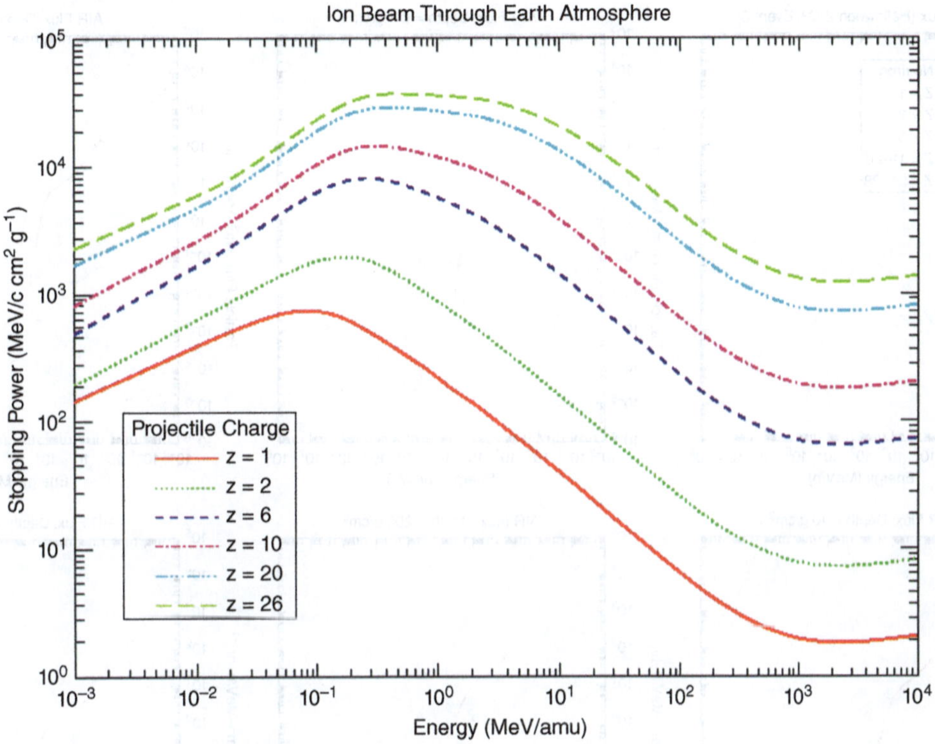

Figure 4.12 Stopping power versus kinetic energy of incident ions on Earth's atmosphere. The charge number (z) of the ion is specified in the legend.

incident ions lose their kinetic energy before reaching commercial aircraft cruising altitudes. For energies greater than 1 GeV/amu, the charged particle flux densities decrease with a power-law dependence on energy (Figure 4.2). Although secondary charged particles can be produced by nuclear fragmentation reactions, which is represented by the third term in equation (4.18), the stopping range in Figure 4.11 explains why the heavy-ion fluence rates in Figure 4.13 are significantly less than the proton fluence rate. The same is true for the light ions in Figure 4.14.

Two approximations can be made to the total ion-target interaction cross-section in equation (4.18). First, elastic ion-nucleus scattering has been neglected for cosmic ray transport through the atmosphere. Second, the CSDA has been invoked in the representation of atomic ion-electron energy transfer collisions. As a consequence of these two approximations, the coupled Boltzmann transport equations in equation (4.17) can be expressed, alternatively, as

$$\bar{B}\left[\Phi_j(\mathbf{x},\Omega,E)\right]=\sum_k\iint\sigma_{jk,r}(\Omega,\Omega',E,E')\times\Phi_k(\mathbf{x},\Omega',E')\,d\Omega' dE', \tag{4.22}$$

where

$$\bar{B}\left[\Phi_j(\mathbf{x},\Omega,E)\right]\equiv\left[\Omega\bullet\nabla-\frac{1}{A_j}\frac{\partial}{\partial E}S_j(E)+\sigma_{j,r}(E)\right]\Phi_j(\mathbf{x},\Omega,E). \tag{4.23}$$

The $\bar{B}\left[\Phi_j\right]$ in the above equations denotes a differential operator acting on the directional fluence.

The differential operator in equation (4.22) can be inverted using the method of characteristics in order to transform the integro-differential equation into a Volterra-type integral equation (Wilson, 1977). As a result, the integral equation for cosmic ray transport is given by

$$\Phi_j(\mathbf{x},\Omega,E)=\frac{S_j(E_\gamma)P_j(E_\gamma)}{S_j(E)P_j(E)}\Phi_j(\Gamma_{\Omega,\mathbf{x}},\Omega,E_\gamma)+\sum_k\int_E^{E_\gamma}\frac{A_jP_j(E')}{S_j(E)P_j(E)}dE'\int_{E'}^{\infty}dE''\int d\Omega'\sigma_{jk,r}(\Omega,\Omega',E',E'')\times\Phi_k\left[\mathbf{x}+\left(R_j(E)-R_j(E')\right)\Omega,\Omega',E''\right]. \tag{4.24}$$

In the above equation, $\Gamma_{\Omega,\mathbf{x}}$ is a position vector of a point on the boundary surface, and E_γ is given by

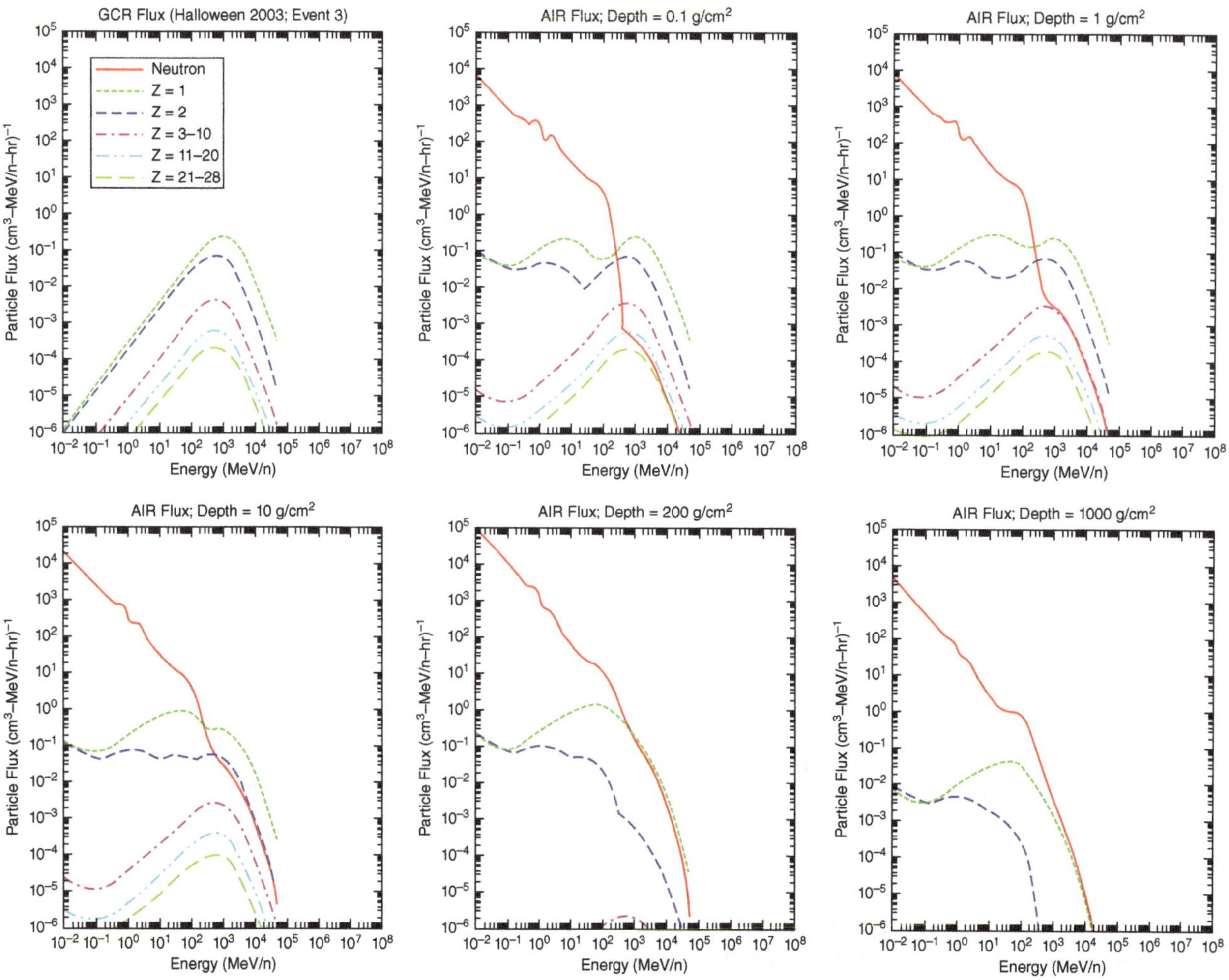

Figure 4.13 Event-averaged GCR spectral fluence rates at zero vertical geomagnetic cutoff rigidity during the Halloween 2003 solar-geomagnetic storm (29 October 2003 [2100 UT] to 1 November [0000 UT]). The panels show the fluence rates at different atmospheric depths. The typical cruising altitudes for commercial aircraft correspond to an atmospheric depth of roughly 200 g/cm². The fluence rates from different charge groups have been summed together to reduce the number of lines.

$$E_\gamma = R_j^{-1}\left[R_j + \Omega\bullet\left(\mathbf{x} - \Gamma_{\Omega,\mathbf{x}}\right)\right]. \tag{4.25}$$

The R_j^{-1} operator in equation (4.25) is the inverse operation of obtaining the energy given the range using the range-energy relation in equation (4.21). The expression for the integral cosmic ray transport equation in (4.24) was made compact by introducing the total nuclear survival probability, which is defined by

$$P_j(E) \equiv \exp\left[-A_j \int_0^E \frac{\sigma_{j,r}(E')dE'}{S_j(E')}\right]. \tag{4.26}$$

The first term in equation (4.24) describes the attenuation of the directional fluence specified at the boundary as a result of transport through the target medium. For atmospheric cosmic ray transport, an isotropic distribution can be assumed for the directional fluence and the boundary specification can be defined as the charged particle fluence rates that have been transported through the heliosphere and magnetosphere and incident at the top of the neutral atmosphere. These incident ions are attenuated by ionization energy loss ($S(E)$) and nuclear absorption ($P(E)$), as indicated by the first term in equation (4.24). The second term in (4.24) describes the generation of type j particles from projectile-target nuclear fragmentation reactions by type k particles. The second term in equation (4.24) includes the production of type j particles from type k particles at all intervening positions between the boundary point and the position of observation, accounting for the attenuation by ionization energy loss and nuclear absorption in between the point of production of a type j particle and the observation point.

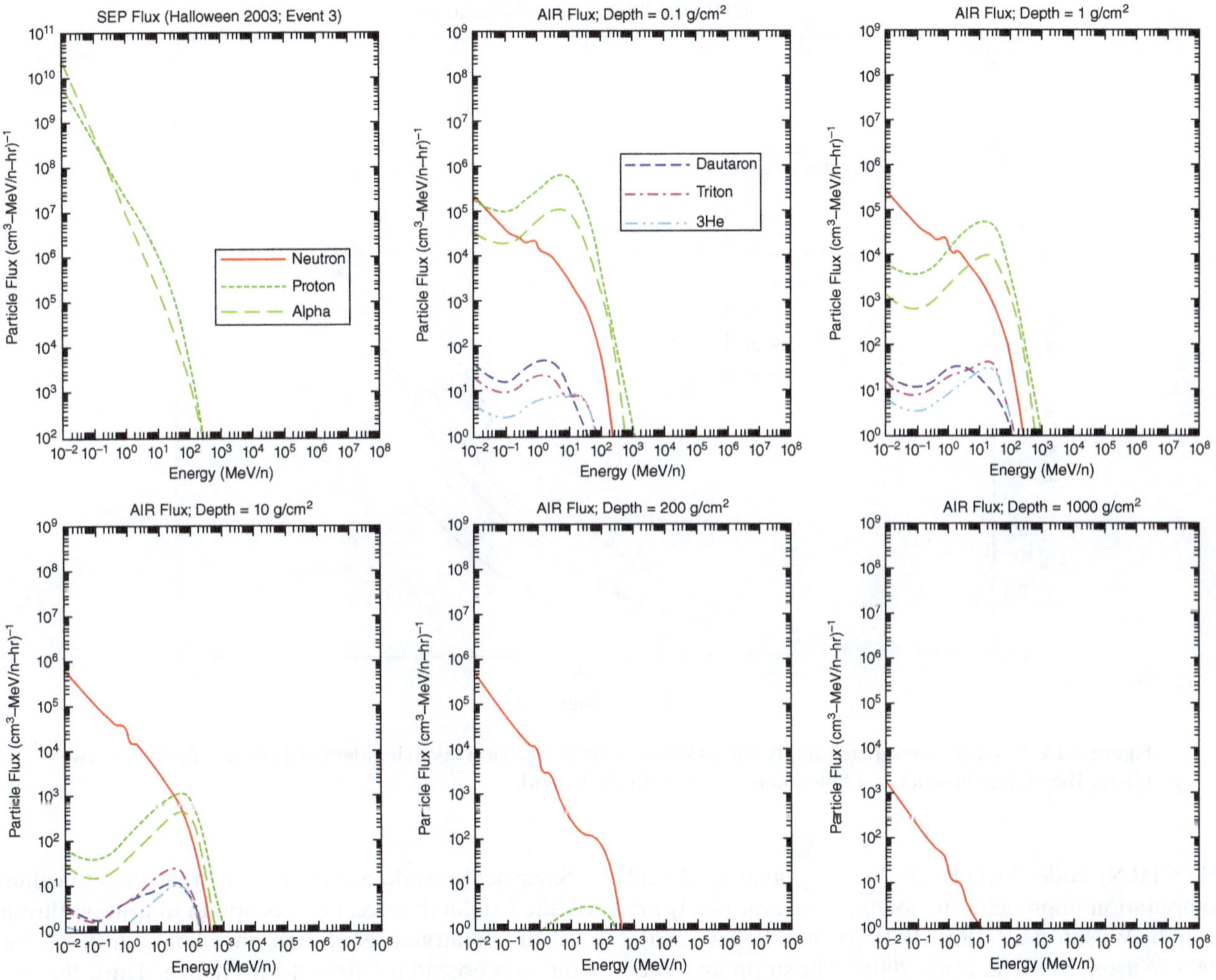

Figure 4.14 Event-averaged SEP spectral fluence rates at zero vertical geomagnetic cutoff rigidity during the Halloween 2003 solar-geomagnetic storm (29 October 2003 [2100 UT] to 1 November [0000 UT]). The panels show the fluence rates at different atmospheric depths. The typical cruising altitudes for commercial aircraft correspond to an atmospheric depth of roughly 200 g/cm^2.

The representation of the relevant total nuclear absorption cross-sections ($\sigma_{j,r}$) and nuclear fragmentation production cross-sections ($\sigma_{jk,r}$) cannot be expressed in a simple, compact form such as the stopping power in equations (4.19) and (4.20). Nevertheless, important insight into the influence of nuclear reactions on the atmospheric transport of cosmic rays can be gained by examining the probability of a nuclear reaction as a function of incident ion kinetic energy. The probability of a nuclear reaction is 1 minus the total nuclear survival probability in equation (4.26) (i.e. $1 - P_j(E)$), which is shown in Figure 4.15. For particles with kinetic energy below 100 MeV/amu, there is a small chance of a nuclear reaction. Recall from Figure 4.2 that the peak of the incident GCR spectrum is between ~200 and 500 MeV/amu. At these energies, one out of every two particles will undergo some kind of nuclear reaction. For kinetic energies greater than 1 GeV/amu, nearly every particle will be subject to some type of nuclear reaction. Combining this discussion with the discussion of Figure 4.11, the only primary incident charged particles that can survive transport through the atmosphere and reach the cruising altitudes of typical commercial aircraft are protons with kinetic energy on the order of 1 GeV or greater. The high-LET particles present at 10–12 km with energies less than 1 GeV/amu are secondary particles created at higher altitudes from nuclear fragmentation reactions, most of which are neutrons.

The coupled cosmic ray integral transport equations in equation (4.24) are solved in the NAIRAS model, for example, using NASA Langley Research Center's deterministic High Charge (Z) and Energy TRaNsport

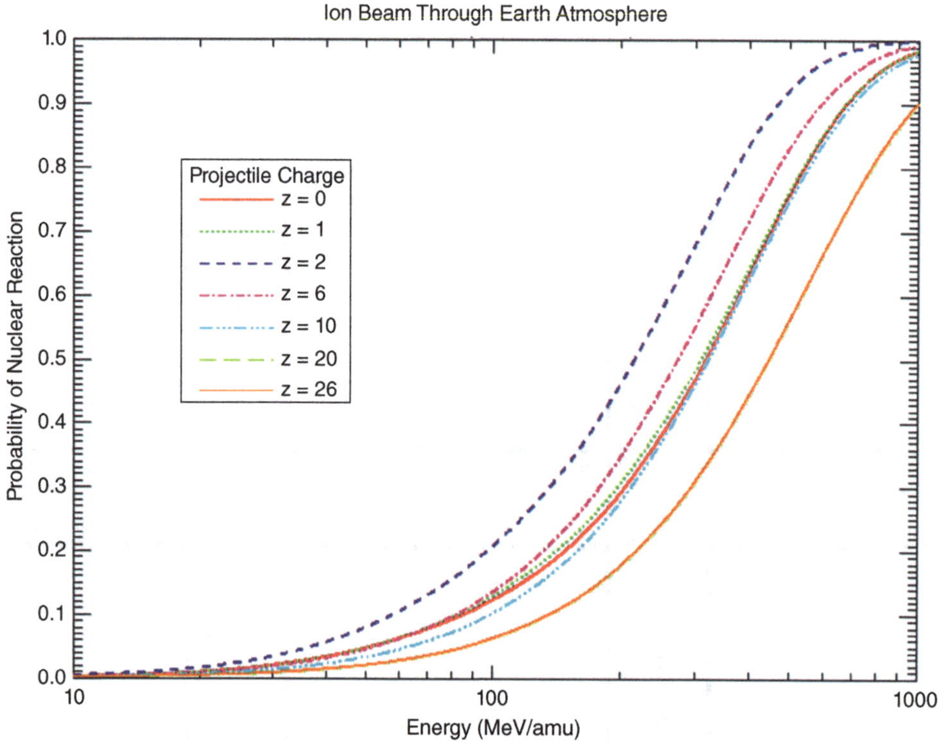

Figure 4.15 Nuclear survival probability versus kinetic energy of an ion beam incident at the top of Earth's atmosphere. The charge number (z) of the ion is specified in the legend.

(HZETRN) code. Details of the early analytical and computation approaches to solving equation (4.24) are given by Wilson et al. (1991, 1997, Wilson, Badavi, et al., 1995; Wilson, Mertens, et al., 2005). The stopping power parameterization used in HZETRN is described by Tai et al. (1997). The nuclear cross-sections for neutron and proton interactions are described extensively in Wilson et al. (1989). The model for calculating the heavy-ion nuclear fragmentation cross-sections are described by Wilson et al. (Wilson, Nealy, et al., 1995; Wilson, Tweed, et al., 2005). HZETRN is used in a wide variety of radiation transport applications: e.g. the calculation of dosimetric quantities for assessing astronaut risk to space radiations on the ISS and the Space Transportation System (STS) Shuttle, including realistic spacecraft and human geometry (Badavi et al., 2005, 2007; Slaba et al., 2009; Wilson et al., 2006). Extensive summaries of HZETRN laboratory and space-flight verification and validation are found in recent reports (Badavi et al., 2007; Nealy et al. 2007; Wilson, Mertens, et al., 2005; Wilson, Tweed, et al., 2005). In the NAIRAS model, there are 59 coupled transport equations in the HZETRN description of GCR transport through the atmosphere. This set includes transport equations for neutrons and GCR nuclear isotopes from protons through nickel (Z = 28, A = 58).

Since neutrons do not interact with the target medium via the Coulomb force, there is not an ionization threshold, or an atomic excitation threshold, or a nuclear Coulomb potential barrier to overcome. Thus, the neutrons are not brought to rest as the charged particles are. Neutrons continue to cascade down in energy through neutron-nucleus interactions and the low-energy neutron fluence continues to build, as is evident in Figure 4.13. The low-energy neutron fluence is quite large, even at a small atmospheric depth of 0.1 g/cm^2. The large low-energy neutron fluence at small atmospheric depth is dominated by backscattered neutrons generated at much larger depths, or at much lower altitudes in the atmosphere. At the atmospheric depth of typical cruising altitudes of commercial aircraft, the heavy ions have largely disappeared, due to a combination of ionization energy loss and nuclear fragmentation reactions into lower-energy lighter particles. Refer to the discussion of Figures 4.11 and 4.15.

Solar particles consist mainly of protons and alpha particles. As a result, only the solution of the six light-particle coupled transport equations defined in HZETRN are required in the description of SEP transport through the atmosphere. Similar to the GCR atmospheric transport properties, the large low-energy neutron fluence at small atmospheric depth for SEP events is due to the large backscattered neutron component, which also originates at larger penetration depths. Furthermore, only nucleon (protons + neutrons) fluences remain at the typical cruising altitude of commercial aircraft, for the same reasons as previously described.

4.2.7. Radiation Dosimetry

Radiation can result in several injury pathways within human tissues by ionizing, dissociating, or exciting atoms and molecules. Pathways include (1) free radical production, (2) chemical bond breakage, (3) new chemical bond production and cross-linkages between macromolecules, and (4) molecular damage related to regulation of vital cell processes, such as deoxyribonucleic acid (DNA), ribonucleic acid (RNA), and proteins (UNSCEAR, 2000; Tobiska et al., 2017). High linear energy transfer (LET; i.e. the energy lost by a particle per unit track length) radiations are typically more harmful to living tissues per unit dose (energy deposited per unit of target mass) than low-LET radiations. Photons, muons, and electrons comprise low-LET radiations, while neutrons, alpha particles, and heavier ions are high-LET radiations. Protons and charged pions can be considered low-LET radiation but interact more like high-LET radiation such that they are treated separately in dosimetry (ICRP, 2008). Although cells can repair damage from low ionizing radiation doses, higher doses can result in cell death. At extremely high doses, an organ's cell population can rapidly drop and cells may not be replaced quickly enough such that the tissue fails to function normally (IARC, 2000; UNSCEAR, 2000). These latter effects may be observed when absorbed doses exceed 100 mGy. However, such doses or corresponding dose rates have not been observed from cosmic radiation in the atmosphere. Instead, epidemiological studies in occupational groups conducted over several decades usually focus on long-term effects of radiation-associated cancer (Grajewski et al., 2011).

The energy deposited in a target medium by the radiation field of particle j is the dose, which is given by

$$D_j(\mathbf{x}) = K\int_{\Omega}\int_0^{\infty} S_j(E)\Phi_j(\mathbf{x},\Omega,E)\,d\Omega dE. \quad (4.27)$$

In the above equation, $S_j(E)$ is the target stopping power for particle j (Mev/g/cm²), and K is a unit conversion factor (1.602×10^{-10}) to convert dose to units of Gray (1 Gy = J/kg). The target stopping power is given by equations (4.19) and (4.20), which is shown in Figure 4.12 for representative cosmic ray ions incident on Earth's atmosphere. Radiation health risk and the probability of biological damage depend not only on the absorbed dose but also on the particle type and energy of the radiation causing the dose. This is taken into account by weighting the absorbed dose by a factor related to the quality of the radiation. The weighted absorbed dose has been given the name *dose equivalent* by the ICRP (1991). The unit of dose equivalent is the Sievert (Sv). Dose equivalent in tissue T from particle j ($H_{j,T}(\mathbf{x})$) is defined in terms of the tissue LET dependent quality factor, Q, such that

$$H_{j,T}(\mathbf{x}) = \int_L Q(L)D_j(\mathbf{x},L)dL, \quad (4.28)$$

where L is LET, which can be approximated by the stopping power in units of keV/um; $D_j(\mathbf{x}, L)$ is the spectral dose distribution from particle j in terms of LET, and $Q(L)$ is the tissue LET-dependent quality factor.

The relationship between the probability of biological damage and dose equivalent is found to also depend on the organ or tissue irradiated. A further dosimetric quantity, called the effective dose, is defined to include the relative contributions of each organ or tissue to the total biological detriment caused by radiation exposure. The effective dose ($E(\mathbf{x})$) is the sum of weighted dose equivalents in all the organs and tissues in the human body, such that

$$E(\mathbf{x}) = \sum_T\sum_j w_T H_{j,T}(\mathbf{x}). \quad (4.29)$$

The organ/tissue weighting factors are given in the ICRP 60 report (ICRP, 1991). A computationally efficient approach is to calculate the effective dose rates directly from the particle spectral fluence rates using precomputed fluence-to-effective-dose conversion coefficients. For example, the NAIRAS model uses neutron and proton conversion coefficients tabulated by Ferrari et al. (1997a, 1997b), and the ARMAS-derived data use fitted functions (Matthiä et al., 2014; Tobiska et al., 2016). The effective dose contributions from the other ions are obtained by scaling the proton fluence-to-effective-dose conversion coefficients by Z_j^2/A_j, according to stopping power dependence on charge and mass in equations (4.19) and (4.20). All recommended ICRP radiation exposure limits are defined in terms of effective dose.

Figure 4.16 shows the proton and neutron fluence-to-effective-dose conversion coefficients and the event-averaged GCR and SEP spectral fluence rates computed during the Halloween 2003 SEP event. The fluence rates were computed at zero vertical geomagnetic cutoff rigidity at an atmospheric depth of 200 g/cm², which is the depth corresponding to typical cruising altitudes of 10–12 km for commercial aircraft. Above 20 MeV, protons make a larger contribution to effective dose per unit fluence compared to neutrons. Below 20 MeV, the reverse is true. The fluence rates are shown in Figure 4.16 as a product of the fluence rates times the energy. This is a convenient representation on a log-log scale since the spectral integration

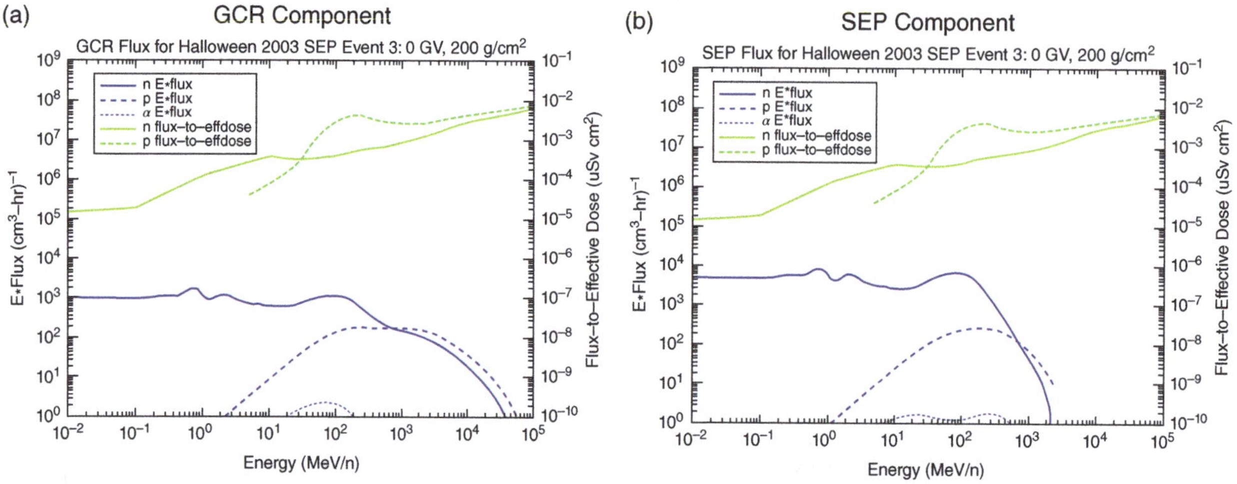

Figure 4.16 The green lines show the fluence-to-effective dose conversion coefficients. The blue lines show the product of the kinetic energy (MeV/amu) times the event-averaged spectral fluence rates for the Halloween 2003 storm period (29 October [2100 UT] to 1 November [0000 UT]). The fluence rates were evaluated at zero vertical geomagnetic cutoff rigidity and at an atmospheric depth of 200 g/cm^2. The lines denote neutron quantities; dashed lines denote proton quantities; dotted lines denote alpha quantities. All quantities are shown for (a) GCR component and (b) SEP component.

with respect to log-energy, which is performed in order to obtain the effective dose rate, is proportional to energy multiplied by the fluence rate. In this representation, neutrons dominate below about 1 GeV. The peak in energy times the proton fluence rate is slightly larger for the SEP component as compared to the GCR component. At 100 MeV and below, the energy times the neutron fluence rate is nearly an order of magnitude greater for the SEP component compared to the GCR component.

The GCR and SEP normalized spectral and accumulated spectral effective dose rates are presented in Figure 4.17 for the event-averaged Halloween 2003 SEP event. Similar to the previous figure, the effective dose rate quantities were computed at zero vertical geomagnetic cutoff rigidity at an atmospheric depth of 200 g/cm^2. The spectral effective dose rates are normalized with respect to the peak in the spectrum. The peaks in both the proton and neutron spectral effective dose rates occur between 100 and 200 MeV, which is true for both GCR and SEP contributions. Neutrons make the largest contributions to effective dose at energies below the peak in the spectrum. Protons make the largest contribution to effective dose at energies above the peak in the spectrum. The relative spectral contribution of protons and neutrons to effective dose holds for both the GCR and SEP components. Half of the total effective dose rate comes from spectral contributions at energies less than 100 MeV for the GCR component and at energies less than roughly 20 MeV for the SEP component. Half of the neutron effective dose comes from spectral contributions at energies less than about 30 MeV for the GCR component, which is similar for the SEP component. Recall that the incident GCR and SEP proton fluence rates spectra are very different. As a result, half of the proton effective dose rates come from spectral contributions at energies less than 100 MeV for the SEP component and at energies less than about 500 MeV for the GCR component.

4.2.8. Avionics

High-energy neutrons (E > 10 MeV) and very-low-energy thermalized neutrons can cause single-event-effect (SEE) errors in avionics (Dyer & Lei, 2001; Dyer et al., 2003a; Dyer & Truscott, 1999; Mutuel, 2016; Normand et al., 1994, 2006). The high-energy neutrons directly interact with silicon (Si) nuclei in electronics through nuclei recoils to produce excess charge carriers. The very-low-energy neutrons are created by scattering from atmospheric constituents as well as aircraft materials, fuel, and passengers and reach thermal equilibrium in an energy range of 0.02–0.2 eV. They are then absorbed by boron (particularly ^{10}B), often found in Si-based aircraft electronics. After absorption, a γ-ray (480 keV) is produced along with an alpha particle (~4 MeV) and a lithium ion. The charged alpha particle is then available to interact with semiconductor structures and may cause a SEE. Higher Z elements near the silicon layers (e.g. tungsten connectors) can increase the SEE effect (Tobiska et al., 2015). SEE in avionics is a probabilistic phenomenon, and its mitigation is mainly achieved through improved engineering processes (IEC-62396-1). The existing certification is for quiet cosmic ray conditions

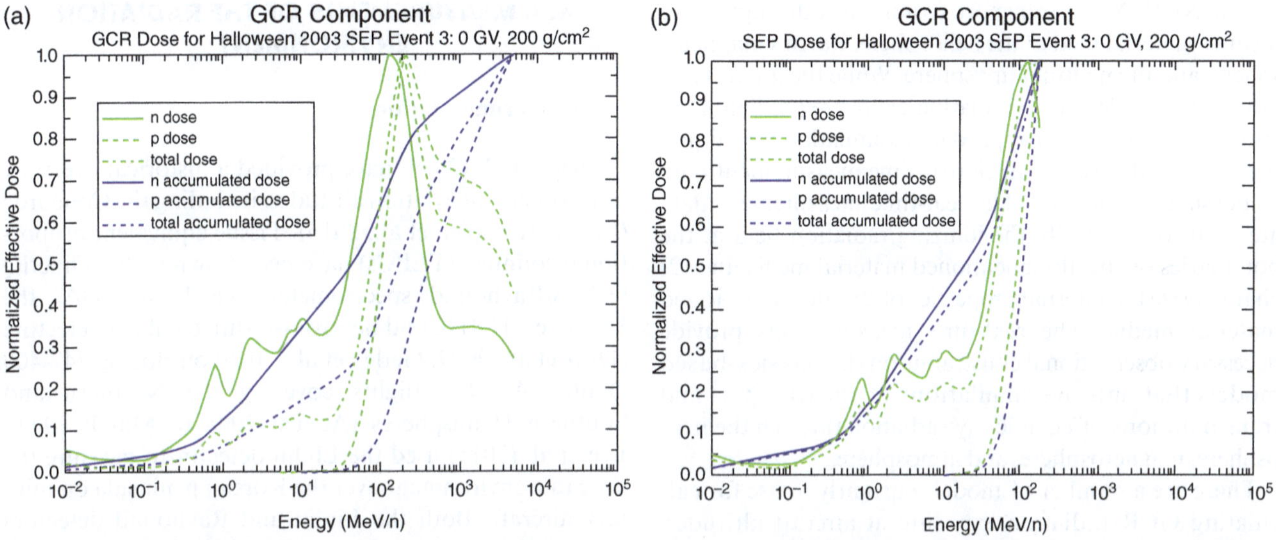

Figure 4.17 Normalized spectral effective dose rates evaluated during the Halloween 2003 storm period (29 October [2100 UT] to 1 November [0000 UT]) at zero vertical geomagnetic cutoff rigidity and at an atmospheric depth of 200 g/cm^2. The green lines show the spectral effective dose rate normalized to the peak in the spectrum. The blue lines are normalized to the spectrally integrated effective dose rate and show the accumulated effective dose rate as the spectral effective dose rate is integrated over energy. The solid lines denote neutron quantities; dashed lines denote proton quantities; dash-dot lines denote total quantities.

only; active or extreme space weather conditions are not considered.

4.2.9. Models

Many models are capable of reporting the aviation radiation environment. They include AIR (Johnston, 2008), AVIDOS (Latocha et al., 2009, 2014), CARI-7 (Copeland, 2017), EPCARD.NET (Mares et al., 2009), FDOSCalc (Wissmann et al., 2010), FREE (Felsberger et al., 2009), KREAM (Hwang et al., 2014), NAIRAS (Mertens et al., 2013), PANDOCA (Matthiä et al., 2014), PARMA/EXPACS (Sato et al., 2008; Sato, 2015), and PC-AIRE (McCall et al., 2009). While all of these models are based on or compared with flight data, the drivers needed for individual models vary. Empirical models like PC-AIRE and FDOSCalc are built from empirical functions fit to in-flight measurement databases representing the GCR environment only. PCAIRE, for example, used the TEPC instrument for assembling its database and is driven by date/time and flight path. These inputs are correlated with GCR-related dose estimates based on the solar cycle, altitude, and magnetic latitudes/longitudes. SEP-related dose can be estimated in postevent analyses. Physics-based models like CARI-7, EPCARD.NET, NAIRAS, and PANDOCA start from the particle spectrum (SEP or GCR local interstellar spectrum, based on measurements) and then propagate the particles through Earth's magnetosphere and atmosphere (or even heliosphere for GCRs). Joyce et al. (2014) utilized Monte Carlo simulations of showers coupled to CRaTER measurements (Spence et al., 2010; Schwadron et al., 2012) in deep space to estimate dose rates through the Earth's atmosphere at a range of different altitudes down to aviation heights. Tobiska et al. (2018) provided analytical functions to represent the GCR and GCR+REP contributions within the ARMAS flight database, driven by L-shell, NOAA geomagnetic level, and flight altitude.

The NAIRAS model and ARMAS data have been compared extensively (Tobiska et al., 2018). NAIRAS is an operational model for predicting aircraft radiation exposure from galactic and solar cosmic rays (Mertens et al., 2007, 2008, 2009; Mertens, Kress, et al., 2010). The real-time exposure rate graphical and tabular data products from the operational prototype stream live from the NAIRAS public web site at https://sol.spacenvironment.net/~nairas/index.html. NAIRAS provides data-driven, global, real-time predictions of atmospheric ionizing radiation exposure rates on a geographic 1 × 1 degree latitude and longitude grid from the surface of the Earth to 100 km with a vertical resolution of 1 km. The real-time, global predictions are updated every hour. Improvements have adopted, as far as possible, the meteorological weather forecasting paradigm of combining physics-based forecast models with data assimilation techniques, i.e. an activity being produced through the NASA RADIAN project (**RADI**ation environment using **A**RMAS data in the **N**AIRAS model). Physics-based models are utilized within NAIRAS to transport cosmic rays through three distinct material media: the heliosphere, Earth's magnetosphere, and the neutral atmosphere. While the quantity of observations relevant to radiation exposure predictions is currently too sparse to apply data assimilation techniques per se, nevertheless, as much real-time measurement data as possible are utilized. The real-time measurement data are used to (1) specify the ionizing radiation field at the boundaries of the aforementioned material media and (2) characterize the internal properties of the aforementioned material media. The real-time measurements provide necessary observational constraints on the physics-based models that improve simulations of the transport and transmutations of cosmic ray radiation through the heliosphere, magnetosphere, and atmosphere.

There are a number of models currently in use for calculating GCR radiation exposure at aircraft altitudes. The CARI model utilizes a database of transport calculations generated by the deterministic LUIN code for a wide variety of geographic locations, altitudes, and solar activity levels (O'Brien et al., 2003). The EPCARD model is based on a similar approach but uses the Monte Carlo FLUKA code for the transport calculations (Schraube et al., 1999). PC-AIRE is a semiempirical model based on fits to measurement data (Lewis et al., 2002). Other aircraft radiation exposure models are described in the European Radiation Dosimetry Group report (Lindborg et al., 2004). Currently, the above models calculate SEP atmospheric radiation exposure poststorm on a case-by-case basis, although PC-AIRE incorporated low-Earth orbit measurements to develop a simple extrapolation to SEP events (Lewis et al., 2002). Copeland et al. (2008) calculated adult and conceptus aircraft exposure rates for 170 SEP events for years 1986–2008 using the Monte Carlo MCNPX transport code.

The main differences that distinguish the NAIRAS model from the others include dynamical solar wind–magnetospheric interactions and the accompanying geomagnetic effects that govern the transport of cosmic rays through the magnetosphere. These are included in real time in the NAIRAS radiation exposure calculations (Kress et al., 2004, 2010; Mertens, Kress, et al., 2010). Furthermore, the physics-based deterministic HZETRN code is used in transporting cosmic rays through the atmosphere. NAIRAS v2 includes the entire EM cascade in its calculations. The HZETRN transport calculations are continuously updated using real-time measurements of boundary condition specifications of the space radiation environment and of atmospheric density versus altitude (Mertens et al., 2007, 2008). Finally, both GCR and SEP atmospheric radiation exposure predictions are included in real time (Mertens et al., 2009; Mertens, Kress, et al., 2010).

4.3. MEASUREMENTS OF THE RADIATION ENVIRONMENT

4.3.1. Instrumentation

Meier et al. (2009) have provided a historical overview of dosimetry at aviation altitudes, including the Concorde (Meier, Trompier, et al., 2016). Tissue equivalent proportional counters (TEPCs) have been flown with a QinetiQ QDOS/Rayhound spectrometer, the Liulin 4SA, the Eberline FH 41B, and neutron-sensitive bubble detectors (Dyer et al., 2009; Getley et al., 2010) on Boeing 747-400 Qantas Airways flights covering the Northern and Southern Hemispheres (August 2008 to March 2009). Lee et al. (2015) used the Liulin detector to measure the dose rate environment over the Korean peninsula on military aircraft. Both the Liulin and Rayhound detectors compared favorably with the TEPC and are within the generally accepted tolerances of 20% (Dyer et al., 2009).

ARMAS has built an extensive database of over a half million 10-second global data records from 0–107 km at all magnetic latitudes and most representative magnetic longitudes. The time frame of ARMAS is a half solar cycle from 2013 to 2020. This system uses the Teledyne micro dosimeter capturing absorbed dose in silicon (Tobiska et al., 2016, 2018; Gersey et al., 2020).

An extensive aviation radiation database has been compiled using the Liulin instrument (Ploc et al., 2011; Spurny & Dachev, 2002) on commercial Czech Airlines aircraft since 2001 (Ploc et al., 2013). The dose rates contain dependencies on flight altitude, geographical position, and solar activity to demonstrate three phenomena: (1) the solar cycle effect upon GCR-induced radiation with higher dose rates seen during solar cycle 23 minimum, (2) increased geomagnetic cutoff rigidity shielding while moving from high to low geomagnetic latitudes, and (3) an ability to measure ground-level events while using a continuous monitoring system (Spurny & Dachev, 2001).

Several flight databases have been created during the past decades capturing a wide range of dosimetric measurements. Measurements on aircraft (Dyer et al., 2003a, 2003b, 2005, 2009; Federico et al., 2015; Getley et al., 2010; Iles et al., 2004; Lee et al., 2015; Meier et al., 2009; Meier, Trompier, et al., 2016; Ploc et al., 2013; Tobiska et al., 2016, 2018), balloons (Mertens, 2016; Mertens et al., 2016; Tobiska et al., 2018), suborbital vehicles (Tobiska et al., 2018), and the ISS (Dachev, 2017; Dachev et al., 2015, 2017a, 2017b) have provided valuable information. Some of the instruments involved in these campaigns include the following devices.

Tissue Equivalent Proportional Counter (TEPC)

A standard instrument to monitor the radiation dose received by aircraft crews and passengers is the TEPC. The

TEPC instrument is a defacto industry standard and measures energy deposition in simulated volumes that are comparable to those of a living cell (Braby et al., 1994; Gersey, et al., 2002). The TEPC uses materials and gases that are essentially equivalent to human tissue in chemical composition. Measurements taken with the TEPC are used to calculate radiation dose, radiation quality factor, and dose equivalent. TEPCs have been well characterized in a variety of controlled radiation environments such as particle beam lines and this has provided confidence in their use for operations (Badhwar et al., 1992; Gersey et al., 2002; Guetersloh et al., 2004). Energy deposition spectra produced by a TEPC can be used to calculate absorbed dose in tissue and to estimate the average quality factor, Q, of radiation during a flight. A TEPC active volume is a right cylinder where the walls are fabricated from tissue equivalent A-150 plastic. The right cylinder is enclosed in stainless steel walls. The active volume is then enclosed in an aluminum cylinder, which is referred to as a "chamber." The active volume of the TEPC is filled with low-pressure propane gas. An anode wire runs the length of the cylinder and is kept at a potential of 640 volts relative to the cylinder walls. When a high-energy particle enters the TEPC active volume, charge is collected at the anode wire and processed by a preamplifier. The signal then moves to a shaping amplifier, is converted into a digital pulse height via an analog to digital converter and stored on a flash-ROM card. This information is then translated into lineal energy (the energy deposited in the active volume by a single energy deposition event) is extracted in units of keV/micrometer. An LET spectrum is created from these measurements and is used to calculate dose equivalent and an average quality factor, Q, of radiation in tissue.

Automated Radiation Measurements for Aerospace Safety (ARMAS)

The difficult task of continuous radiation environment monitoring, reporting, and modeling has not yet been achieved on regional or global scales. There are no existing routine dose measurements at aviation altitudes, either for tissue-relevant or avionics applications. Because of this, the ARMAS program began developing a calibrated, real-time, global monitoring, reporting, and modeling capability of the aviation radiation environment in 2012 (Tobiska et al., 2016).

By the end of 2020 ARMAS has obtained real-time radiation measurements from the ground to 107 km for 781 flights consisting of 713348 10-second measured absorbed dose (silicon) and derived effective dose rate records. The ARMAS monitoring system (Tobiska et al., 2016, 2018) consists of two components: (1) a flight instrument that measures the environment absorbed dose in silicon on an aircraft and (2) a real-time data stream from the aircraft to the ground, which is then processed to level 4–derived effective dose rates for location and time. Measurements are made using the Teledyne micro dosimeter UDOS001 (μDos) in combination with a microprocessor, a GPS chip, an iridium transceiver or a Bluetooth transmitter, and associated electronics. The μDos chip is sensitive to heavy ions (Fe^{+}), alphas, protons, neutrons, electrons, and γ-rays, especially above 1 MeV, based on extensive ground beam line testing (Tobiska et al., 2016, 2019; Gersey et al., 2020). All these components are mated to a printed circuit board and housed in a milled aluminum case. Once the absorbed dose (Si) is measured within the aircraft, it is relayed to the ground via iridium satellite link or aircraft WiFi.

The existing state-of-the-art for real-time total ionizing dose measurements is the ARMAS Flight Module (FM) instrument, now it its seventh generation. It is manufactured by Space Environment Technologies and is built for use by research aircraft and suborbital commercial space travel vehicles. The FM uses 5 VDC power from a vehicle-supplied connector passing through a DC-DC converter or with a COTS battery. Total power is 0.5 W (peak), 0.05 W (average), and 0.05 W (standby). It also contains a redundant micro SD data logger. The 1/8-inch milled aluminum unibody housing contains the electronics. Level 0 engineering units related to absorbed dose in silicon (rad) are integrated in 10-s time-resolved data packets that include UT time tag, latitude, longitude, and altitude. Level 1 dose and Level 2 dose rates in silicon and tissue, as well as Level 3 dose equivalent and ambient dose equivalent rates are provided through an iPhone and iPad ARMAS app paired with the FM7. The data stream has real time with 10-second time resolution, no latency, and an ability to store the data inside the app or downlink the data to the ground if the smart device has WiFi connectivity. Level 4 effective dose rates are processed by ground servers and distributed for the benefit of aerospace users. They are available at the ground with 1-minute latency or less. Figure 4.5 (top panel) shows an example of radiation weather (colored dots) vs. NAIRAS results as well as daily record (black plus and diamond) along with the ARMAS mean statistical database estimate (black asterisk). Final data are archived at the website URL https://sol.spacenvironment.net/ARMAS_Archive/.

Liulin

The Liulin-6SA1 dosimeter spectrometer is well known in the radiation measurement community (Dachev, 2013; Dachev et al., 2015; Spurny & Dachev, 2003; Stassinopoulous et al., 2002) and was used on the RaD-X balloon campaign (Mertens, 2016). It is manufactured by the Space Research and Technology Institute of the Bulgarian Academy of Sciences in Sofia (Dachev, 2013; Dachev et al., 2015). The Liulin contains a silicon detector, a charge-sensitive preamplifier, and two microcontrollers.

The energy deposition of charged particles is measured by pulse height analysis. The pulse amplitudes, following preamplifier processing, is proportional to the energy deposition in the silicon and thus to the dose and LET in the ratio 240 mV/MeV. The amplitudes are digitized by an A-to-D converter resulting in a 256-channel spectrum of the energy deposition. The Liulin detects energy in the range of 0.0813–20.8 MeV, corresponding to LET in the silicon detector in the range of 0.27–69.3 keV/μm for perpendicular charged particle flight through the 300 μm thick silicon detector. LET in tissue is lower.

Active Tissue Equivalent Dosimeter (ATED)

The Active Tissue Equivalent Dosimeter (ATED) is a radiation detector built by Oklahoma State University for measuring the LET spectrum, absorbed dose, and biologically weighted dose equivalent of ionizing radiation. The instrument is a gas-filled TEPC designed to simulate a 3 μm^3 biological cell; the sensitivity of the detector to ionizing radiation is similar to that of human tissue. A prototype ATED was extensively tested at the Japanese HIMAC calibration facility and at several proton particle accelerators. Currently, the ATED is being flown on the ISS. ATED occupies <5000 cm^3 with a mass of <2.5 kg. Total power consumption is <25 Watts. All components except the power cable are housed inside an aluminum enclosure. The pressurized detector head is a spherical ionization cavity. A large diameter is used in order to increase the cross-sectional area of the detector to increase the probability of detecting low fluxes of HZE particles. The spherical shell of the proportional counter is made from acrylic ($C_5O_2H_8$), and the inner surface of the shell is made conductive by coating it with colloidal graphite. The detector head is pressurized to 2300 Pa. ATED requires less than 25 Watts of external electrical power. High voltage needed to bias the proportional counter is provided internally by a DC to DC converter. The preamplifier and amplifier circuitry remain unchanged from the original prototype. The signal from the ionization cavity is first impedance matched by the preamplifier located inside the detector head. The signal is then sent to an operational amplifier and then to a pulse-shaping amplifier before being sent to the spectrometer.

4.3.2. Campaigns

The NASA Radiation Experiment (RaD-X) balloon and aircraft campaign (Meier, Matthiä, et al., 2016; Mertens, 2016; Norman et al., 2016; Tobiska et al., 2016) collected dosimetric measurements at seven altitudes from 8 to 32 km over 18 h near Fort Sumner, New Mexico, during solar and geomagnetically quiet conditions. The measurements started at commercial aviation altitudes and extended beyond the Regener-Pfotzer maximum. They were taken at similar cutoff rigidities, enabling dosimetric statistical uncertainty to be reduced below 5%. One finding was that protons comprise the dominant contribution from cosmic ray primaries between 21 and 27 km (Norman et al., 2016). In addition, cosmic ray heavy-ion primaries are discernible at altitudes >32 km.

Recently, the ARMAS database was compared with the NAIRAS model database as part of the RADIAN project. While ARMAS data were first validated with the RaD-X measurements (Tobiska et al., 2016), the more extensive ARMAS database was encapsulated as polynomial fit equations (Tobiska et al., 2018). Using equations based on altitude, L-shell, and geomagnetic conditions, an effective dose rate deriving from a GCR environment can be calculated for any location on the planet. A subset of the ARMAS database was represented by a second polynomial fit equation for the GCR plus probable relativistic energetic particle, creating bremsstrahlung γ-ray–induced effective dose rates. This subset was contained within a narrow band of L-shells ($2 \leq L \leq 6$) combined with altitudinal and geomagnetic dependencies. These databases can help predict the statistically expected radiation environment for regional air traffic management, for airport overflight operations, and for individual aircraft route planning (Tobiska et al., 2018).

4.4. REGULATORY ACTIVITIES

4.4.1. International

The Council of the EU adopted Directive 96/20/EURATOM on 13 May 1996. Article 42 of the EU directive imposes requirements relating to the assessment and limitation of aircrew cosmic ray radiation exposure (AMS, 2007). EU member states were required to implement the directive by 13 May 2000 through national legislation. Consistent with "best practice" radiation protection procedures, which is to keep all radiation exposures as low as reasonably achievable (i.e. the ALARA principle), the EU has also adopted an "action level" of 6 mSv/yr. For those likely to exceed 6 mSv/yr, individual record keeping and medical surveillance is required of the aircraft operators (Dyer & Lei, 2001). For exposures less than 6 mSv/yr, only monitoring is recommended and the actual implementation of these recommendation varies among the EU Member States (Meier et al., 2009). The EU directive recommendation is that individual exposure be assessed by the aircraft operators if one is likely to exceed 1 mSv/yr, that workers be educated on radiation health risks, and that work schedules be adjusted to ensure that the 6 mSv/yr level is not exceeded (EURADOS, 1996).

4.4.2. United States

While there are no US FAA regulations relevant to the radiation environment hazard management at aviation altitudes, the US Space Weather Strategy and Action Plan (SWAP) (National Science and Technology Council, 2015) and White House executive order "Coordinating Efforts to Prepare the Nation for Space Weather Events" (The White House, 2016) supported the development and validation of an operational aviation radiation monitoring and forecasting capability. In particular, SWAP sought to (1) identify real-time monitoring requirements of the charged particle radiation environment to protect the health and safety of crew and passengers during space weather events; (2) define the scope and requirements for a real-time reporting system conveying situational awareness of the radiation environment to orbital, suborbital, and commercial aviation users during space weather events; and (3) develop or improve models for the real-time assessment of radiation levels at commercial flight altitudes. These executive branch initiatives have made their way into Congress where the House and Senate passed the PROSWIFT bill enacted into law on October 21, 2020 that embodies the elements of SWAP.

4.5. CONCLUSIONS

Air safety improvement from the effects from space weather, particularly ionizing radiation exposure, is a concern faced by aircrew, high-altitude pilots, frequent flyers, and commercial space travelers. This chapter describes the physics of atmospheric radiation, including its primary sources of GCRs and SEPs, as well as an important newly proposed third source, i.e. the secondary radiation environment caused by precipitating radiation belt charged particles.

The GCR primary radiation processes are described, including their transport through the heliosphere and their modulation by the solar wind and changes in the IMF, the modulation connected with the Sun's polar field, their arrival at Earth, and the importance of magnetic shielding by the Earth's geomagnetic field. The SEP primary radiation differential energy spectrum for shock acceleration is described mathematically, and the magnetic shielding by the Earth's geomagnetic field is reinforced. Radiation belt precipitated charged particles, a newer topic in this discipline, is described and data discussed from a new, very large database of atmospheric radiation measurements made by the ARMAS system. The RB source is confined to a band of higher magnetic latitudes mapping to L-shells 2–7.

The transport of charged particles, from any of the three sources, through the magnetic shielding from the Earth's main and magnetospheric magnetic fields is described and the concept of geomagnetic cutoff rigidity is thoroughly explained, both conceptually and mathematically. The transport of radiation through the atmosphere from precipitating charged particles below the mesopause is presented from a theoretical perspective, including the interactions of neutrons and charged particles with target species in the atmosphere. Radiation dosimetry is described, including the concepts of dose quantities as well as quality factor.

Additional discussions are presented, including an introduction on the effects of radiation on avionics, the models capable of reporting the aviation radiation environment, the measurements that have been made of the radiation environment in the atmosphere by specific instruments, and introductory remarks on the regulatory activity internationally and in the US related to the effects of radiation faced by aircrew, high-altitude pilots, frequent flyers, and commercial space travelers.

REFERENCES

American Meteorological Society (AMS) (2007). Integrating space weather observations and forecasts into aviation operations [Technical report]. American Meteorological Society Policy Program and SolarMetrics.

Artamonov, A. A., Mishev, A. L., & Usoskin, I. G. (2016). Model CRAC:EPII for atmospheric ionization due to precipitating electrons: Yield function and applications. *J. Geophys. Res. Space Physics*, *121*, 1736–1743. doi:10.1002/2015JA022276

Badavi, F. F., Nealy, J. E., de Angelis, G., Wilson, J. W., Clowdsley, M. S., Luetke, N. J., et al. (2005). Radiation environment and shielding model validation for CEV design. Space 2005, No. AIAA 2005-6651. Long Beach, CA: Am. Inst. of Aeronaut. and Astronaut.

Badavi, F. F., Tramaglina, J. K., Nealy, J. E., & Wilson, J. W. (2007). Low Earth orbit radiation environments and shield model validation for ISS. *Space 2007*, No. AIAA 2007-6046. Long Beach, CA: Am. Inst. of Aeronaut. and Astronaut.

Badhwar, G. D., Konradi, A., Hardy, A., & Braby, L. A. (1992). *Nucl. Tracks Radiat. Meas.*, *20*, 13–20.

Badhwar, G. D., & O'Neill, P. M. (1991). An improved model of galactic cosmic radiation for space exploration missions. *22nd International Cosmic Ray Conference* (OG-5.2-13, pp. 643–646).

Badhwar, G. D., & O'Neill, P. M. (1992). An improved model of galactic cosmic radiation for space exploration missions. *Nuclear Tracks Radiat. Meas.*, *20*, 403–410.

Badhwar, G. D., & O'Neill, P. M. (1993). Time lag of twenty-two year solar modulation. *23nd International Cosmic Ray Conference* (vol. 3, pp. 535–539).

Badhwar, G. D., & O'Neill, P. M. (1994). Long term modulation of galactic cosmic radiation and its model for space exploration. *Adv. Space Res.*, *14*, 749–757.

Badhwar, G. D., & ONeill, P. M. (1996). Galactic cosmic radiation model and its applications. *Adv. Space Res.*, *17*, 7–17.

Blum, L. W., Halford, A., Millan, R., Bonnell, J. W., Goldstein, J., Usanova, M., et al. (2015). Observations of coincident EMIC wave activity and duskside energetic electron precipitation on 18–19 January 2013. *Geophys. Res. Lett.*, *42*, 5727–5735. doi:10.1002/2015GL065245

Braby, L. A., Badhwar, G. D., Conroy, T. J., Elegy, D. C., & Brackenbush, L. W. (1994). *Radiat. Protection Dosimetry*, *52*, 423–426.

Brandt, S. (1999). *Data analysis, statistical and computational methods for scientists and engineers*. New York: Springer-Verlag.

Chilingarian, A., Chilingaryan, S., & Reymers, A. (2015). Atmospheric discharges and particle fluxes. *J. Geophys. Res. Space Physics*, *120*, 5845–5853. doi:10.1002/ 2015JA021259

Clem, J., Clements, D. P., Esposito, J., Evenson, P., Huber, D., LHeureux, J., et al. (1996). Solar modulation of cosmic electrons. *Astrophys. J.*, *464*, 507.

Clilverd, M. A., Rodger, C. J., Millan, R. M., Sample, J. G., Kokorowski, M., McCarthy, M. P., et al. (2007). Energetic particle precipitation into the middle atmosphere triggered by a coronal mass ejection. *J. Geophys. Res.*, *112*, A12206. doi:10.1029/2007JA012395

Clucas, S. N., Dyer, C. S., & Lei, F. (2005). The radiation in the atmosphere during major solar particle events. *Adv. Space Res.*, *36*, 1657–1664.

Copeland, K. (2017). CARI-7A: Development and validation. *Radiat. Prot. Dosim.* https://doi.org/10.1093/rpd/ncw369

Copeland, K., Sauer, H. H., Duke, F. E., & Friedberg, W. (2008). Cosmic radiation exposure on aircraft occupants on simulated high-latitude flights during solar proton events from 1 January 1986 through 1 January 2008. *Adv. Space Res.*, *42*, 1008–1029.

Dachev, T. P. (2013). Profile of the ionization radiation exposure between the Earth surface and free space. *J. Atmos. Sol. Terr. Phys.*, *102*, 148–156.

Dachev, T. P. (2017). Relativistic electron precipitation bands in the outside radiation environment of the international space station. *Journal of Atmospheric and Solar-Terrestrial Physics*, *177*, 247–256. https://doi.org/10.1016/j.jastp.2017.11.008

Dachev, T. P., Bankov, N. G., Tomov, B. T., Matviichuk, Y. N., Dimitrov, P. G., Häder, D.-P., & Horneck, G. (2017a). Overview of the ISS radiation environment observed during the ESA EXPOSE-R2 mission in 2014–2016. *Space Weather*, *15*, 1475–1489. https://doi.org/10.1002/ 2016SW001580

Dachev, T. P., Bankov, N. G., Tomov, B. T., Matviichuk, Y. N., Dimitrov, P. G., Häder, D.-P., & Horneck, G. (2017b). Overview of the ISS radiation environment observed during the ESA EXPOSE-R2 mission in 2014–2016. *Space Weather*, *15*, 1475–1489. https://doi.org/10.1002/ 2016SW001580

Dachev, T. P., Semkova J. V., Tomov B. T., Matviichuk Y. N., Dimitrov P. G., & Koleva, R. T. (2015). Overview of the Liulin type instruments for space radiation measurement and their scientific results. *Life Sci. Space Res.*, *4*, 92–114. doi:10.1016/j.lssr.2015.01.005

Dyer, C., Hands, A., Lei, F., Truscott, P., Ryden, K. A., Morris, P., et al. (2009). Advances in measuring and modeling the atmospheric radiation environment. *IEEE Trans. Nucl. Sci.*, *56*(8).

Dyer, C., & Lei, F. (2001). Monte Carlo calculations of the influence on aircraft radiation environments of structures and solar particle events. *IEEE Trans. Nucl. Sci.*, *48*(6), 1987–1995.

Dyer, C. S., Lei, F., Clucas, S. N., Smart, D. F., & Shea, M. A. (2003a). Solar particle enhancements of single event effect rates at aircraft altitudes. *IEEE Transactions on Nuclear Science*, *50*(6), 2038–2045.

Dyer, C. S., Lei, F., Clucas, S. N., Smart, D. F., & Shea, M. A. (2003b). Calculations and observations of solar particle enhancements to the radiation environment at aircraft altitudes. *Advances in Space Research*, *32*(1), 81–93.

Dyer, C., Lei, F., Hands, A., Clucas, S., & Jones, B. (2005). Measurements of the atmospheric radiation environment from cream and comparisons with models for quiet time and solar particle events. *IEEE Trans. Nucl. Sci.*, *52*(6), 2326–2331.

Dyer, C. S., & Truscott, P. (1999). Cosmic radiation effects on avionics. *Radiat. Prot. Dosim.*, *86*(4), 337–342.

Ellison, D. C., & Ramaty, R. (1985). Shock acceleration of electrons and ions in solar flares. *Astrophys. J.*, *298*, 400–408.

European Radiation Dosimetry Group (EURADOS) (1996). Exposure of air crew to cosmic radiation. A report of EURADOS Working Group 11, *EURADOS Report 1996.01*. In McAulay, I. R, Bartlett, D. T, Dietze, G. et al. (Eds.), *European Commission Report Radiation Protection 85*. Luxembourg: Office for Official Publications of the European Communities.

Fedder, J.G.L.J.A. & Mobarry, C. M. (2004). The Lyon-Fedder-Mobarry (LFM) global MHD magnetospheric simulation code. *J. of Atmos. and Solar-Terrestrial Phys.*, *66*(15–16), 1333–1350.

Federico, C. A., Gonçalez, O. L., Caldas, L. V. E., Pazianotto, M. T., Dyer, C., Caresana, M., & Hands, A. (2015). Radiation measurements onboard aircraft in the South Atlantic region. *Radiation Measurements*, *82*, 14e20–14e20. https://doi.org/10.1016/j.radmeas.2015.07.008

Felsberger, E., O'Brien, K., Kindl, P. (2009). IASON-FREE: Theory and experimental comparisons. *Radiat. Prot. Dosim.*, *136*(4), 267–273. https://doi.org/10.1093/rpd/ncp128

Ferrari, A., Pelliccioni, M., & Pillon, M. (1997a). Fluence to effective dose conversion coefficients for neutrons up to 10 tev. *Radiat. Prot. Dos.*, *71*(3), 165–173.

Ferrari, A., Pelliccioni, M., & Pillon, M. (1997b). Fluence to effective dose and effective dose equivalent conversion coefficients for protons from 5 mev to 10 tev. *Radiat. Prot. Dos.*, *71*(2), 85–91.

Forsyth, C., Rae I. J., Murphy K. R., Freeman M. P., Huang C.-L., Spence H. E., et al. (2016). What effect do substorms have on the content of the radiation belts? *J. Geophys. Res. Space Physics*, *121*, 6292–6306. doi:10.1002/2016JA022620

Friedberg, W., & Copeland, K. (2003). What aircrews should know about their occupational exposure to ionizing radiation ionizing (DOT Report No. DOT/FAA/AM-03/16). Oklahoma City, OK: Federal Aviation Administration, Civil Aerospace Medical Institute.

Friedberg, W., & Copeland, K. (2011). Ionizing radiation in Earth's atmosphere and in space near Earth (DOT Report No. DOT/FAA/ AM-11/9). Oklahoma City, OK: Federal Aviation Administration, Civil Aerospace Medical Institute.

Gaines, E. E., Chenette, D. L., Imhof, W. L., Jackman, C. H., & Winningham, J. D. (1995). Relativistic electron fluxes in May 1992 and their effect on the middle atmosphere. *J. Geophys. Res.*, *100*(D1), 1027–1033.

Gaisser, T. (1990). *Cosmic rays and particle physics*. Cambridge University Press.

Gersey, B., T. B. Borak, S. Guetersloh, C. Zeitlin, J. Miller, L. Heilbronn, T. Murakami, and Y. Iwata (2002), *Radiat. Res.*, *157*, 350–360.

Gersey, B. "B.", Tobiska, W. K.,Atwell, W., Bouwer, D., Didkovsky, L., Judge, K., et al. (2020). Beamline and flight comparisons of the ARMAS flight module with the tissue equivalent proportional counter for improving atmospheric radiation monitoring accuracy. Space Weather, 18, e2020SW002599. https://doi.org/ 10.1029/2020SW002599

Getley, I. L., Bennett, L.G.I., Lewis, B. J., Bennett, B., Dyer, C. S., Hands, A.D.P., & Duldig, M. L. (2010). Evaluation of new cosmic radiation monitors designed for aircrew exposure assessment. *Space Weather*, *8*, S01001. https://doi.org/10.1029/2009SW000492

Gopalswamy, N. (2004). A global picture of CMEs in the inner heliosphere. In G. Poletto & S. Suess (Ed.), *The sun and the heliosphere as an integrated system* (ASSL Series, p. 201). Boston: Kluwer.

Grajewski, B., Waters, M. A., Yong, L. C., Tseng, C.-Y., Zivkovich, Z., & Cassinelli II, R. T. (2011). Airline pilot cosmic radiation and circadian disruption exposure assessment from logbooks and company records. *Ann. Occup. Hyg.*, *55*(5): 465–475.

Guetersloh, S. B., Borak, T. B., Taddei, P. J., Zeitlin, C., Heilbronn, L., Miller, J., et al. (2004). *Radiat. Res.*, *161*, 64–71.

Hajra, R., Tsurutani, B. T., Echer, E., Gonzalez, W. D., Brum, C.G.M., Vieira, L.E.A., & Santolik O. (2015). Relativistic electron acceleration during HILDCAA events: Are precursor CIR magnetic storms important? *Earth Planets Space*, *67*, 109. doi:10.1186/s40623-015-0280-5

Hajra, R., Tsurutani, B. T., Echer, E., Gonzalez, W. D., & Santolik, O. (2015a). Relativistic (E > 0.6, >2.0, and >4.0 MeV) electron acceleration at geosynchronous orbit during high-intensity, long-duration, continuous AE activity (HILDCAA) events. *Astrophys. J.*, *799*, 39–46. doi:10.1088/0004-637X/799/1/39

Heinrich, W., Roesler, S., & Schraube, H. (1999). Physics of cosmic radiation fields. *Radiat. Prot. Dosim.*, *86*, 253–258.

Hwang, J., Dokgo, K., Choi, E., Kin, K.-C., Kim, H.-P., & Cho, K.-S. (2014). Korean radiation exposure assessment model for aviation route dose. In KREAM, KSS Fall meeting, Jeju, Korea, 29–31 October.

Iles, R.H.A., Jones, J.B.L., Taylor, G. C., Blake, J. B., Bentley, R. D., Hunter, R., et al. (2004). Effect of solar energetic particle (SEP) events on the radiation exposure levels to aircraft passengers and crew: Case study of 14 July 2000 SEP event. *Journal of Geophysical Research*, *109*, A11103. https://doi.org/10.1029/2003JA010343

Ingraham, J. C., Cayton T. E., Belian R. D., Christensen R. A., Friedel R. H. W., Meier M. M., et al. (2001). Substorm injection of relativistic electrons to geosynchronous orbit during the great magnetic storm of March 24, 1991. *J. Geophys. Res.*, *106*(A11), 25759–25776. doi:10.1029/2000JA000458

International Standards Organization, IS 21348 (2007). Space environment (natural and artificial) – Process for determining solar irradiances. Geneva: ISO.

International Standards Organization, IS 16695 (2013). Geomagnetic Reference Models. Geneva: ISO.

International Agency for Research on Cancer (IARC) (2000). Ionizing radiation: Part 1. X- and γ-radiation and neutrons. In *IARC Work. Group Eval. Carcinog. Risks Hum.* (vol. 75). Lyon, France: IARC and World Health Organization.

International Commission on Radiological Protection (ICRP) (1991). *ICRP Publication 60, 1990 Recommendations of the International Commission on Radiological Protection*, *21*(1–3). Pergamon Press.

International Commission on Radiological Protection (ICRP) (2008). *ICRP Publication 103, 2007 Recommendations of the International Commission on Radiological Protection.* ISBN 0-7020-3048-1. Elsevier.

International Standards Organization (ISO) (2013). IS:16695: Space environment (natural and artificial)—Geomagnetic reference models. Geneva.

Johnston, C. O. (2008). A comparison of EAST shock-tube radiation measurements with a new radiation model. AIAA Paper 2008-1245.

Joyce, C. J., Schwadron, N. A., Wilson, J. K., Spence, H. E., Kasper, J.C., Golightly, M., et al. (2014). Radiation modeling in the Earth and Mars atmospheres using LRO/CRaTER with the EMMREM Module. *Space Weather*, *12*, 112–119. https://doi.org/10.1002/2013SW000997

Kahler, S. W. (2001). Origin and properties of solar energetic particles in space. In P. Song, H. J. Singer, & G. L. Siscoe (Eds.), *Space weather*. Washington, DC: American Geophysical Union.

Katsiyannis, A. C., Dominique, M., Pierrard, V., Rosson, G. L., Keyser, J. D., & Berghmans, D. (2018). The detection of ultra-relativistic electrons in low Earth orbit. *J. Space Weather Space Clim.*, *8*, A01.

Kim, H., Lessard, M. R., Engebretson, M. J., & Young, M. A. (2011). Statistical study of Pc1–2 wave propagation characteristics in the high-latitude ionospheric waveguide. *J. Geophys. Res.*, *116*, A07227. doi:10.1029/2010JA016355

Kress, B. T., Hudson, M. K., Perry, K. L., & Slocum, P. L. (2004). Dynamic modeling of geomagnetic cutoff for the 23–24 November 2001 solar energetic particle event, *Geophys. Res. Lett.*, *31*(L04808). doi:10.1029/2003GL018599

Kress, B. T., Mertens, C. J., & Wiltberger, M. (2010). Solar energetic particle cutoff variations during the 28–31 October 2003 geomagnetic storm. *Space Weather*, *8*(S05001). doi:10.1029/2009SW000488

Langlais, B., & Mandea, M. (2000). An IGRF candidate geomagnetic field model for epoch 2000 and a secular variation model for 2000–2005, *Earth Planets Space*, *52*, 1137–1148.

Latocha, M., Beck, P., B€utikofer, P., Thommesen, H. (2014). AVIDOS 2.0—Current developments for the assessment of radiation exposure at aircraft altitudes caused by solar cosmic radiation exposure. European Space Weather Week, Liege, 17–21 November. http://stce.be/esww11

Latocha, M., Beck, P., & Rollet, S., 2009. AVIDOS—A software package for European accredited aviation dosimetry. *Radiat. Prot. Dosim.*, *136* (4), 286. https://doi.org/10.1093/rpd/ncp126

Lee, J., Nam, U.-W., Pyo, J., Kim, S., Kwon, Y.-J., Lee, J., et al. (2015). Short-term variation of cosmic radiation measured by aircraft under constant flight conditions. *Space Weather*, *13*, 797–806. https://doi.org/10.1002/2015SW001288

Lewis, B. J., Bennett, G. I., Green, A. R., McCall, M. J., Ellaschuk, B., Butler, A. & Pierre, M. (2002). Galactic and solar radiation exposure to aircrew during a solar cycle. *Radiat. Prot. Dosim.*, *102*(3), 207–227.

Lindborg, L., Bartlett, D. T., Beck, P., McAulay, I. R., Schnuer, K., Schraube, H., et al. (2004). Cosmic radiation exposure of aircraft crew: Compilation of measured and calculated data. A report of EURADOS Working Group 5, European Radiation Dosimetry Group, Luxembourg: Office for the Official Publications of the European Communities.

Mares, V., Maczka, T., Leuthold, G., & Ruhm, M. (2009). Air crew dosimetry with a new version of EPCARD. *Radiat. Prot. Dosim.*, *136* (4), 262–266. https://doi.org/10.1093/rpd/ncp129

Matthiä, D., Meier, M. M., & Reitz, G. (2014). Numerical calculation of the radiation exposure from galactic cosmic rays at aviation altitudes with the PANDOCA core model. *Space Weather*, *12*, 161. https://doi.org/ 10.1002/2013SW001022

McCall, M. J., Lemay, F., Bean, M. R., Lewis, B. J., & Bennett, L. G. (2009). Development of a predictive code for aircrew radiation exposure. *Radiat. Prot. Dosim.*, *136* (4), 274–281. https://doi.org/10.1093/rpd/ncp130

McIlwain, C. E. (1961). Coordinates for mapping the distribution of magnetically trapped particles. *J. Geophys. Res.*, *66*, 3681–3691.

Meier, M., Hubiak, M., Matthiä, D., Wirtz, M., & Reitz, G. (2009). Dosimetry at aviation altitudes. *Radiat. Prot. Dos.*, *136*(4), 251–255.

Meier, M. M., Matthiä, D., Forkert, T., Wirtz, M., Scheibinger, M., Hübel, R., & Mertens, C. J. (2016). RaD-X: complementary measurements of dose rates at aviation altitudes. *Space Weather*, *14*. https://doi.org/10.1002/2016SW001418

Meier, M. M., Trompier, F., Ambrozova, I., Kubancak, J., Matthi, D., Ploc, O., Santen, N., Wirtz, M. (2016). CONCORD: Comparison of cosmic radiation detectors in the radiation field at aviation altitudes. *J. Space Weather Space Clim.*, *6*, A24. https://doi.org/10.1051/swsc/2016017

Mertens, C. J. (2016). Overview of the radiation dosimetry experiment (RaD-X) flight mission. *Space Weather*, *14*, 874–898. https://doi.org/10.1002/2016SW001407

Mertens, C. J., Gronoff, G. P., Norman, R. B., Hayes, B. M., Lusby, T. C., Straume, T., et al. (2016). Cosmic radiation dose measurements from the RaD-X flight campaign. *Space Weather*, *14*, 874–898. https://doi.org/10.1002/2016SW001407

Mertens, C. J., Kress, B. T., Wiltberger, M., Blattnig, S. R., Slaba, T. S., Solomon, S. C., & Engel, M. (2010). Geomagnetic influence on aircraft radiation exposure during a solar energetic particle event in October 2003. *Space Weather*, *8*(S03006). doi:10.1029/2009SW000487

Mertens, C. J., Moyers, M. F., Walker, S. A., & Tweed, J. (2010). Proton lateral broadening distribution comparisons between grntrn, mcnpx, and laboratory beam measurements. *Adv. Space Res.*, *45*, 884–891.

Mertens, C. J., Kress, B. T., Wiltberger, M., Tobiska, W. K., Grajewski, B., & Xu, X. (2012). Atmospheric ionizing radiation from galactic and solar cosmic rays. In M. Nenoi (Ed.), *Current topics in ionizing radiation research*. ISBN: 978-953-51-0196-3. InTech.

Mertens, C. J., Meier M. M., Brown S., Norman R. B., & Xu X. (2013). NAIRAS aircraft radiation model development, dose climatology, and initial validation. *Space Weather*, *11*, 603–635. doi:10.1002/swe.20100

Mertens, C. J., Tobiska, W. K., Bouwer, D., Kress, B. T., Wiltberger, M. J., Solomon, S. C., & Murray, J. J. (2009). Development of nowcast of atmospheric ionizing radiation for aviation safety (NAIRAS) model. *1st AIAA Atmosheric and Space Environments Conference* (AIAA 2009-3633). San Antonio, TX: Am. Inst. of Aeronaut. and Astronaut.

Mertens, C. J., Wilson, J. W., Blattnig, S. R., Kress, B. T., Norbury, J. W., Wiltberger, M. J., et al. (2008). Influence of space weather on aircraft ionizing radiation exposure. *46th Aerospace Sciences Meeting and Exhibit* (AIAA 2008-0463). Reno, NV: Am. Inst. of Aeronaut. and Astronaut.,

Mertens, C. J., Wilson, J. W., Blattnig, S. R., Solomon, S. C., Wiltberger, M. J., Kunches, J., et al. (2007). Space weather nowcasting of atmospheric ionizing radiation for aviation safety. *45th Aerospace Sciences Meeting and Exhibit* (AIAA 2007-1104). Reno, NV: Am. Inst. of Aeronaut. and Astronaut.

Mewaldt, R. A., Cohen, C.M.S., Labrador, A. W., Leske, R. A., Mason, G. M., Desai, M. I., et al. (2005). Proton, helium, and electron spectra during the large solar particle events of October–November 2003. *J. Geophys. Res.*, *110*(A09S10). doi:10.1029/2005JA011038

Miyashita, Y., Hosokawa, K., Hori, T., Kamide, Y., Yukimatu, A. S., Fujimoto, M., et al. (2008). Response of large-scale ionospheric convection to substorm expansion onsets: A case study. *J. Geophys. Res.*, *113*, A12309. doi:10.1029/2008JA013586

Mutuel, L. H. (2016). Single event effects mitigation techniques report (TC-15/62). Department of Transportation/Federal Aviation Administration, February 2016.

National Science and Technology Council (2015). National Space Weather Action Plan, October 2015.

Nealy, J. E., Cucinotta, F. A., Wilson, J. W., Badavi, F. F., Dachev, T. P., Tomov, B. T., et al. (2007). Pre-engineering spaceflight validation of environmental models and the 2005 HZETRN simulations code. *Adv. Space Res.*, *4*, 1593–1610.

Norman, R. B., Mertens, C. J., & Slaba, T. C. (2016). Evaluating galactic cosmic ray environment models using RaD-X flight data. *Space Weather*, *14*, 764–775. https://doi.org/10.1002/2016SW001401

Normand, E., Oberg, D. L., Wert, J. L., Ness, J. D., Majewski, P. P., Wender, S., & Gavron, A. (1994). Single event upset and charge collection measurements using high energy protons and neutrons. *IEEE Trans. Nucl. Sci.*, *41*(6), 2203–2209.

Normand, E., Vranish, K., Sheets, A., Stitt, M., & Kim, R. (2006). Quantifying the double-sided neutron SEU threat, from low energy (thermal) and high energy (>10 MeV) neutrons. *IEEE Trans. Nucl. Sci.*, *53*(6), 3587–3595.

O'Brien, K., O'Brien, K., Smart, D. F., Shea, M. A., Felsberger, E., Schrewe, U., et al. (2003). World-wide radiation doseage calculations for air crew members. *Adv. Space Res.*, *31*(4), 835–840.

O'Neill, P. M. (2006). Badhwar-ONeill galactic cosmic ray model update based on advanced composition explorer (ACE) energy spectra from 1997 to present. *Adv. Space Res.*, *37*, 1727–1733.

Parker, E. N. (1965). The passage of energetic charged particles through interplanetary space. *Planet. Space Sci.*, *13*, 9–49.

Ploc, O., Ambrozova, I., Kubancak, J., Kovar, I., & Dachev, T. P. (2013). Publicly available database of measurements with the silicon spectro- meter Liulin onboard aircraft. *Radiation Measurements*, *58*, 107–112. https://doi.org/10.1016/j.radmeas.2013.09.002

Ploc, O., Pachnerova Brabcova, K., Spurny, F., Malusek, A., & Dachev, T. (2011). Use of energy deposition spectrometer Liulin for individual monitoring of aircrew. *Radiation Protection Dosimetry*, *144*, 611e614.

Reames, D. V. (2013). The two sources of solar energetic particles. *Space Sci Rev*, *175*, 53. doi:10.1007/s11214-013-9958-9

Regener, E., & Pfotzer, G. (1935). Vertical intensity of cosmic rays by threefold coincidences in the stratosphere. *Nature*, *136*, 718.

Reitz, G., Schnuer, K., & Shaw, K. (1993). Editorial: Workshop on radiation exposure of civil aircrew. *Radiat. Prot. Dosim.*, *48*, 3.

Rodger, C. J., Clilverd, M. A., Thomson, N. R., Gamble, R. J., Seppälä, A., Turunen, E., et al. (2007). Radiation belt electron precipitation into the atmosphere: Recovery from a geomagnetic storm. *J. Geophys. Res.*, *112*, A11307. doi:10.1029/2007JA012383

Sato, T. (2015). Analytical model for estimating terrestrial cosmic ray fluxes nearly anytime and anywhere in the world: Extension of PARMA/EXPACS. *PLoS One.*, *10*(12), e0144679.

Sato, T., Yasuda, H., Niita, K., Endo, A., & Sihver, L. (2008). Development of PARMAPHITS-based analytical radiation model in the atmosphere. *Radiat. Res.*, *170*, 244.

Schraube, H., Mares, V., Roesler, S., & Heinrich, W. (1999). Experimental verification and calculation of aviation route doses. *Radiat. Prot. Dosim.*, *86*(4), 309–315.

Schwadron, N. A., Baker, T., Blake, B., Case, A. W., Cooper, J. F., Golightly, M., et al. (2012). Lunar radiation environment and space weathering from the Cosmic Ray Telescope for the Effects of Radiation (CRaTER). *J. Geophys. Res. Planets*, *117*. https://doi.org/ 10.1029/2011JE003978. E00H13

Shea, M. A., & Smart, D. F. (2012). Space weather and the ground-level solar proton events of the 23rd solar cycle. *Space Sci. Rev.*, *171*, 161–188.

Simpson, J. A. (1983). Elemental and isotopic composition of the galactic cosmic rays. *Annu. Rev. Nucl. Part. Sci.*, *33*, 323–382. https://doi.org/10.1146/annurev.ns.33.120183.001543

Slaba, T. C., Qualls, G. D., Clowdsley, M. S., Blattnig, S. R., Simonsen, L. C., Walker, S. W., & Singleterry, R. C. (2009). Analysis of mass averaged tissue doses in max, fax, and cam, and caf. *Technical Report TP-2009-215562*. NASA.

Smart, D. F., & Shea, M. A. (1994). Geomagnetic cutoffs: A review for space dosimetry calculations, *Adv. Space Res.* *14*(10), 10787–10796.

Smart, D. F., & Shea, M. A. (2005). A review of geomagnetic cutoff rigidities for earth-orbiting spacecraft, *Adv. Space Res.*, *36*, 2012–2020.

Spence, H. E., Case, A., Golightly, M. J., Heine, T., Larsen, B. A., Blake, J. B., et al. (2010). CRaTER: The Cosmic Ray Telescope for the Effects of Radiation Experiment on the Lunar Reconnaissance Orbiter Mission. *Space Sci. Rev.*, *150*(1–4), 243–284.

Spurny, F., & Dachev, T. (2001). Measurements during an intense solar flare, GLE 60. *Radiation Protection Dosimetry*, *95*, 273e275.

Spurny, F., & Dachev, T. (2002). On board aircrew dosimetry with a semiconductor spectrometer. *Radiation Protection Dosimetry*, *100*, 525e528.

Spurny, F., & Dachev, T. S. (2003). Long-term monitoring of the onboard aircraft exposure level with a Si-diode based spectrometer. *Adv. Space Res.*, *32*(1), 53–58.

Stassinopoulos, E. G., Stauffer, C. A., Dachev, T. P., Brucker, G. J., Tomov, B. T., & Dimitrov, P. G. (2002). The Liulin-3M radiometer for measuring particle doses in space and on aircraft (NASA/TM-2002-210003).

Störmer, C. (1965). *The polar aurora*, Oxford at the Clarendon Press.

Tai, H., Bichsel, H., Wilson, J. W., Shinn, J. L., Cucinotta, F. A., & Badavi, F. F. (1997). Comparison of stopping power and range databases for radiation transport study. *Technical Report TP-3644*. NASA.

Tang, C. L., Zhang, J.-C., Reeves, G. D., Su, Z. P., Baker, D. N., Spence, H. E., et al. (2016). Prompt enhancement of the Earth's outer radiation belt due to substorm electron injections. *J. Geophys. Res. Space Physics*, *121*, 11826–11838. doi:10.1002/ 2016JA023550

Tascione, T. F. (1994). *Introduction to the space environment*, 2nd ed. Malabar, FL: Krieger Publishing Co.

Tobiska, W. K., Atwell W., Beck P., Benton E., Copeland K., Dyer C., et al. (2015). Advances in atmospheric radiation measurements and modeling needed to improve air safety. *Space Weather*, *13*, 202–210.

Tobiska, W. K., Bouwer, D., Smart, D., Shea, M., Bailey, J., Didkovsky, L., et al. (2016). Global real-time dose measurements using the Automated Radiation Measurements for Aerospace Safety (ARMAS) system. *Space Weather*, *14*, 1053–1080.

Tobiska, W. K., Didkovsky, L., Judge, K., Weiman, S., Bouwer, D., Bailey, J., et al. (2018). Analytical representations for characterizing the global aviation radiation environment based on model and measurement databases. *Space Weather*, 16. https://doi.org/10.1029/2018SW001843

Tobiska, W. K., Meier, M.M., Matthiae, D., & Copeland, K. (2017). Characterizing the variation in atmospheric radiation at aviation altitudes. In N. Buzulukova (Ed.), *Extreme EVENTS in geospacer* (pp. 453–471). Elsevier, ISBN: 9780128127001.

Tobiska, W. K., Didkovsky, L., Judge, K., Bouwer, D., Wieman, S., Gersey, B., et al. (2019). ARMAS flight system for operational aerospace radiation measurements. In 49th International Conference on Environmental Systems, 7–11 July 2019, Boston, Massachusetts, ICES-2019- 406. Retrieved from https://ttu-ir.tdl.org/handle/2346/84649

Toffoletto, F. R., Sazykin, S., Spiro, R. W., Wolf, R. A., & Lyon, J. G. (2004). RCM meets LFM: Initial results of one-way coupling. *J. of Atmos. and Solar-Terrestrial Phys.*, *66*(15–16), 1361–1370.

Townsend, L. W., Stephens, D. L., Hoff, J. L., Zapp, E. N., Moussa, H. M., Miller, T. M., et al. (2006). The carrington event: Possible doses to crews in space from a comparable event. *Adv. Space Res.*, *38*, 226–231.

Townsend, L. W., Zapp Jr., E. N., D. L. S., & Hoff, J. L. (2003). Carrington flare of 1859 as a prototypical worst-case solar energetic particle event. *IEEE Trans. Nucl. Sci.*, *50*(6): 2307–2309.

Tsurutani, B. T., Gonzalez, W. D., Gonzalez, A.L.C., Guarnieri, F. L., Gopalswamy, N., Grande, M., et al. (2006). Corotating solar wind streams and recurrent geomagnetic activity: A review. *J. Geophys. Res.*, *111*, A07S01. doi:10.1029/2005JA011273

Tsurutani, B. T., Hajra, R., Tanimori, T., Takada, A., Bhanu, R., Mannucci, A. J., et al. (2016). Heliospheric plasma sheet (HPS) impingement onto the magnetosphere as a cause of relativistic electron dropouts (REDs) via coherent EMIC wave scattering with possible consequences for climate change mechanisms. *J. Geophys. Res. Space Physics*, *121*, 10130–10156. doi:10.1002/2016JA022499

Tsyganenko, N. A. (1989). Determination of magnetic current system parameters and development of experimental geomagnetic field models based on data from IMP and HEOS satellite. *Planet Space Sci.*, *37*, 5–20.

Tsyganenko, N. A. (2002). A model of the near magnetosphere with dawn-dusk asymmetry: 1. mathematical structure. *J. Geophys. Res.*, *107*(A8), 1179. doi:10.1029/2001JA000219

Tsyganenko, N. A., & Sitnov, N. I. (2005). Modeling the dynamics of the inner magnetosphere during strong geomagnetic storms. *J. Geophys. Res.*, *110*, A03208. doi:10.1029/2004JA010798

Tylka, A. J., Cohen, C.M.S., Dietrich, W. F., Lee, M. A., Maclennan, C. G., Mewaldt, R. A., et al. (2005). Shock geometry, seed populations, and the origin of variable elemental composition at high energies in large gradual solar particle events. *ApJ.*, *625*, 474–495.

Tylka, A. J., & Lee, M. A. (2006). Spectral and compositional characteristics of gradual and impulsive solar energetic particle events, in solar eruptions and energetic particles. In N. Gopalswamy, R. Mewaldt, & J. Torsi (Eds.), *Solar eruptions and energetic particles*. Geophysical Monograph 165. Washington, DC: American Geophysical Union.

United Nations Scientific Committee on the Effect of Atomic Radiation (UNSCEAR) (2000). Sources and effect of ionizing radiation. United Nations Scientific Committee on the Effect of Atomic Radiation UNSCEAR 2000 Report to the General Assembly, with Scientific Annexes, Vol. II, Annex G.

Van Allen, J. A. (1968). Particle description of the magnetosphere. In R. L. Carovillano, J. F. McClay, & H. R. Radoski, *Physics of the magnetosphere*, Vol. 10. New York: Springer-Verlag.

Wang, W., Wiltberger, M., Burns, A. G., Solomon, S. C., Killeen, T. L., Maruyama, N., & Lyon, J. G. (2004). Initial results from the coupled magnetosphere-ionosphere-thermosphere model: Thermosphere-ionosphere responses. *J. of Atmos. and Solar-Terrestrial Phys.*, *66*, 1425–1441.

The White House (2016). Executive Order—Coordinating efforts to prepare the nation for space weather events (13744). 13 October 2016.

Wilson, J. W. (1977). Analysis of the theory of high-energy ion transport. *Technical Report TN D-8381*, NASA.

Wilson, J. W. (2000). Overview of radiation environments and human exposures. *Heath Phys.*, *79*(5), 470–494.

Wilson, J. W., Badavi, F. F., Cucinotta, F. A., Shinn, J. L., Badhwar, G. D., Silberberg, R., et al. (1995). HZETRN: Description of a free-space ion and nucleon transport and shielding computer program. *Technical Report TP 3495*, NASA.

Wilson, J. W., Cucinotta, F. A., Golightly, M. J., Nealy, J. E., Qualls, G. D., Badavi, F. F., et al. (2006). International space station: A testbed for experimental and computational dosimetry. *Adv. Space Res.*, *37*, 1656–1663.

Wilson, J. W., Joes, I. W., Maiden, D. L., & Goldhagan, P. (Eds.) (2003). Analysis, results, and lessons learned from the June 1997 ER-2 campaign (NASA CP-2003-212155). NASA Langley Research Center.

Wilson, J. W., Mertens, C. J., Goldhagan, P., Friedberg, W., Angelis, G. D., Clem, J. M., et al. (2005). Atmospheric ionizing radiation and human exposure. Technical Report TP-2005-213935, NASA.

Wilson, J. W., Miller, J., & Cucinotta, F. A. (Eds.) (1997). *Shielding strategies for human space exploration*. NASA Conference Publication 3360, NASA Johson Space Center.

Wilson, J. W., Nealy, J. E., Cucinotta, F. A., Shinn, J. L., Hajnal, F., Reginatto, M., & Goldhagen, P. (1995). Radiation safety aspects of commercial high-speed flight transportation. NASA Tech. Pap. 3524.

Wilson, J. W., Townsend, L. W., Nealy, J. E., Chung, S. Y., Hong, B. S., Buck, W. W., et al. (1989). Bryntrn: A baryon transport model. Technical report, NASA.

Wilson, J. W., Townsend, L. W., Schimmerling, W., Khandelwal, G. S., Khan, F., Nealy, J. E., et al. (1991). Transport methods and interactions for space radiation. Technical Report RP-1257, NASA.

Wilson, J. W., Tweed, J., Walker, S. A., Cucinotta, F. A., Tripathi, R. K., Blattnig, S., & Mertens, C. J. (2005). A benchmark for laboratory exposures with 1 a GeV iron ions. *Adv. Space Res.*, *35*, 185–193.

Wissmann, F., Reginatto, M., & Möller, T. (2010). Ambient dose equivalent at flight altitudes: a fit to a large set of data using a Bayesian approach. *J. Radiol. Prot.*, *30*, 513–524.

Xapsos, M. A., Barth, J. L., Stassionopoulos, E. G., Messenger, S. R., Walters, R. J., Summers, G. P., & Burke, E. A. (2000). Chracterizing solar proton energy spectra for radiation effects applications. *IEEE Trans. Nucl. Sci.*, *47*(6), 2218–2223.

Xu, W., Marshall R. A., Fang X., Turunen E., & Kero A. (2018). On the effects of bremsstrahlung radiation during energetic electron precipitation. *Geophysical Research Letters*, *45*, 1167–1176. https://doi.org/10.1002/2017GL076510

5

Remaining Issues in Upper Atmosphere Satellite Drag

Jeffrey P. Thayer[1], W. Kent Tobiska[2], Marcin D. Pilinski[3], and Eric K. Sutton[4]

ABSTRACT

The greatest concentration of spacecraft and space debris resides in an orbital band between 100 and 2000 km altitude, classified as low-Earth orbit (LEO). Of these space objects, those below about 1000 km all experience an appreciable drag force due to the presence of Earth's atmosphere. It has been a common approach to use LEO spacecraft orbital behavior to extract upper atmosphere properties and then reapply this atmosphere description to predict future spacecraft behavior using orbit propagators. This approach leads to mutual coupling between spacecraft response and atmosphere change that results in convoluted uncertainties. To unravel such interdependencies, a physical description of the LEO space environment and how it changes due to external influences is required independently of the physical description for atmosphere-spacecraft interactions. The main challenge lies in attributing how much of the orbital perturbation is the result of changes in the upper atmosphere versus changes in a satellite's ballistic coefficient. Today, greater fidelity in space environment specification is required for LEO spacecraft operations, collision avoidance procedures, and space environment forecasting. This chapter highlights the remaining issues by segmenting the LEO drag environment into a lower and upper register, and describing separately the LEO environment and gas-surface interaction in each register. In addition, advances in solar and geomagnetic drivers for improved density specification are described.

5.1. INTRODUCTION

Upper atmospheric drag represents a scientific, engineering, and societal issue where spacecraft in low-Earth orbit (LEO) experience uncertain and variable forces due to rapid and appreciable changes in the rarefied gas. These upper atmosphere changes are challenging to describe and to predict, requiring LEO operating spacecraft to be regularly monitored. The LEO domain is popular among satellite providers as it gives access to space operations using only modest launch capability, provides remote sensing of Earth at high spatial resolution, gives astrophysical instruments an unobstructed view of space void of atmospheric distortion and absorption, and offers a set of orbits known as sun-synchronous repeat orbits that provide a near-constant alignment of the orbital plane and the projected direction of the Sun, giving optimum and repeatable illumination conditions for imaging Earth's surface.

[1] *Ann and H. J. Smead Aerospace Engineering Sciences Department, University of Colorado at Boulder, Boulder, Colorado, USA*

[2] *Space Environment Technologies, Pacific Palisades, California, USA*

[3] *Laboratory for Atmospheric and Space Physics, University of Colorado at Boulder, Boulder, Colorado, USA*

[4] *Space Weather Technology, Research, and Education Center, University of Colorado at Boulder, Boulder, Colorado, USA*

Space Physics and Aeronomy Collection Volume 5: Space Weather Effects and Applications, Geophysical Monograph 262, First Edition.
Edited by Anthea J. Coster, Philip J. Erickson, and Louis J. Lanzerotti.

DOI: 10.1002/9781119815600.ch5

The public expectation of uninterrupted satellite services with sustained high-fidelity performance is a constant challenge for operators of LEO spacecraft, not to mention the human spaceflight activity in this domain. The variable atmospheric drag forces and torques on the spacecraft can alter its orbit and attitude. Furthermore, a continuous account must be made of the large population of space objects (also altered by atmospheric drag) that includes over 17,000 objects larger than 10 cm and growing. Today, the LEO domain, often defined between 100 and 2000 km, contains almost 74% of the on-orbit catalog population and 34% of its overall mass in just 0.4% of the volume extending from the Earth's surface to geostationary orbit (Klinkrad, 2017). A new addition to LEO activity is the planned large satellite constellations, or *megaconstellations*, promising continuous, global communication and internet services. Such proposed constellations in LEO include SpaceX with 4000 satellites, Samsung with 4200 satellites, and OneWeb with 720 satellites (Radtke et al., 2017). The benefit of operating at lower altitudes is the much lower signal delay times and less power required to transmit a signal. However, operating in this altitude range requires a considerable number of satellites to provide complete on-ground coverage. The presence of the megaconstellation in LEO makes the need for detailed descriptions of the LEO environment even more critical.

Upper atmosphere drag involves the behavior of a multiconstituent rarefied gas in the thermosphere and exosphere, and its interaction with the surfaces of resident space objects in LEO. This interaction represents the largest nongravitational perturbation to Keplerian orbits below approximately 500 km altitude, causing an overall decrease in orbital energy and decaying the orbit towards re-entry. At LEO altitudes, satellite drag is the most variable orbital perturbation and the one that is the most difficult to predict. To fully account for and predict atmospheric drag perturbations on LEO objects, a physical description of the global gas state and the spacecraft's gas-surface interaction is required. The requirements for satellite drag modeling have become more acute as the LEO object population continues to increase; the current, and future, situation requires continuous precise orbit determination, conjunction assessment, and collision avoidance actions that involve not only the spacecraft response to atmospheric drag but the drag response of all the other space objects in or passing through LEO. Inherently, there are errors in the system that make the position of space objects uncertain, and these uncertainties become cumulative when trying to predict an object's future position and velocity in a changing atmosphere environment. Atmospheric drag in LEO is highly dependent on the state of the neutral gas population that makes up the Earth's thermospheric/exospheric mass density (100–1000 km altitude). It has been well established that atmospheric drag, and in particular variability in atmospheric neutral density, is the primary error source for LEO objects (Bowman, Tobiska, Marcos, Huang, et al., 2008; Hejduk & Snow, 2018; Storz et al., 2005). Thus, improved estimates of atmospheric density will positively impact the operations and performance of these critical space assets. Space weather events, and longer-term space climate, can alter the state of the upper atmosphere and produce variability that can exceed several hundred percent of the mean state at some LEO altitudes. Figure 5.1 illustrates the major pathways by which space weather couples to the upper atmosphere, leading to changes in upper atmosphere gas properties, impacting the forces on the satellite, and resulting in the assessment of orbital properties.

The magnitude of satellite drag forces is proportional to the gas mass density and the square of the magnitude of relative velocity between the satellite and the atmosphere. The drag experienced by an orbiting object also depends on the satellite area, the shape of the satellite, the temperature and composition of the atmosphere, the atmosphere mean-free path relative to the size of the satellite, and the scattering dynamics of atmospheric molecules from the spacecraft surfaces. Scattering dynamics stochastically describe the velocity, energy, and momentum of molecules scattered from the surface and are themselves a function of incident molecular energy, incident angle, composition, and a variety of surface properties. These properties (area, shape, etc.) are often subsumed in an object's ballistic coefficient, essentially a sensitivity factor unique to each space object that gauges the level of responsivity of the space object to a change in density and/or relative velocity. The challenge lies in the interdependencies that exist between the changes in the state of the space environment that directly impact drag accelerations and indirectly affect the ballistic coefficient. Thus, the net drag acceleration depends on the state of the space environment and the sensitivity of the space object response to changes in that state.

The following sections address remaining issues in the LEO drag environment, specified here to be the region where atmosphere drag has an appreciable influence on satellites, from 100 to 1000 km. Section 5.2 provides background information and places in context some current capabilities while expressing the societal need for more research. Section 5.3 addresses the physics behind the LEO drag problem and establishes two specific registers within the LEO drag environment: lower (100–500 km) and upper (500–1000 km), where the physics of the problem differs. Section 5.4 covers the lower register and the issues related to rarefied gas dynamics treated as a

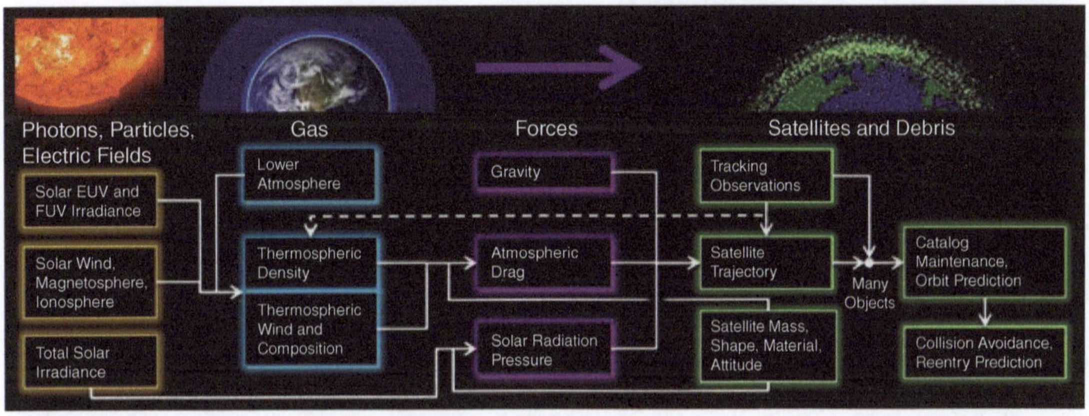

Figure 5.1 Major pathways by which space weather couples to satellite aerodynamic forces. (Source: J. Emmert, NRL.)

continuum and gas-surface interactions under free molecular flow with high accommodation. Section 5.5 covers the upper register and the issues related to rarefied gas dynamics transitioning from a continuum to a kinetic regime, and gas-surface interactions transitioning from an oxygen to a helium and hydrogen environment with low accommodation. Section 5.6 addresses the significant advancements in the last 40 years related to the improvement of density specification from solar and geomagnetic drivers.

5.2. BACKGROUND

Since the beginning of the space age, atmospheric drag has been studied to discern its impact on LEO objects. Observations of changes in LEO objects have revealed large-scale, long-term behavior in upper atmosphere density (e.g. King-Hele & Quinn, 1965). It has been a common approach to use observed changes in LEO spacecraft orbits as instruments for estimating space environment properties, particularly gas mass density and winds, by assuming knowledge, or stability, of the ballistic coefficient. The ballistic coefficient assumptions made when constructing or calibrating atmospheric models based on orbital observations are implicitly entered into those descriptions of the atmospheric mass density. This derived description of the space environment is then reapplied to space objects for nowcasting and forecasting behavior using spacecraft orbit propagators. A "fitted" ballistic coefficient is often computed as part of this process since the ballistic coefficient assumptions made in constructing/calibrating any given atmospheric model are usually not known to the model user (e.g. Hoots & Roehrich, 1980; Gaposchkin & Coster, 1988; Marcos et al., 1998; Storz et al., 2005; Vallado et al., 2006). To determine a fitted coefficient, empirical or physics-based descriptions of the LEO space environment are assumed to be "known" properties to solve for the most recent ballistic coefficient given the observed drag acceleration. These observations, along with others of varying degrees of direct thermosphere state specification, have led to the development of empirical models. Emmert (2015) has provided an extensive review of approaches to describe and observe thermosphere mass density spanning the period from 2000 to 2014 and identified mass density features that have been elucidated using these various approaches over the prescribed time span. The approaches described cover measuring techniques, empirical models that provide a synthesis of historical data, and physical models that simulate the environment by solving fluid equations. Figure 5.2 provides a climatological view of the thermosphere state as described by NRLMSISE-00® empirical model (shortened to MSIS; Picone et al., 2002). The largest uncertainties driving change to this climatological view of the LEO drag environment are solar, lower atmosphere, and geomagnetic drivers. Progress to date in solar and geomagnetic drivers for the LEO drag environment is discussed in section 5.6.

He et al. (2018) have also assessed several empirical/semiempirical thermosphere models to assess their ability to reproduce spatial variations and their ability to capture complex features in thermosphere mass density. The models included the Jacchia model (Jacchia, 1977), High Accuracy Satellite Drag Model (HASDM; Storz et al., 2005), MSIS, Jacchia-Bowman (JB2008; Bowman, Tobiska, Marcos, Huang, et al., 2008), Drag Temperature Model (DTM-2013; Bruinsma, 2015), and Global Average Mass Density Model (GAMDM; Emmert et al., 2008). The ISO International Standard 14222 on Earth's upper atmosphere (2013) recommends using

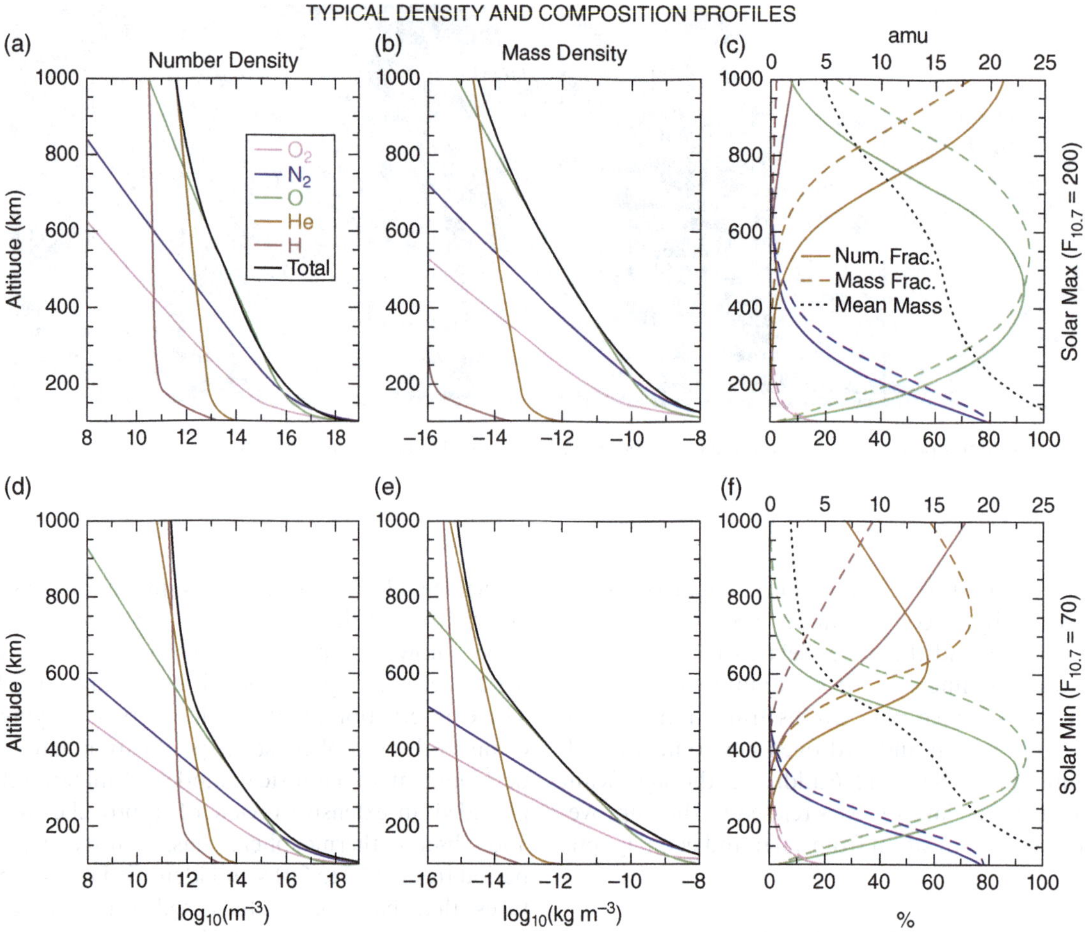

Figure 5.2 (a) Global average number density profiles at solar maximum ($F_{10.7}$ = 200 solar flux units [1 sfu = 10^{-22} W m^{-2} Hz^{-1}]) from the NRLMSISE-00® empirical model. Shown are profiles of O_2 (purple), N_2 (blue), O (green), He (orange), H (red), and the total number density (black). (b) Corresponding mass density profiles for each species. (c) Number mixing ratios (solid lines) and mass mixing ratios (dashed lines) for each species. Also shown is the mean molecular mass in atomic mass units (dotted black line), referenced to the upper axis of the plot. (d)–(f) Same as (a)–(c), but for solar minimum ($F_{10.7}$ = 70) conditions. (Source: Emmert, 2015. Reproduced with permission of Elsevier.)

MSIS for relative constituent abundances and JB2008 for mass densities related to satellite drag. Marcos et al. (2006) led the first conclusive studies that identified the Jacchia-Bowman series of models, starting with JB2006, as the most significant advance in empirical modeling since the 1960s. This was due to several factors, including the use of three new solar indices and proxies tying source regions on the Sun to energy absorption and species altitudes in the Earth's upper atmosphere (removing the dependence on the F10.7 proxy), the use of Dst as a geomagnetic driver during storm periods (removing the dependence on the Ap proxy), and the inclusion of semiannual hemispheric variability. In JB2008, the 1-sigma uncertainties at 400 km for a given epoch dropped from 15% to 8% compared with Jacchia and MSIS models. The use of JB2008 in HASDM, and the assimilation of calibration satellite data into the initial JB2008 density solutions, further dropped the comparable HASDM density uncertainties to less than 5%.

HASDM has the unique quality of including calibration satellites to solve for thermospheric neutral density in near real time by incorporating the observed drag effects on a set of low-perigee inactive payloads and debris. Many different calibration satellites with different orbits may be exploited to recover a dynamically varying global density field. The greater the number of calibration satellites, the better the accuracy. Typically, 75 calibration satellites are used. HASDM is an example of a type of data/model assimilation. This method improves the accuracy of current epoch and near-term parameters compared to modeling alone. Because HASDM is able to mitigate uncertainty in atmosphere effects at each epoch, by resolving the differences between calibration satellites and the core JB2008 model, real-time nowcasting specification of the LEO drag environment has improved significantly. Forecasting is then achieved with the JB2008 model, driven by the solar and geomagnetic drivers described in section 6, with uncertainties growing significantly a few hours postepoch. The calibration satellites operated by the Air Force are regularly monitored for changes in their orbital properties, and a subset of objects are replaced on an annual basis. The HASDM results are then implemented into orbit propagators for all other space assets.

In more recent times with the advancement of scientific-grade accelerometers sensitive to nano-G forces, the LEO spacecraft has become a refined instrument for deriving high-resolution behavior in thermosphere neutral mass density and neutral winds. This approach has been profoundly demonstrated by the CHAMP and GRACE missions (Doornbos et al., 2010; Sutton, 2008; Sutton et al., 2007). (The GOCE mission also provided unprecedented sensitivity for mass density and winds but performed such precision through sensitive thruster operations to create a free-flying spacecraft at altitudes near 250 km.) These available data sets have led to improvements in empirical and physics-based models, either as a data source or a validation procedure. Recent work by Mehta and Linares (2017) invokes techniques to exploit the capabilities of both empirical and physics-based models using proper orthogonal decomposition. High-resolution GPS data from LEO spacecraft have also been introduced as a means to extract density using GPS accelerometry (van den IJssel & Visser, 2010) at a time cadence higher than conventional ground-based tracking approaches can provide but lower than high-precision accelerometers (Ren et al., 2018).

The increasing congestion of the space environment is well known even outside of the space industry. Figure 5.3 provides a current and future snapshot of LEO objects provided by The Aerospace Corporation. Allowed to continue unabated, increases in space debris will eventually prevent the most useful Earth orbital bands from being safely occupied, thus threatening the commercial, civil, and military uses of space on which modern society relies. The emerging approach to remediation involves both reducing existing space debris (active debris removal) and preventing future debris-producing events (conjunction assessment and collision avoidance operations). Conjunction assessment is the process of determining close approaches between two objects, assessing the probability of collision, and providing this information to satellite owners/operators (Hejduk & Snow, 2018). Collision avoidance involves weighing the risk posed by the close

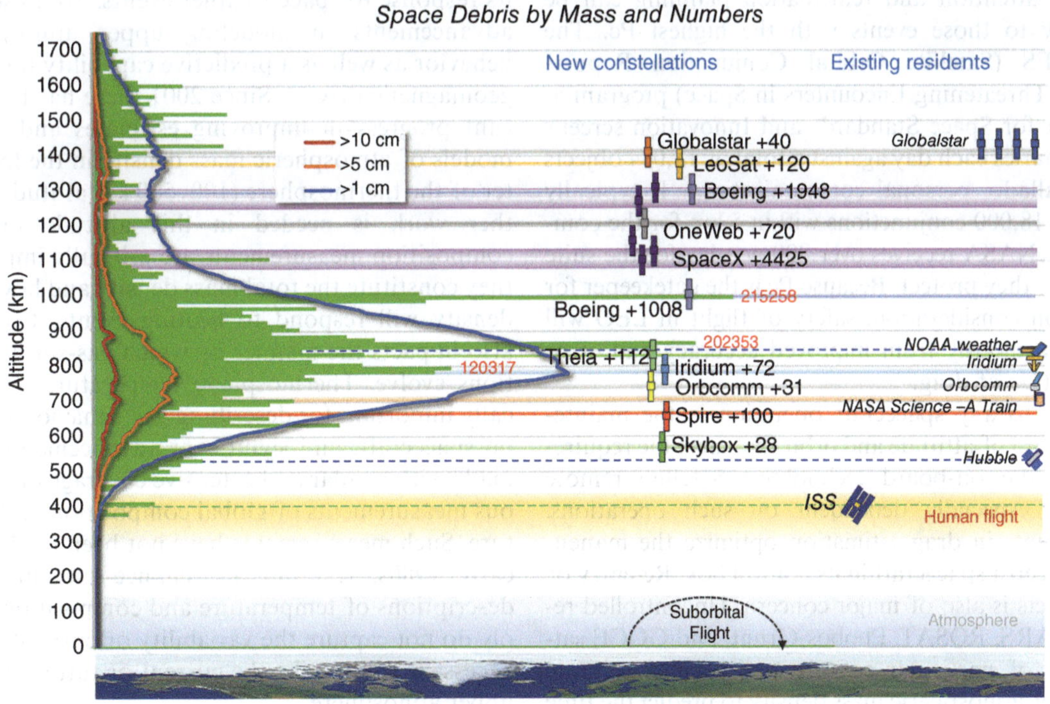

Figure 5.3 LEO object population densities and future potential constellations. (Source: Dr. Ted Muelhaupt of Aerospace Corporation.)

approach against the cost and risk of a maneuver. The latter used to be practiced only to protect high-value space assets from a catastrophic collision that would prematurely end its mission life. Today, conjunction risk assessment and remediation activities are seen as necessary to preserve the space environment for all space assets (Hejduk & Snow, 2018; Bussy-Virat et al., 2018).

Industry has embraced the probability of collision (Pc) as the standard parameter for conjunction risk assessment; to calculate this parameter requires state estimates of the two satellites and estimates of the uncertainties at the time of the satellites' closest approach. The greatest encumbrance to accurate Pc calculations today is the lack of reliable estimates of atmospheric drag and, at least as important, a reliable quantification of the uncertainty of the drag estimates that are produced. Drag variability produces position errors for both satellites. These errors change the time of closest approach, which in turn can often change the Pc severely enough to cause the event to be incorrectly categorized (i.e. as not high risk when it is in fact high risk). Because the Pc calculation considers the estimation errors as part of the calculation, an accurate atmospheric drag error estimate can increase or decrease the conjunction risk depending on the quality of the atmospheric density determination.

A highly accurate calculation of Pc is important because remediation efforts must be reserved for only the most risky conjunctions. The growing debris environment now produces so many conjunction notifications that individual attention and remediation planning can be given only to those events with the highest Pc. The SOCRATES (Satellite Orbital Conjunction Reports Assessing Threatening Encounters in Space) program at the Center for Space Standards and Innovation screens 3,541 primaries each day against a total of 15,084 objects (David Vallado, personal communication). It typically finds over 18,000 conjunctions within 5 km for the coming 7 days. NASA receives over 800 per day for the suite of satellites they protect. Because Pc is the gatekeeper for remediation consideration, safety of flight in LEO will benefit enormously from improved accuracy and error quantification of drag.

Inherent to any spacecraft operation is the maintenance/control of attitude and orbit that meet the requirements of the on-board technology. Satellite remote sensing is especially dependent on such operations. Improvements in drag estimation optimize the maneuvers that keep a spacecraft in its control box. Re-entry of space objects is also of major concern. Uncontrolled reentries (UARS, ROSAT, Phobos-Grunt, and GOCE satellites) are of particular concern and require accurate estimates of atmospheric mass density to predict the time and place of surface impact. Estimated times of reentry typically span roughly a week with +/- 3-day accuracy, and the estimated area of impact is often described to occur over broad regions of Earth. Improved estimates of the atmospheric drag environment would significantly improve re-entry predictions.

Various forms of energy from space weather drivers (solar soft X-ray [XUV], extreme ultraviolet [EUV], and far ultraviolet [FUV] flux, including the temporal effects of solar flares, coronal mass ejections, and corotating interaction regions) and lower atmosphere events (gravity, planetary, and tidal wave propagation) significantly perturb the upper atmosphere state over a variety of temporal and spatial scales. The broad range of external forcing on the upper atmosphere and its variability results in a highly dynamic, continuously changing state. These energy inputs and subsequent processes need to be adequately represented and are required for the assessment of physics-based models under different influences of space weather and lower atmosphere events. Physics-based models used for data–model comparisons include the Thermosphere-Ionosphere-Electrodynamics General Circulation Model (TIE-GCM; Richmond et al., 1992; Roble et al., 1988), Coupled Thermosphere-Ionosphere-Plasmasphere electrodynamics model (CTIPe; Fuller-Rowell et al., 1996), and the Global Ionosphere-Thermosphere Model (GITM; Ridley et al., 2006); see Emmert (2015) for more details.

The pathway for ensuring continuous and secure operations of LEO spacecraft technology is to improve the estimate of the upper atmosphere state and predict its response to space weather events. To do so requires advancements in modeling upper atmosphere gas behavior as well as a predictive capability for solar and geomagnetic drivers. Since 2001 there has been significant progress in improving estimates and advancing models of atmospheric mass density in the lower register of the thermosphere (100–500 km altitude), but further work is needed in this altitude range. Gas composition measurements are critically important, as they constitute the total mass density and how the mass density will respond to heating events. Composition also impacts the manner by which gas-surface interactions evolve. Thermosphere temperature is also critically important for describing gas behavior and impact on spacecraft, and significant advancements could be made with regular and extensive coverage of simultaneous measurements of global composition and temperature. Such measurements have not been made since the early 1980s, resulting in reliance on climatological descriptions of temperature and composition that simply do not capture the variability or correlated activity necessary to describe the current or future state of the upper atmosphere.

A key deficiency that requires much more investigation is nowcasting and predicting atmospheric drag in the upper register of the LEO drag environment (500–1000 km altitude). In fact, atmospheric drag estimates are more coveted by operators in this domain than at lower altitudes because of the larger number of operational space assets and debris (see Figure 5.3). For upper atmosphere research, this altitude domain includes the challenging transition from gas continuum to particle dynamics, as the thermosphere gives way to the exosphere and is highly sensitive to space weather and the changing boundary conditions imposed by the underlying layers. For the engineering community, the diverse and dynamic population of light ion and neutral constituents in this altitude range makes estimation of spacecraft drag coefficients especially challenging. Finally, for the operator community, conjunction assessment and collision avoidance are daily issues in this LEO altitude range. Improvements in nowcasting and forecasting atmospheric drag in the lower register will be important for expanding the models into the higher register, making the 500–1000 km altitude range particularly challenging and in need of improvements in all aspects of the atmospheric drag problem.

5.3. THE PHYSICS OF ATMOSPHERIC DRAG ON SPACECRAFT

The physics of atmospheric drag on spacecraft has similarities to atmospheric drag described by low-speed aerodynamicists on aircraft in that forces exerted on the body in a moving gas produce a drag force defined to be parallel to the velocity of the gas relative to the body. The projected aerodynamic forces in the drag direction produce a scalar force per unit mass, or drag acceleration, on the body that is proportional to the dynamic pressure of the gas. A proportionality factor, B, can be applied such that the drag acceleration is

$$a_D = B\frac{1}{2}\rho v_r^2. \tag{5.1}$$

This illustrates that the value of B or its inverse, often called the ballistic coefficient parameter b, represents a measure of sensitivity for a body to be perturbed by the dynamic pressure of the gas. It is common to apply dimensional analysis to express B in terms of a drag coefficient, C_D, multiplied by an area-to-mass ratio, and express the drag acceleration as

$$a_D = C_D\frac{A_{ref}}{m}\frac{1}{2}\rho v_r^2. \tag{5.2}$$

Here A_{ref} is more generally expressed as a reference area of the body, m is the mass of the body, ρ is the mass density of the gas, and v_r is the velocity of the gas relative to the body. The reference area is a value assigned to represent the cross-sectional footprint of the satellite in the plane normal to v_r. The mass represents that of the entire spacecraft including fuel. The mass density is that of the multiconstituent, rarefied upper atmosphere gas. The relative velocity is the vector addition of the inertial velocity of the spacecraft, the inertial velocity of the corotating atmosphere, and the velocity due to winds in an Earth-fixed atmosphere. Accounting for signs, the velocity of the gas relative to the body, v_r, is $-(v_{sat} - v_{corot} - v_{wind})$, where the outside negative sign indicates that the relative velocity is opposite to the body velocity. The corotation and wind velocities represent the bulk velocity of the gas, but thermal motion of the gas particles may also need to be included when considering atmospheric drag in LEO. This thermal component is often accounted for in the estimate of the C_D and becomes particularly important for spacecraft bodies with elongated surfaces and high-altitude portions of LEO orbits (Sutton, 2009).

Much of the physics of the gas and body interaction is contained within the nondimensional drag coefficient and, for continuum aerodynamics, is often experimentally retrieved by performing a series of wind tunnel tests that measure the drag force with all other properties known. This is not viable for bodies in space and, furthermore, the physics of the gas and body interaction in LEO is quite different than for the troposphere and aircraft. Consequently, various modeling and measurement approaches are employed to estimate C_D, or the combined ballistic coefficient, while in orbit. Often, spacecraft are modeled as surface segments by which individual drag coefficients are determined and then summed to determine the total drag coefficient of the spacecraft (Pilinski & Argrow, 2013). Furthermore, the individual species of the thermosphere interact differently with the spacecraft surface and should be addressed separately, and these accumulated effects are captured in a mass-weighted sum. Thus, gas-surface interactions must be properly defined and physically accurate to describe spacecraft drag.

A common physical description for the upper atmosphere gas employs continuum gas dynamics, where average values of macroscopic upper atmosphere properties are determined and predicted with appropriately justified mathematical models for describing the conservation of mass, momentum, and energy of the gas within the LEO environment. The reason for the continuum mechanics in this context is that the mean free paths, while large relative to the size of a satellite, are much smaller than the characteristic size of most atmospheric features. Such continuum atmospheric models provide a total mass density and mass-averaged wind globally, among many other properties, that can be applied to equation (5.2). Gas-surface interactions invoke a different physical description where

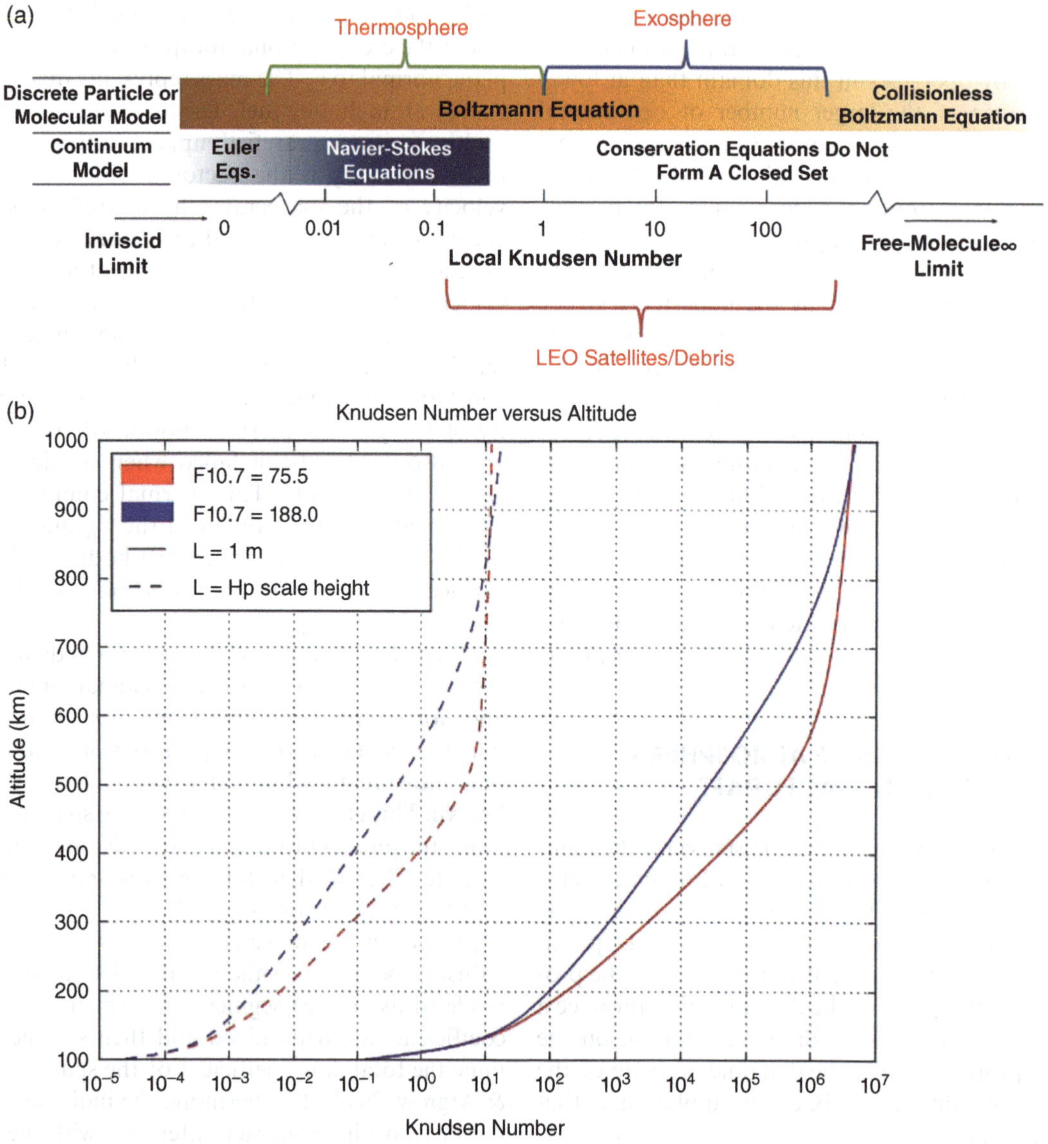

Figure 5.4 (a) Flow regimes according to the Knudsen number for gas dynamics and gas-surface interactions (adopted from Bird, 1994). (b) The graph indicates the Knudsen number for a 1-meter spacecraft (solid line) and for atmospheric gas behavior using density scale height (H_p) as a representative scale length under solar maximum and minimum conditions. (Source: Bird, 1994. Reproduced with permission of Oxford University Press.)

microscopic properties of the gas are needed to account for the energy and momentum exchange between the gas particles and the spacecraft surface. In this case, gas kinetic theory is the more appropriate framework by which the drag coefficient can be modeled.

Figure 5.4 summarizes the physical description of these two different aspects to atmospheric drag (upper atmosphere gas description and gas-surface interactions) in terms of the Knudsen number. The Knudsen number (Kn) is the ratio of the mean free path between collisions to a representative gradient scale length of the gas, or scale length of a body, and is a useful dimensionless number to evaluate approximations to gas flow regimes. For Kn less than 1, the gas is adequately described by continuum gas dynamics. For Kn greater than 1, the gas description requires more sophistication, and above about 10 a particle trajectory approach employing gas kinetics must be applied (Bird, 1994). The Knudsen number for describing the thermosphere gas mathematically using transport equations relates the mean free path (determined using MSIS and hard-sphere, neutral-neutral collisions) to the scale length of the macroscopic gradient in mass density, considered as the absolute value of the density scale height, $H_\rho = -\rho/(d\rho/dz)$, also determined using MSIS. This is displayed as an altitude plot versus Knudsen number. A value of Kn < 0.1 satisfies the application of Navier-Stokes equations to the thermosphere gas (i.e. continuum aerodynamics). Often, Kn = 1 is used to indicate the transition between the thermosphere and the exosphere, or the exobase, but this transition is a more extended domain where further approximations to the heat flux vector and momentum flux tensors could be required before ultimately replacing the continuum mathematical model with collision-less gas kinetic theory (Kn > 10).

There are several scenarios where the Knudsen number can exceed 1 in describing a gas. One is at the point where gradients in gas variables become so steep that smaller and smaller volumes of the flow are required to describe its characteristics. In this case, one must consider whether there are a sufficient number of molecules or atoms within the volume to appropriately describe the gas macroscopic properties. In other words, the gradients of the macroscopic properties become so steep that their scale length is of the same order as the average distance traveled by the molecules and atoms between collisions. This can occur in regions of shocks or sharp boundaries. Another case is where the gradients are sufficiently large but the gas becomes so tenuous that the mean free path exceeds the gradient scale. This is commonly experienced in transitioning from the thermosphere to exosphere, and the gas must be treated more by its free molecular flow than by its continuum flow. For the entire LEO domain, the upper atmosphere gas description will need to span continuum and free molecular flow regimes.

The Knudsen number for gas-surface interactions of a spacecraft with a physical dimension of 1 meter versus altitude is also plotted in Figure 5.4, again using MSIS to determine the mean free path. Here it is clear that values exceed 1 in the lower thermosphere and that LEO objects are generally in the free-molecular flow regime for most of the thermosphere and above. Under this flow condition the probability that two molecular collisions occur simultaneously is very small. Furthermore, it is assumed the incident and reflected molecules do not interact with each other near the surface (see Sentman 1961b, section V.A). The dynamical processes reduce to addressing the successive collisions among a gas atom/molecule and the object. It is in this regime of free-molecular flow that gas-surface interactions are modeled and used in estimating the drag coefficient of the spacecraft.

In the early stages of the space age, a judicious selection of a constant drag coefficient value of 2.2 was employed to carry out many calculations of upper atmosphere density behavior derived from orbit decay. It was recognized then that this value will likely vary over altitude and geophysical conditions (Cook, 1965), but the required precision was not as stringent as it is today. An example of drag coefficient variability is shown in Figure 5.5, where a SL6 cylindrical rocket body (satellite 16125) shows clear change in the drag coefficient derived from the ballistic coefficient (black triangles) as the

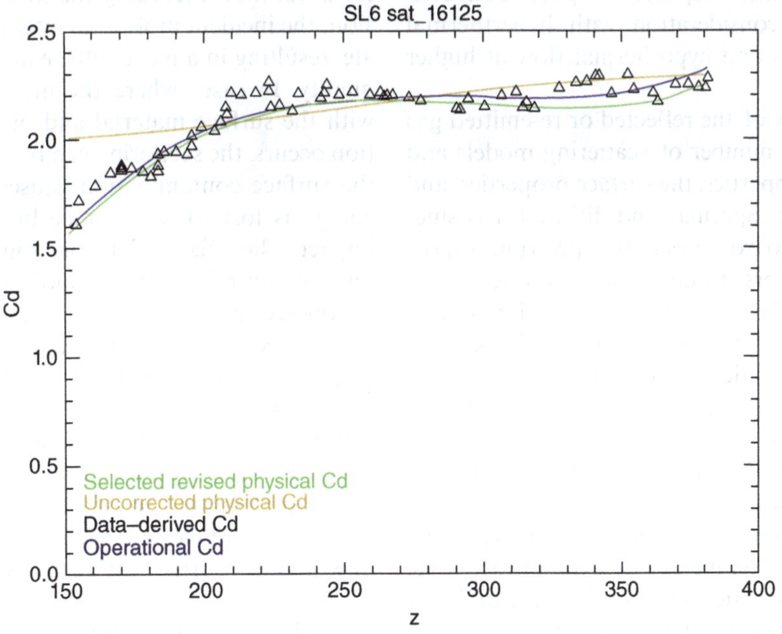

Figure 5.5 The altitude-variable drag coefficient for a SL6 cylindrical rocket body (satellite 16125) that changed from 2.3 to 1.5 as the object moved through atmosphere regimes with different species compositions. (Source: Tobiska and Bowman, unpublished work.)

altitude changes from an atomic oxygen region above 200 km to a mixed atmosphere below 200 km. The maximum C_D was 2.3 and the minimum was 1.5. In this figure, polynomial fits to the data are shown using assumptions of corrections to a physical C_D in the green line (Moe & Moe, 2005), an uncorrected physical C_D as the gold line, and a predicted C_D for operational purposes as the blue line.

Thus, density models derived from such observations inherently include uncertainties introduced by the lack of accuracy in the drag coefficient. As in any collisional exchange, the fundamental issue in describing the drag coefficient resides in the manner by which momentum and energy are exchanged through the interaction process. More specifically, account must be made of how momentum and kinetic energy of the gas changes when interacting with the satellite body. Analytical solutions of free molecular flow for simple shapes have had an extensive history, and a review of these techniques was recently given by Prieto et al. (2014). The force per unit area on the satellite surface is equal to the change in the momentum flux of the incident and reflected gas. Descriptions for the incident momentum flux have been organized based on the molecular speed ratio, s, which is the relative speed divided by the gas species most probable speed. A speed ratio greater than 5 describes the flow as a collimated beam of molecules (hyperthermal flow), and the frontal area is most significantly impacted. A speed ratio less than 5 requires the inclusion of thermal motion and its impact on all satellite surfaces (hypothermal flow). Over the extent of the thermosphere/exosphere both conditions may require consideration with hyperthermal flow at lower altitudes and hypothermal flow at higher altitudes.

The momentum flux of the reflected or re-emitted gas can be described by a number of scattering models and depends on the gas properties, the surface properties, and the angle of incidence. Specular and diffuse (or cosine) reflection represent two extremes in the spectrum of possible scattering behaviors. In describing the re-emission of an incident particle, a specular re-emission is one where the angle of the re-emitted particle equals the angle of incidence and the kinetic energy of the particle is conserved, also described as an elastic collision. A diffuse re-emission is normally associated with one where the surface adsorbs the incident molecules; these molecules equilibrate thermally with the surface and are then re-emitted with a random distribution of speed and direction associated with the temperature of the surface. This process is a form of inelastic collision where the diffusely reflected molecules are completely accommodated to the surface temperature before being re-emitted. A number of more complex scattering models are able to represent behavior spanning from specular to diffuse and have been applied to the computation of aerodynamic forces on satellites. These include Schamberg's model for quasi-specular reflection (Pilinski & Argrow, 2013) as well as versions of the drifting Maxwellian distribution such as the Cercignani Lampis Lord (CLL) model (Walker et al., 2014). These models represent the distribution of scattered velocities as a "scattering lobe" of variable width centered on either a sub-specular (closer to the surface normal direction), specular, or super-specular (closer to the surface tangent direction) vector. For a flat plate facing into the flow, a more specular reflection with kinetic energy conserved produces the greatest change in momentum and thus will increase the drag on the spacecraft over a diffuse reflection. However, for a flat plate parallel to the flow, a more diffuse scattering will produce the greatest change in momentum exchange as specular reflection results in a zero change of tangential momentum.

A thermal accommodation coefficient, α, is often used to allow for a range of energy exchanges between the gas and the surface. An $\alpha = 1$ is fully accommodated with the re-emitted molecule having a temperature equal to the surface, and an $\alpha = 0$ is one where the reflected molecule maintains its original temperature, i.e. kinetic energy is conserved upon reflection. An additional but related element to the re-emitted direction and accommodation of the incident molecules is adsorption. Adsorption is the process by which satellite surfaces become covered by atmospheric molecules that are trapped close to the surface. In this case, where the atoms "stick" to the surface for a duration exceeding the time of a gas-surface collision, the incident molecules interact with the gas adsorbate, resulting in a more diffuse and fully accommodating quality. In cases where the incident molecule interacts with the surface material and only partial accommodation occurs, the scattering can become less diffuse. Due to the surface contamination caused by adsorption, more energy is lost to the surface by incident molecules on impact. The effects of adsorption are a broader angular distribution of re-emitted molecules, which is closer to the diffuse case, and a higher accommodation of the incident molecules to the surface (Moe & Moe, 2005). These effects generally lead to an overall decrease in the drag coefficient.

Both continuum (atmospheric dynamics) and free molecular flow regimes (atmosphere-spacecraft interactions) are required to address upper atmosphere drag. Much more research is necessary in upper atmosphere flow and gas-surface interactions to adequately predict, at the necessary levels of fidelity, the actual effects on LEO objects. It will be common throughout the next sections of the chapter to move in and out of these flow regimes when describing upper atmosphere behavior and gas-surface interactions on LEO objects. It is important to consider the approach of treating the upper atmosphere gas description separately from the physics associated with determining the drag coefficient. To facilitate the discussion, the LEO drag environment will be segmented into two altitude regimes: a lower register from approximately 100–500 km altitude and an upper register from approximately 500–1000 km. The lower register represents a region where most thermosphere measurements have been made, external influences deposit their energy, atmospheric drag is the dominant nongravitational force with high accommodation, the thermosphere gas transitions to predominantly atomic oxygen (as shown in Figure 5.2), and the thermosphere gas is well represented by using continuum gas dynamics. The upper register represents a region where the most populous LEO resident space objects reside, atmospheric drag and radiation pressure compete in magnitude (though drag forces cause far more uncertainty), gas-surface interactions tend towards lower accommodation coefficients and possibly less diffuse scattering, the thermosphere gas is transitioning to an exosphere where particle motions take on ballistic trajectories, the gas composition transitions to predominantly helium and hydrogen, plasma density becomes more important, and much fewer measurements of the gas properties are available. Segmenting by geocentric altitude is favorable for identifying LEO domains but not for describing the thermosphere gas, whose more natural vertical coordinate is pressure. However, this is a common issue when discussing atmospheric drag as several different coordinate systems are often employed to represent instrument frames, observing frames, and model frames, and some caution is required when combining all elements in the satellite drag problem.

5.4. THE LOWER REGISTER (100–500 KM ALTITUDE)

5.4.1. Rarefied Gas Dynamics

In this lower register, continuum mixed gas dynamics applies for describing the mass, momentum, and energy conservation of the thermosphere. Numerical models solving these equations typically produce gas properties including mass density, composition, mass-averaged wind, and mean temperature on a 3-D global grid and short time steps (TIEGCM, CTIPe, GITM). Here, sufficient collisions between gaseous species regularly take place over represented scale lengths, i.e. low Kn, that these macroscopic properties can be determined and applied to describe gas behavior. This statistical approach involves integrations over velocity space of Boltzmann's equation. The procedure of multiplying the species velocity distribution function by powers of velocity and then integrating over all velocities is called taking velocity moments (Bittencourt, 2004). Taking higher-order velocity moments leads to the constitutive equations of mass continuity, momentum, and energy as well as higher-order moment equations. Closure of the equations can only be achieved through approximation, where the higher-order equations are expressed in terms of lower-order macroscopic properties. The transport equations for the thermospheric gas are generally known as the Navier-Stokes equations in continuous gas dynamics. This form of continuum gas dynamics is applicable if the velocity distribution function can be closely approximated by a Maxwellian and if the higher-order moment equations describing the momentum flux tensor and thermal flux tensor are expressed in terms of temperature, mass density, and wind (i.e. lower-order macroscopic variables). In other words, the momentum flux tensor is described by pressure along its diagonal and by viscosity, or stress tensor, for its off-diagonal terms. The thermal flux tensor is reduced to the heat conduction equation. This is an example of the closure issue in taking moments of the Boltzmann equation and setting approximations (see Schunk & Nagy, 2009).

The Navier-Stokes equations form the foundations for many of the global circulation models used by the upper atmosphere community to describe the behavior of the thermosphere in the lower register. In most of these models, the natural coordinate system is latitude and longitude, with pressure (or log pressure) as the vertical component (GITM is the exception using geopotential height for its vertical component). In pressure coordinates, geopotential height is a 4-D variable described by the three coordinate components and time. Geopotential height is used in description of the upper atmosphere to reduce the complexity of describing gravitational forces on a gas in a pressure coordinate system and to simplify numerical solutions. Geopotential height will deviate by a few percentage points from geocentric height as distance from the surface increases. However, geopotential height will vary appreciably on a pressure surface, and consequently the upper bound of the lower register (defined here as 500 km) will experience a range of pressure surfaces depending on the temperature and compositional makeup of the gas. Only in a very deep solar minimum will the continuum mechanics approximation be called into question near this upper boundary. Thus, much research has been applied to these models in order to describe the thermospheric gas within this lower register. A LEO object in circular orbit propagated through the model output will experience a range of pressure levels in a single orbit and, to first order, those levels will also have varying mass density, winds, and constitutive species' mixing ratios.

The lower bound of this register is marked by the mesopause, around 100 km, using the temperature gradient as a thermal fiducial, but this region can be identified dynamically as the location where turbulent mixing of the gas (eddy diffusion) equals the molecular diffusion, i.e. the turbopause. This description proves to be more relevant as turbulent mixing is highly influential on the distribution of mass density throughout the vertical column. Strong mixing processes force the variety of species to remain well mixed below the turbopause, i.e. a fixed mixing ratio similar to that observed in the lower atmosphere. The mixing results in a uniform scale height for all species that are not chemically active and is dictated by the local temperature and mean molecular weight of the gas mixture. Above the turbopause, the more efficient molecular diffusion process will work to separate individual species, each having a scale height described by their own molecular weight. However, several factors are at play that can impact a species seeking to reach diffusive equilibrium after experiencing the turbopause. One is that each species has its own turbopause height and the other is that the turbopause height can change depending on the momentum flux divergence of wave activity in the region. A change in height of the turbopause alters the number density of an individual species at the altitude by which it begins to experience molecular diffusion. This means that a light species undergoing turbulent mixing (light is defined as having a molecular weight less than the mean molecular weight of the surrounding gas composition), such as atomic oxygen or helium, will change faster with height in the turbulent mixing region than if it were allowed to diffuse on its own in that region (oxygen is also influenced by chemistry near this boundary so a more complex distribution results). Once molecular diffusion takes over at a higher altitude, the lighter species concentration has already decreased more significantly than if the turbopause action occurred at a lower altitude. The opposite process occurs with a heavy species, such as argon, where the species changes more slowly with height through the turbopause region. This leads to more argon at higher altitudes if the turbopause were to extend to a higher altitude.

Another effect that challenges molecular diffusion above the turbopause, and also effects the columnar mass density, is mixing caused by 3-D thermospheric circulations within the multiconstituent gas (Fuller-Rowell, 1998). This process can be effective in altitudes of 100–250 km where the time constant for circulation mixing can be competitive with molecular diffusion. Indeed, this is reflected in the time delays for the effects of energy inputs by different solar indices captured by the JB2008 model and described in section 5.6. Circulation processes seek to maintain mass conservation in the system, resulting in vertical and horizontal transport of mass throughout the thermosphere. Under this condition, a vertical wind in a multiconstituent gas is described as the mass-average wind needed to produce the vertical mass flux that satisfies the horizontal mass flux created in regions of divergent/convergent motion. It does not explicitly describe the behavior of individual species but only the mean mass motion of the gas, which is closely represented by the local major species. Species in the region whose molecular weight deviates from the mean molecular weight of the gas mixture results in a species distribution that differs from major gas behavior. For instance, in the lower thermosphere the vertical mass flux is dictated by the majority species, i.e. molecular nitrogen. As mass density decreases with height, the mass-averaged vertical wind must increase with height to conserve mass. The maintenance of mass conservation in the presence of vertical flows causes an adjustment process in the gas pressure so that the gas maintains hydrostasis. The major gas in the region then behaves closely to its own diffusive equilibrium. However, other species, particularly minor species in the region that, by definition, are not involved in maintaining hydrostatic equilibrium, demonstrate behavior that can deviate significantly from diffusive equilibrium. Corollary to turbopause action, vertical motions due to circulation result in vertical structure that deviates from diffusive equilibrium. The amount of deviation differs by species, and that difference depends on their molecular weight relative to the local mean molecular weight of the multiconstituent gas in the region. Species lighter than the mean molecular weight, such as helium, will experience a steeper change in concentration with height in a region of mass-averaged, upward vertical winds, and a shallower change in concentration with height in a region of downward winds, compared to its distribution under a purely molecular diffusion process. The opposite response occurs for species that are heavier than the mean molecular weight.

Essentially, upward circulations try to remix the non-major gas species, i.e. oppose molecular diffusion, so that all species follow the scale height of the mass-averaged mixture. Downward circulations work to separate non-major gas species, i.e. enhance molecular diffusion, so that species separate more efficiently. The variation in mean molecular weight with height in the lower register is nonlinear and tends to follow the transition from molecular nitrogen to atomic oxygen. Consequently, the response of each species to 3-D motion and molecular diffusion within the multiconstituent gas will vary throughout the lower register. The process of molecular diffusion increases with decreasing collision frequency, such that, in the upper portions of the lower register the species will finally reach diffusive equilibrium. However, as at the turbopause, the species concentration in the upper register will depend on its starting concentration at the lower altitude where molecular diffusion begins to dominate over circulation. This boundary lies somewhere in the middle of the lower register and is surely a dynamic and extended domain that is presently not well defined. These competitive processes of turbulence, circulation, and molecular diffusion are very much at play through most of the lower register. The processes are altitude dependent and strongly affect the mixing ratio of species through the column and the overall mass density experienced by a satellite.

Summary of the remaining issues in rarefied gas dynamics in the lower register:

• Turbopause height with location, season, year, and species

• 3-D thermosphere circulation globally beyond climatology

• Thermosphere temperature and composition distribution globally with height.

5.4.2. Gas-Surface Interactions

For gas-surface interactions (GSI), free molecular flow is a valid assumption throughout most of the thermosphere when considering most spacecraft; very large spacecraft in the lower thermosphere and vehicles at re-entry altitudes <~120 km altitude must consider continuum-rarefied transitional flow regimes when assessing spaceflight dynamics (e.g. Turansky & Argrow, 2014). Here, free molecular flow will be assumed and gas-surface interactions will be applied from a kinetic point of view where the momentum and energy exchange between

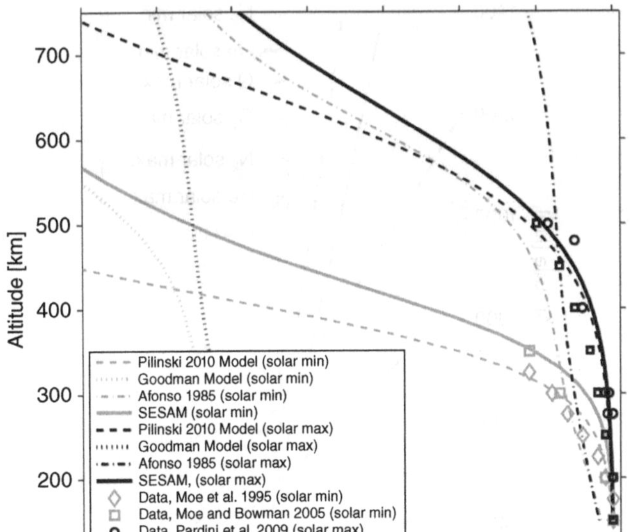

Figure 5.6 Various methods of estimating energy accommodation compared with SESAM (Source: Pilinski et al., 2013. Reproduced with permission of American Institute of Aeronautics and Astronautics, Inc.)

gaseous molecules/atoms and satellite surfaces must be evaluated at the microscopic level. The inherent assumption is that the impinging gas molecules/atoms interact with the surface before interacting with emitted molecules/atoms, thus allowing the impinging population to be evaluated independently from the outgoing population.

The description and observation of GSI in the lower register has been most developed given the availability of measurements and objects for analysis at altitudes below 500 km (Bowman & Moe, 2005; Moe & Moe, 2005; Moe et al., 1995). Pardini et al. (2010) has further evaluated physical drag coefficients for solar maximum conditions in this lower register. Pilinski et al. (2010, 2013) extended this work by developing an energy accommodation model, called the Semi-Empirical Satellite Accommodation Model (SESAM), that is consistent with the physical accommodation estimates while being applied to all levels of solar activity. Figure 5.6, from Pilinski et al. (2013), shows the energy accommodation values for LEO satellites along with SESAM (solid grey and black lines) and several other models.

This approach enables the computation of drag coefficients for changing solar activity and altitudes for spacecraft in low-eccentricity orbits. Estimates of the drag coefficients are further complicated by spacecraft geometries. Commonly, a panel method is applied to represent satellite geometries. In general, the panel method consists of the application of aerodynamic relations, such as Sentman's equations (Sentman, 1961a, 1961b), assuming diffuse scattering and a range of energy accommodation coefficients, to simplified geometries constructed from multiple panels with different orientations. A limited number of flat panels describe the entire structure of the satellite. Normal vectors and areas of each panel provide individual drag coefficients for each panel, and then a weighted sum is performed to achieve the net drag coefficient (Pilinski & Argrow, 2013). The energy accommodation coefficient has been introduced in Sentman's equation, and consequently in the panel method, by Moe et al. (2004) and Sutton (2009) as well as Doornbos (2011) for accelerometer data processing. However, the inability to model multiple reflections and shadowing effects introduces systematic errors for more complex geometries, for example the CHAMP and GRACE satellites.

Pilinski et al. (2011) and more recently Mehta et al. (2017), and references therein, have developed sophisticated numerical techniques to physically model the drag coefficient for satellites with complex geometries. They applied those techniques to the CHAMP and GRACE satellites using test particle Monte Carlo schemes (Mehta et al., 2014) and found sensitivities to gas-surface interactions and complex geometries can cause an average bias of 14%–18% for CHAMP and 10%–24% for GRACE satellites. March et al. (2019) has also addressed the complex geometry issue and has applied direct simulation Monte Carlo techniques (Bird, 1994) to CHAMP, GRACE, GOCE, and SWARM satellites resulting in more consistent density data among the four satellite missions.

Central to these studies is the manner by which the energy accommodation coefficient is estimated. In this lower register, atomic oxygen is a significant portion of the total density and a dominant species in the surface adsorption process. Adsorption, whereby monoatomic layers of gaseous atoms form on satellite surfaces, facilitates the accommodation of impinging atoms. Upon impact, the atom may maintain its kinetic energy ($\alpha = 0$), take on the temperature of the surface ($\alpha = 1$), or take some value in between. For altitudes below 500 km, the adsorbed atomic oxygen shields the incoming flow from surface properties, resulting in a GSI with high energy accommodation ($\alpha > 0.85$) and a re-emission process that is largely diffuse, essentially the Sentman model. This combination of interaction processes typically results in a surface drag coefficient that is lower than if kinetic energy were conserved upon reflection ($\alpha = 0$). The speed ratio in this lower register is often greater than 5, making the flow hyperthermal and resulting in the frontal surfaces of the spacecraft to be the dominant contributor to the overall spacecraft drag coefficient (Cook, 1965); however, elongated spacecraft and contraction of the thermosphere can reduce the speed ratio in the lower register.

Summary of the remaining issues in gas-surface interactions in the lower register:

- Modeling of spacecraft geometries and analytic versus computational methods of drag coefficients
- Assessing forces and torques on spacecraft to extract density and winds (accelerometers, GNSS tracking, radar tracking)
- Assessing errors in energy accommodation and scattering properties

5.5. THE UPPER REGISTER (500–1000 KM)

The starting altitude of the upper register at 500 km can be identified by the gas Knudsen number exceeding 1. As shown in Figure 5.4, this occurrence is defined to be the transition from the thermosphere to the exosphere. This transition is more gradual, but 500 km is a reasonable altitude to consider the start of such a transition. A correlated, and more observable transition, is the altitude by which helium and oxygen concentrations become equal. O/He transition dynamics are very important in affecting the total mass density of the LEO space environment and gas-surface interactions. The altitude of the He/O transition is lowest in solar minimum, occurring around 500 km. This is illustrated in Figure 5.7 in terms of partial pressures for each of the main species of the thermosphere for both solar maximum and solar minimum. This transition will prove useful in the discussion of rarefied gas behavior in sections 5.5.1 and 5.5.2. Typically, LEO is defined up to 2000 km, but here a limit is placed on the upper register description to 1000 km, as the neutral gas and drag behavior above this altitude is poorly known.

Other starting altitudes could be chosen for the upper register based on different considerations. For example, a common altitude threshold used in LEO deorbiting schemes concerned with orbital debris is 650 km altitude (FCC, 2018). Satellites deployed below 650 km will typically re-enter Earth's atmosphere within 25 years, even absent of any propulsive or other special deorbit capabilities. Thus, the collision risks presented by such satellites are generally lower, even if the satellites fail on-orbit and are unable to perform any active deorbiting maneuvers. Above this approximate 650 km threshold, a satellite that is not actively deorbited will remain in orbit for significantly longer periods of time. Satellites in this higher orbit require propulsive deorbit capabilities, but those missions may involve greater risk from an orbital debris perspective due to the possibility of a satellite failure that

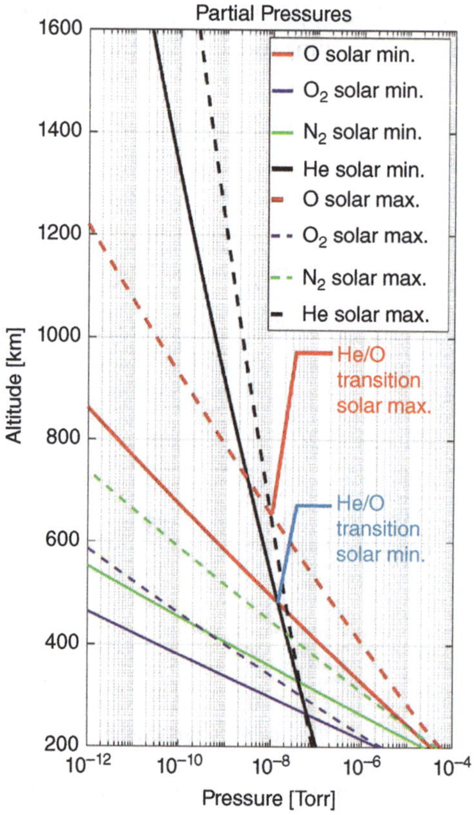

Figure 5.7 Partial pressures for thermosphere species as determined by MSIS.

can leave objects residing in LEO for much longer than 25 years. The most populous geocentric altitude range for LEO satellites and debris lies between 600 and 900 km. Many of the planned megasatellite constellations are targeting this upper register and require significant planning for deorbiting, as well as contingency plans for failed satellites that cannot be actively deorbited and other mission-generated debris.

5.5.1. Rarefied Gas Dynamics

In this upper register, failure to meet the conditions for Navier-Stokes flow requires more sophisticated descriptions and a greater expansion of terms. An example of this behavior is in transitioning through the exobase where gas velocity distributions begin to deviate more significantly from assumed near-Maxwellian behavior. Deviations from Maxwellian behavior can be accommodated by taking higher-order velocity moments, but that level of sophistication quickly creates challenges in providing sufficient descriptions. This leads to kinetic theory/Monte Carlo approaches, which have parallels to the free molecular flow descriptions used for gas-surface interactions. Here, collisional processes that describe gas behavior in the lower register give way to collision-less processes, leading to ballistic trajectories and gaseous

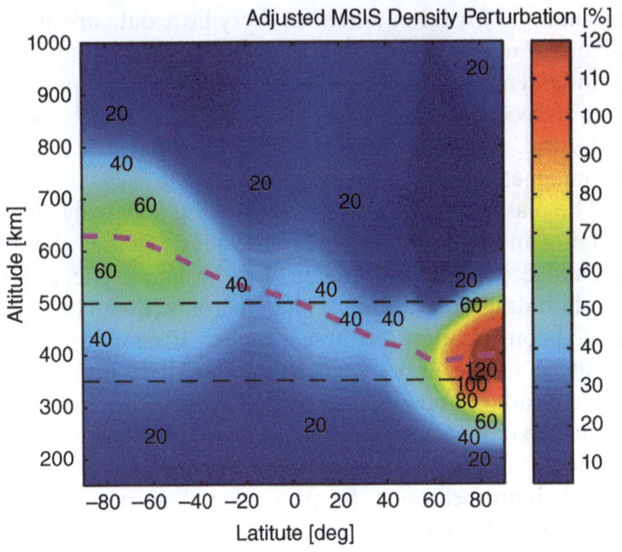

Figure 5.8 Altitudinal and latitudinal variations in mass density change during the 5–6 December 2008 geomagnetic storm estimated by MSIS after CHAMP and GRACE data adjustment. The black dashed lines represent the CHAMP and GRACE altitudes at that time. The pink dashed line indicates the maximum change in mass density and corresponds closely to where oxygen and helium have the same concentration. (Source: Liu et al., 2014. Reproduced with permission of John Wiley & Sons.)

escape as light and thermally energized species are able to overcome gravity. Inherently, the lighter and more thermally energetic species of atomic oxygen, helium, and hydrogen occupy this domain, but their relative concentrations and behavior are impacted by lower-register dynamics. The processes from below and above alter the manner by which the species change with altitude. Consequently, the total mass density can vary appreciably in regions of gas species transitions under different geophysical conditions. A particular example given here is the dynamic behavior of the oxygen/helium transition in the upper register.

The thermosphere-neutral constituents redistribute with altitude and latitude during geomagnetic activity in different ways that depend on the severity of the storm and the relative concentrations and molecular weights of the species that make up the multiconstituent gas. Much research has been dedicated to studying the behavior of the O/N_2 ratio during geomagnetic storms (e.g. Fuller-Rowell et al., 1990; Prölss, 1976; Rishbeth et al., 1985) in the lower register. Only recently has it been demonstrated that the helium/oxygen transition can dominate changes in mass density in the upper register during active geomagnetic times (e.g. Liu et al., 2014; Thayer et al., 2012). Furthermore, helium behavior will play a critical role in describing mass density structure and changes for altitudes extending above the thermosphere (Sutton et al., 2015).

While a major species in this upper register, helium behaves as a minor species in the lower register. This is an important distinction because a minor species may distribute itself in nonintuitive ways since, by definition, it is minor to the overall description of the gas. The description of the gas is determined by the major species that are required to follow the rules established by the constitutive equations; a minor species does not provide feedback to the major species. One well-known example that was actually revealed through spacecraft orbital behavior in the 1960s is the existence of the winter hemisphere helium bulge. In the cold and contracted winter thermosphere, helium abundance is found to be 5–10 times greater than in the summer hemisphere. This has been proven to be caused by major gas circulations imposed to satisfy mass conservation with the minor helium species responding to this circulation to form the wintertime helium bulge. The helium bulge creates an increase in mass density that can be easily detected by LEO spacecraft orbital behavior. Recently Bruinsma and Forbes (2010) and Thayer et al. (2012) revealed the winter helium bulge in GRACE data. The phenomenon has even been detected on Mars by the Maven mission (Elrod et al., 2017). For CHAMP no evidence of such behavior was detected, as CHAMP's altitude was about 150 km below GRACE, leaving it in an oxygen-rich environment. Consequently, GRACE during the solar minimum period 2008–2010 was subject to oxygen and helium composition changes along the orbit and mass density structures that deviated from diffusive equilibrium.

Upon the occurrence of a geomagnetic storm, composition and mass density changes occur. Figure 5.8 is an indication of how mass density can change with altitude during a geomagnetic storm. The black dashed line indicates the altitudes for the GRACE and CHAMP satellites in December 2008. The dashed pink line indicates where helium and oxygen are nearly equal in concentration. The mass density response to geomagnetic activity is the combined influence of variations in temperature and composition change. The altitude structure produced in Figure 5.8 is due to the different manner by which oxygen and helium respond to geomagnetic activity. This leads to a complex response at GRACE altitudes due to the dynamics of the helium/oxygen transition region. The density perturbation from quiet to active conditions is positive for all latitudes, indicating the mass density increases from quiet to active conditions, as expected, but with significant variability with latitude and altitude in the amount of change. From the summer hemisphere through the equatorial region, mass density perturbations at GRACE altitudes exceed those at CHAMP altitudes. This is indicative of mostly temperature influencing the response with little composition change as it remains primarily oxygen over this latitude region. Northward of about 40°N, the perturbation in mass density at GRACE altitudes increases less than at lower latitudes and less than at CHAMP altitudes. This response is indicative of composition change offsetting the temperature, resulting in the density perturbation in the winter hemisphere being less than that in the summer hemisphere at GRACE altitudes. In the winter polar region at GRACE altitudes, helium concentrations dominate during geomagnetic quiet times, while oxygen dominates during geomagnetic active times, which makes the mean molecular weight change significant in this region.

The mass density change with altitude maximizes near the altitude where oxygen and helium have the same concentration. This transition region is indicated by the pink dashed line in the figure. Above that altitude, the mass density change is suppressed as helium and oxygen dynamics lead to competing processes affecting the mass density: helium by transport and oxygen by thermal expansion. This is a very climatological view of the mass density change and much more investigation is required. The above description illustrates the tradeoff between temperature and mean molecular weight changes, such that the corresponding density perturbation due to geomagnetic storm varies with latitude and altitude. Note that besides temperature and mean molecular weight change, the change in the vertical gradient of mean molecular weight can also modify the density response to a geomagnetic storm, as described by Thayer et al. (2012). Thus, understanding the dynamics of oxygen and helium are paramount to understanding mass density change in this upper register. Temperature does play a role in determining how rapid the composition changes with altitude (as can be seen when contrasting altitude changes between southern summer mass density change and northern winter mass density change), as well as in how their relative scale heights change during geomagnetic activity. However, helium is strongly tied to transport processes, and this can have unexpected effects on composition and mass density distribution. The extreme solar minimum conditions accentuate this effect as the atmosphere is more contracted, and thus, transition regions occur over shorter altitude ranges and have greater variability. Under less extreme conditions, this transition region would be several hundred kilometers higher in altitude but still quite dynamic.

There remain many challenges in specifying the neutral state in this upper register. At greater altitudes, helium and hydrogen become the major species and plasma densities begin to approach those of the neutral species. Investigators have come to rely on the climatological empirical model of MSIS to provide composition information of the upper thermosphere. However, almost all of the measurements that define MSIS come from the lower register and are extrapolated to the upper register assuming diffusive equilibrium. Very little data are available to justify how high this approximation holds. Fundamental questions pervade the most basic properties of species relative concentrations, species global distribution, plasma properties, and behavior during geomagnetic activity.

The plasma environment in this region is equally challenging and strongly coupled to the neutral gas as charge exchange processes dominate. Furthermore, effects on satellite drag due to the plasma environment become an added complication for these altitudes. Plasma ion composition in the upper register measured by incoherent scatter radars and satellites show the pressing need for updated studies of topside mass-resolved plasma concentrations, and these can yield indirect information on neutral helium behavior. In particular, helium ions are dynamic and can reach more than 50% of total ion populations at solar maximum (Bailey & Sellek, 1990; Gonzalez et al., 2004; Heelis et al., 1990), reflecting the variations of neutral helium itself. Craven et al. (1995) noted that observed He^{+} from the Atmosphere Explorer spacecraft were lower than modeled predictions by over a factor of 2 in a wide range of conditions, a difference that was ascribed to fundamental uncertainty in the neutral helium concentration. Further observations of upper register He^{+} combined with modeling studies remain

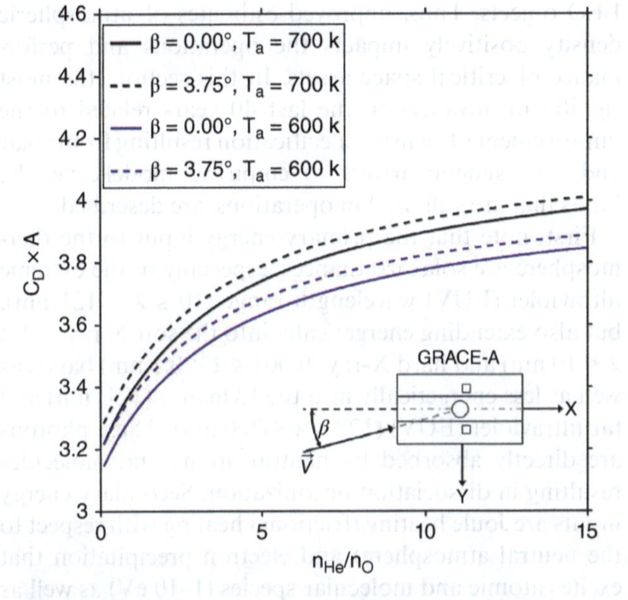

Figure 5.9 GRACE-A panel model C_DxA as a function of He/O ratio at two values of atmospheric temperature and angle of attack (β). (Source: Thayer et al., 2012. Reproduced with permission of John Wiley & Sons.)

extremely valuable, as they can derive important indirect information on neutral helium through correct interhemispheric plasma transport modeling (e.g. Lie-Svendsen et al., 1992) using CTIP-e, TIE-GCM, and other advanced models.

Summary of the remaining issues in rarefied gas dynamics in the upper register:

- Lower register specification of thermosphere gas
- Composition behavior in oxygen, helium, and hydrogen
- Physics-based simulations of thermosphere-exosphere transitions
- Exosphere gas dynamics
- Plasma environment

5.5.2. Gas-Surface Interactions

The description of gas-surface interactions becomes acutely more uncertain in this upper register as the gas becomes more tenuous, changes in gas composition occur, and the drag force weakens to where solar radiation pressure, Earth-reflected solar radiation pressure, and IR-emitted Earth radiation pressure must also be included in order to isolate the influence of atmospheric drag. The adsorbed atomic oxygen surface coverage on spacecraft surfaces along with the associated changes in energy accommodation become strongly altitude dependent in this upper register, further complicating the description of drag (Pilinski et al., 2010, 2013). Pardini et al. (2006) demonstrate the variability for a single drag coefficient model caused by changes in the ambient environment and solar conditions. The discrepancies between various aerodynamics models, caused largely by a variety of GSI assumptions within those models, is as big as the whole range of C_D values predicted by some of the C_D models and is largest in the exospheric region above 500 km. Note also that aerodynamic models such as the Semi-Empirical Satellite Accommodation Model are constrained by aerodynamic data but that this data has thus far been limited to altitudes below approximately 500 km (Pilinski et al., 2013). The discrepancies in atmospheric neutral densities between various atmospheric models computed along the paths of two satellites flying at 800 km and 600 km can reach as much as 100%. It has long been understood that a significant number of these model discrepancies are caused by assumptions made about satellite drag coefficients during the construction of those models.

Effectively, both adsorbed surface coverage and energy accommodation decrease as the molar mass decreases, and this occurs with increasing altitude. An upper altitude threshold postulated to maintain near full accommodation and diffuse emission is about 700 km during solar maximum conditions (Moe & Bowman, 2005; Pardini et al., 2010). In these instances, analytic approaches to determine the drag coefficient, such as by Sentman (1961a), have been applied at these altitudes and below. This altitude will be even lower under solar minimum conditions. Above that altitude, the surface contamination decreases, the energy accommodation coefficients decrease, and the angular distribution of reemitted molecules may have an increasingly important quasi-specular component (Saltsburg et al., 1967). Furthermore, the flow becomes more hypothermal as the speed ratio decreases, requiring the inclusion of thermal motion of the impinging gas in the analysis. All of these effects lead to an increase in the drag coefficient with height, but little observational data are available to evaluate these processes sufficiently.

A change in thermosphere composition from the chemically reactive and heavier atomic oxygen to the inert and much lighter helium impacts the drag coefficient. Figure 5.9 illustrates this point by computing the drag coefficient: the area product for the GRACE-A satellite using the Sentman (1961a) model, which assumes near-full accommodation and diffuse re-emission. This analysis was to evaluate the presence of helium during the deep solar minimum of 2008 at GRACE altitudes near 479 km. During this period, the MSIS model was performing poorly and a sensitivity test of different exospheric temperatures and He/O ratios was employed to assess their impact on the drag coefficient. An increase in the amount of helium relative to oxygen increases the drag coefficient for a given temperature. This is due to a decrease in the speed ratio as the most probable thermal speed increases with the presence of a lighter species. The lower speed ratio causes more atoms to reach the long side of the satellite and increase the momentum exchange between the spacecraft and the atmosphere. An increase in He/O ratio from 5 to 10 increased the drag coefficient by about 20%. An increase in temperature also produces an increase in the drag coefficient again associated with a decrease in the speed ratio.

A spacecraft with elongated surfaces in a low-eccentricity orbit, such as GRACE, in the upper register can experience significant changes in its drag coefficient within a single orbit due to varying composition structure. As described in section 5.5.1, for solstice conditions, the winter high latitudes are inundated with helium while oxygen dominates in the summer high latitudes. A change in He/O from less than 1 in the summer hemisphere to more than 50 in the winter hemisphere is possible within a single orbit (e.g. Liu et al., 2014). Thus, one quarter of its orbit will experience a change in drag coefficient that, if not properly accounted for, would introduce a bias in retrieving mass density. Poor representation of composition and temperature in the upper register, in addition to the changing conditions of gas-surface interactions, can therefore limit the accuracy of physically modeled drag coefficients.

A further consideration is the role of charged aerodynamics in the upper register. Here, plasma concentrations approach neutral concentrations, creating a direct plasma-surface interaction effect that can rival neutral gas-surface interactions. Furthermore, a greater susceptibility to spacecraft charging at these altitudes creates greater potentials and introduces an indirect interaction due to field effects. Recent research into plasma effects on satellite drag has been performed by Capon et al. (2018, 2019). The approach taken was to construct a surrogate model to describe the variation of a charged drag coefficient with plasma-scaling parameters to assess its contribution to the overall drag coefficient for LEO objects. This remains a challenging topic and one that will require further evaluation to determine its overall impact on drag.

Summary of the remaining issues in gas-surface interactions in the upper register:

- Introduction of charge drag
- Changes in adsorption and accommodation factors
- Transitions from diffuse to near-specular scattering
- Effects of composition change on drag coefficient modeling

5.6. SOLAR AND GEOMAGNETIC DRIVERS

As mentioned in the introduction to this chaper, the variability in atmospheric neutral density, especially due to space weather events, is the primary error source for LEO objects. Thus, improved estimates of atmospheric density positively impacts the operations and performance of critical space assets. In this section, the most significant advances in the last 40 years related to the improvement of density specification resulting from solar and geomagnetic drivers to empirical models, i.e. the kinds that are still used in operations, are described.

First, note that the primary energy input to the thermosphere are solar irradiances, especially in the extreme ultraviolet (EUV) wavelength range ($10 \leq \lambda < 121$ nm), but also extending energetically into the soft X-ray ($0.1 \leq \lambda < 10$ nm) and hard X-ray ($0.001 \leq \lambda < 0.1$ nm) bands as well as less energetically into the Lyman-α (121. nm) and far ultraviolet (FUV) ($122 \leq \lambda < 200$ nm). These photons are directly absorbed by neutral atoms and molecules resulting in dissociation or ionization. Secondary energy inputs are Joule heating (frictional heating with respect to the neutral atmosphere) and electron precipitation that excites atomic and molecular species (1–10 eV) as well as ionizes the neutrals (0.01–10 keV).

Drivers to thermosphere models use solar irradiances, indices, or proxies as well as geomagnetic indices or proxies. It is important to clarify that proxies and indices are not the same, as identified by the ISO standard for determining solar irradiances (IS 21348). Specifically, a solar irradiance proxy is a measured or modeled surrogate for solar spectral irradiances at specified wavelengths or over a wavelength bandpass, i.e. it may be only empirically related. A solar irradiance index, on the other hand, is a measured or modeled indicator, i.e. expresses an activity level, of actual solar irradiances at specific wavelengths or over a wavelength bandpass. The same is true for a geomagnetic activity index or proxy.

Thus, for physics-based thermospheric density models, solar spectral irradiances themselves are used. Common bandpasses are 1 nm bins or 37 wavelength bins and groups (Hinteregger et al., 1981). Geomagnetic drivers, such as the electron or ion fluxes, are input with an energy spectrum to characterize particle precipitation. Electric fields can be modeled or measured as an important method for characterizing Joule heating.

However, for empirical models, solar and geomagnetic indices or proxies are used to parameterize those energy inputs. This is because empirical models have historically been faster and more efficient to run in operational centers. For example, the $F_{10.7}$ and Ap proxies described below have been used extensively as solar and geomagnetic energy inputs, with their legacies extending back to the 1960s. In this section of the chapter, we describe more useful indices and proxies that have been developed since the early 2000s to better characterize solar inputs and the evolution of geomagnetic storms. For solar energy inputs, it is desirable to have solar indices and proxies that vary differently through time. This strategy of using multiple solar indices has significantly improved the accuracy of density modeling. These are also summarized in the ISO International Standard on Earth upper atmosphere (IS 14222:2013).

5.6.1. Legacy Drivers ($F_{10.7}$ and Ap)

Paetzold (1960) discovered that his analysis of solar data from a 20-cm antenna failed in the summer because of moisture condensation in the antenna. However, he noted that the 10-cm antenna in Ottawa did not have a moisture problem. Thus, using the data from the Ottawa 10-cm antenna, Paetzold discovered annual and semiannual variations in density as determined from early satellite drag data. Jacchia (1965, 1971) used this information to develop correlations with the 10.7-cm solar radio flux, $F_{10.7}$ to define the basis for thermospheric temperatures, which have an effect on densities. Thus, while the 10.7 cm measurements ($F_{10.7}$) are measured on the ground, i.e. the signal is not absorbed by the atmosphere, he found a good proxy for the solar irradiances absorbed in the thermosphere.

Jacchia used $F_{10.7}$ to represent all solar energy that was available for thermospheric heating, and it came to be thought of as an EUV proxy. Thermospheric heating is dominated by the solar chromospheric EUV energy, but it additionally comes from coronal soft X-ray, Lyman-α, and photospheric FUV wavelengths (Banks & Kockarts, 1973).

Geomagnetic activity for empirical models has been represented by the Ap index, which reports planetary geomagnetic amplitude activity (Mayaud, 1980). It is translated from the Kp geomagnetic index that is derived from geomagnetic field measurements made at locations around the world. Official ap values are calculated at the GeoForschungsZentrum Potsdam Adolf-Schmidt-Observatory for Geomagnetism in Germany, where the daily Ap is obtained by averaging the eight 3-hour values of ap for each day. The U.S. Air Force Weather Agency (AFWA) also calculates an estimated ap from a different, smaller set of stations than used for calculating the official ap values. The AFWA ap values are available through products issued by NOAA Space Weather Prediction Center. Daily Ap and 3-hour ap indices were used in early satellite orbit analysis because the time scales of geomagnetically induced variability they represented were consistent with the variations in satellite drag–derived densities (Moe, 1966; Moe & Nebergall, 1969).

5.6.2. Contemporary Drivers (S_{10}, M_{10}, Y_{10}, Dst)

The use of other indices to represent complex thermospheric heating processes has evolved over the years. A range of solar irradiances in a variety of spectral bands emanate from different solar atmosphere layers; all these photons are absorbed by various atomic and molecular species at unique unit optical depth layers in the terrestrial atmosphere. Schmidtke (1976) first postulated developing multiple solar indices to improve thermospheric density models. Tobiska (1988) then applied this concept using both Lyman-α and $F_{10.7}$ to represent chromospheric and coronal solar emissions. This methodology was extended to the development of the JB2006 model (Bowman, Tobiska, Marcos, & Valladares, 2008) using three separate solar indices to map these complex solar irradiances to thermospheric optical depths. JB2008 (Bowman, Tobiska, Marcos, Huang, et al., 2008) further extended this methodology to include an additional solar temperature region (the hot corona) and an additional thermospheric layer, i.e. the 85–100 km mesopause and lower thermosphere.

The first three indices, $F_{10.7}$, $S_{10.7}$, $M_{10.7}$ (Tobiska et al., 2008) were combined with a fourth index, $Y_{10.7}$. Collectively, these were adopted as part of the COSPAR International Reference Atmosphere (CIRA) 2014. It was the ability to insert more physics into thermospheric density drivers in the form of specific solar wavelengths linked to optical depths that enabled JB2008 to substantially reduce thermospheric densities' variation uncertainties (Bowman, Tobiska, Marcos, Huang, et al., 2008; Marcos et al., 2006; Tobiska et al., 2008).

5.6.3. $F_{10.7}$ Proxy

$F_{10.7}$ was first measured by Covington (1948) on a daily basis beginning 14 February 1947. At the present time, the 10.7 cm solar radio flux is measured at the Dominion Radio Astrophysical Observatory (DRAO) in Penticton, British Columbia, by the Solar Radio Monitoring Programme, which is operated jointly by the National Research Council (http://www.nrc-cnrc.gc.ca) and the Canadian Space Agency. Observations of the $F_{10.7}$ flux density values are made at 17, 20, and 23 UTC each day, then processed and transmitted to users automatically (see the DRAO website, https://nrc.canada.ca/node/1101). The 20 UTC observed values (not 1 AU) are archived at the World Data Center and used for the JB2008 historical inputs. The physical units of $F_{10.7}$ are $\times 10^{-22}$ W m^{-2} Hz^{-1} and their numerical value is customarily used without the multiplier, being called solar flux units (sfu). In other words, a 10.7 cm solar radio emission of 150×10^{-22} W m^{-2} Hz^{-1} is simply referred to as $F_{10.7} = 150$ sfu.

$F_{10.7}$ is a useful proxy for the combination of chromospheric, transition region, and coronal solar EUV emissions modulated by bright solar active regions whose energies at Earth are deposited in the thermosphere. Because of the high EUV– $F_{10.7}$ correlation, Jacchia developed empirical equations for exospheric temperatures that varied with solar activity and that could be used in atmospheric density models, e.g. CIRA72 (COSPAR, 1972). $F_{10.7}$ does not actually interact with the Earth's (or other planetary) atmosphere, i.e. the radiation can be measured all the way to the Earth's surface. It originates mostly in the solar cool (low) corona by electrons in thermal free-free (bremsstrahlung) emission in the vicinity of sunspots and in widely distributed areas associated with the hot complexes of solar activity (Tapping & DeTracey, 1990). The dependence on few processes, combined with its localized formation in the cool corona, i.e. a region that is closely coupled with magnetic structures responsible for creating the XUV–EUV irradiances, makes this a good generalized solar proxy for thermospheric heating.

The running 81-day centered smoothed $F_{10.7}$ values, using the moving boxcar method, are referred to as F_{81}. The observed archival daily $F_{10.7}$ values and their 81-day running center-smoothed values, F_{81}, with a 1-day lag are shown in Figure 5.10 and are described in other work (Bowman, Tobiska, Marcos, Huang, et al., 2008; Bowman, Tobiska, Marcos, & Valladares, 2008). The 1-day lag had the best correlation with satellite-derived density residuals. Figure 5.10 shows the solar cycle 23 $F_{10.7}$ index from 1 January 1997 to 1 January 2009.

Linear regression is used with daily $F_{10.7}$ to scale and report all JB2008 solar indices in units of sfu. Days where there are missing data values are not included in the regressions. $F_{10.7}$ is the recognized historical EUV proxy and, by reporting other proxies or indices in sfu, it is very easy to qualitatively identify similarities and differences between them. For solar energy inputs, it is desirable to have solar indices and proxies that vary differently through time. This strategy of using multiple solar indices has significantly improved the accuracy of density modeling as reported by Bowman, Tobiska, Marcos, Huang, et al. (2008).

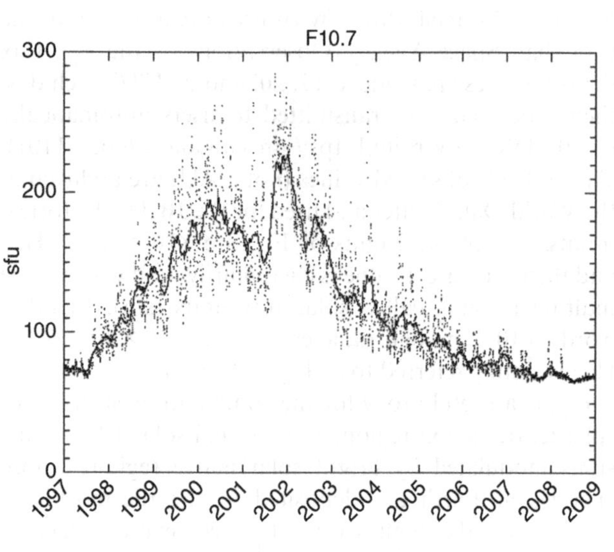

Figure 5.10 $F_{10.7}$ daily and 81-day smoothed values from 1 January 1997 to 1 January 2009.

5.6.4. $S_{10.7}$ Index

As described by Tobiska et al. (2008), the $S_{10.7}$ index is the integrated 26–34 nm solar irradiance that is measured by the Solar Extreme-Ultraviolet Monitor (SEM) instrument on the NASA/ESA Solar and Heliospheric Observatory (SOHO) research satellite. SOHO had an uninterrupted view of the Sun by operating in a halo orbit at the Lagrange Point 1 (L1) on the Earth-Sun line, approximately 1.5 million km from the Earth. SEM was built and operated by University of Southern California's Space Science Center PI team (Judge et al., 2001). SOHO was launched on 2 December 1995, and SEM has been making observations since 16 December 1995. The SEM instrument measures the 26–34 nm solar EUV emission with 15-second time resolution in this first-order broad-band wavelength range.

The $S_{10.7}$ index is an activity indicator of the integrated 26–34 nm solar emission and is created by first normalizing the data, then converting it to sfu via a first-degree polynomial fit with $F_{10.7}$. Spikes from abnormal flares and missing data were excluded from the fitting vectors. Normalization is achieved for the 1 AU adjusted epoch values, denoted as SOHO_SEM26_34, by division of a mean value over a time frame common to multiple datasets. The mean value equals 1.9955×10^{10} photons $cm^{-2}\,s^{-1}$. The common time frame is 16 December 1995 to 12 June 2005, which is generally equivalent to solar cycle 23. The resulting index is called $S_{10.7}$.

In addition to this basic derivation, corrections to $S_{10.7}$ are made as follows. The originally released version of $S_{10.7}$ (v1.8) ranged from 1 January 1996 to 30 December 2005. In versions 3.0–3.9 used by JB2006, the $S_{10.7}$ values between those dates are the original ones derived in v1.8. However, in versions 3.0–3.9 and versions 4+ after 12 June 2005, a slight long-term trend was removed to ensure that similar values at the minima of solar cycles 22 and 23 were achieved. For JB2008 ($S_{10.7}$ values v4.0 and higher), a new derivation was completed as given in equation (5.3), and there may be slight differences of <0.5% compared to earlier versions of $S_{10.7}$. Figure 5.11 shows $S_{10.7}$ and the S_{81} (81-day centered smoothed) values (v4.0) for solar cycle 23 from 1 January 1997 to 1 January 2009. Daily updated values are found at the JB2008 menu link on the Space Environment Technologies (SET) website (http://spacewx.com).

$$S_{10.7} = (-2.90193) + (118.512)\left[\frac{SOHO\_SEM26\_34}{1.9955x10^{10}}\right] \tag{5.3}$$

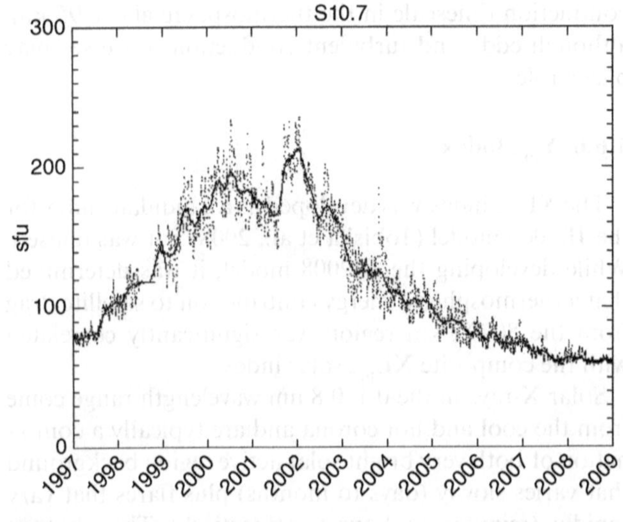

Figure 5.11 $S_{10.7}$ daily and 81-day smoothed values from 1 January 1997 to 1 January 2009.

Chromospheric He II at 30.4 nm and coronal Fe XV at 28.4 nm dominate the broadband SEM 26–34 nm irradiances, but that bandpass includes contributions from other chromospheric, transition region, and coronal lines. When the SOHO SEM and TIMED single event effect (SEE) 26–34 nm integrated data are compared, there are differences in the time series, particularly during active solar conditions. It is possible that the SOHO SEM measurements are slightly contaminated with second-order emissions from the coronal 17.1 nm Fe IX line that have not been removed; however, this topic needs further investigation. The energy in this bandpass principally comes from solar active regions, plage, and network. Once the photons reach the Earth, they are deposited (absorbed) in the terrestrial thermosphere, mostly by atomic oxygen above 200 km. The daily $S_{10.7}$ index and its 81-day running center–smoothed values, S_{81}, with a 1-day lag (the best correlation with satellite density residuals), are also described in Bowman, Tobiska, Marcos, Huang, et al. (2008). A 1-day lag is consistent with the average atomic oxygen thermal conduction timescale in the thermosphere above 180 km.

5.6.5. M10.7 Proxy

Tobiska et al. (2008) describe the development of the $M_{10.7}$ index. It is derived from the Mg II core-to-wing ratio (cwr) that originated from the NOAA series operational satellites, e.g. NOAA-16, -17, -18, which hosted the Solar Backscatter Ultraviolet (SBUV) spectrometer (Heath & Schlesinger, 1986; Viereck et al., 2001). This instrument has the objective of monitoring ozone in the Earth's lower atmosphere and, secondarily, making solar UV measurements. In its discrete operating mode, a diffuser screen is placed in front of the instrument's aperture in order to scatter solar middle ultraviolet radiation near 280 nm into the instrument. Although the NOAA data are from operational satellites, the SORCE/SOLSTICE and ERS-2/GOME research satellites also make the Mg II cwr measurements. In addition, starting with the GOES–16 satellite, Mg II cwr is also measured.

The 280 nm solar spectral band contains photospheric continuum and chromospheric line emissions. The Mg II h and k lines at 279.56 and 280.27 nm, respectively, are chromospheric in origin, while the weakly varying wings or continuum longward and shortward of the core line emission are photospheric in origin. The instruments from all satellites observe both features. On the ground, the ratio of the Mg II variable core lines to the nearly nonvarying wings is calculated. The result is mostly a measure of chromospheric solar active region emission that is theoretically independent of instrument sensitivity change through time. However, long-term changes can occur in the index if instrument wavelength calibrations change in flight or the solar incidence angle into the instrument changes. The daily Mg II cwr, described by Viereck et al. (2001), has historically been provided through the NOAA Space Weather Prediction Center (SWPC). It is likely, however, that the index values at NOAA SWPC are presently uncorrected. SET has developed and provides a corrected, operational Mg II cwr data product (MgIIcwr_SET) available at the "Products" menu link of http://spacewx.com that uses the NOAA-16, -17, -18, SORCE/SOLSTICE, and ERS-2/GOME data sources. Tobiska et al. (2008) describes the detailed processing of the MgIIcwr_SET product. The NOAA data come directly through NOAA NESDIS, and SET uses the DeLand algorithm (Cebula & DeLand, 1998; DeLand & Cebula, 1994) to create the index.

The Mg II cwr is an especially good proxy for some solar FUV and EUV emissions. It well represents photospheric and lower chromospheric solar FUV Schumann-Runge Continuum emission near 160 nm that maps into lower thermosphere heating due to O_2 photodissociation (Bowman, Tobiska, Marcos, Huang, et al., 2008). Since a 160 nm solar FUV emission photosphere index is not produced operationally, the MgIIcwr_SET proxy is used and modified into the $M_{10.7}$ index for comparison with the other solar indices. This derivation is performed in the following manner. A relationship between the long-term (multiple solar cycle) daily MgIIcwr_SET and $F_{10.7}$ is created by making a first-degree polynomial fit to produce a coefficient set that can translate the index into sfu. The result is M*10.7. Next, a correction is added for the decline of solar cycle 23 to account for NOAA 16 instrument degradation that may be related to its diffuser screen illumination geometry changing with time; this cause is

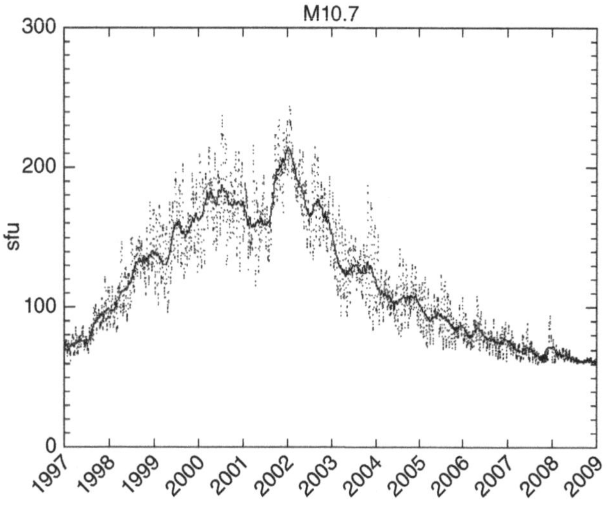

Figure 5.12 $M_{10.7}$ daily and 81-day smoothed values from 1 January 1997 to 1 January 2009.

unconfirmed. The correction is accomplished by using another first-degree polynomial fit between a trend ratio and day number starting 2448542.0 JD (12 October 1991 12:00 UT) near the peak of solar cycle 22. The trend ratio is formed from the 365-day center smoothed M*10.7 divided by the 365-day center smoothed F10.7. Equation (5.4) is the final v4.0 formulation of the $M_{10.7}$ index. The derived $M_{10.7}$ index is reported in sfu, i.e. $F_{10.7}$ units with a lower threshold minimum value set to 60. There may be slight differences of up to 1% compared to earlier versions of $M_{10.7}$.

$$M_{10.7} = \left[-2107.6186 + (8203.0537)(MgIIcwr\_SET)\right] + \left[(M*10.7)\left(1.2890589 + \left(-8.3777235x10^{-5}\right)*x - 1\right)\right] \tag{5.4}$$

The day number is $x = 0, 1, 2, \ldots$, with $x = 0$ equivalent to starting on 2448542.0 JD. Figure 5.12 shows the $M_{10.7}$ index during solar cycle 23 from 1 January 1997 to 1 January 2009. The daily $M_{10.7}$ and its 81-day running center-smoothed values, M_{81}, are shown in Figure 5.12 and are used with a 2-day lag in JB2008 as a proxy for the Schumann-Runge continuum FUV emission (Bowman, Tobiska, Marcos, Huang, et al., 2008). Originally, the JB2006 model, which did not use $Y_{10.7}$, had a lag time for $M_{10.7}$ of 5 days since the index was incorporating a combination of lag times from several energy transfer processes in the lower thermosphere to the mesopause. However, with the addition of the lower altitude (85–100 km), relevant $Y_{10.7}$ index, a shorter lag time was appropriate for $M_{10.7}$, which represents O_2 photodissociation, recombination, conduction, and transport processes at the 95–110 km level. A two-day lag is consistent with the average molecular oxygen dissociation and thermal conduction timescale in the thermosphere above 95 km, although eddy and turbulent conduction processes may play a role.

5.6.6. $Y_{10.7}$ Index

The $XL_{10.7}$ index was developed as a candidate index for the JB2006 model (Tobiska et al., 2008) but was unused. While developing the JB2008 model, it was determined that a thermospheric energy contribution to satellite drag from the 80–95 km region was significantly correlated with the composite $XL_{10.7}$ solar index.

Solar X-rays in the 0.1–0.8 nm wavelength range come from the cool and hot corona and are typically a combination of both very bright solar active region background that varies slowly (days to months) plus flares that vary rapidly (minutes to hours), respectively. The photons arriving at Earth are absorbed in the lower thermosphere to mesopause (85–100 km) by molecular oxygen (O_2) and molecular nitrogen (N_2), where they ionize those neutral constituents to create the ionospheric D-region.

The X-ray spectrometer (XRS) instrument is part of the instrument package on the GOES series operational spacecraft. The GOES/XRS provides the historical through current epoch 0.1–0.8 nm solar X-ray emission data with a 1-minute cadence and as low as 5-minute latency. These data, which are particularly useful for flare detection, are continuously reported by NOAA SWPC at their website, http://www.swpc.noaa.gov/. Tobiska and Bouwer (2005) used the GOES/XRS 0.1–0.8 nm data to develop an index of the solar X-ray active region background, without the flare component, for operational use. This is called the Xb_{10} index and is used to represent the daily energy that is deposited into the mesosphere and lower thermosphere.

While the 0.1–0.8 nm X-rays are a major energy source in these atmospheric regions during high solar activity, they relinquish their dominance to another emission that reaches the same optical depth, i.e. the competing hydrogen (H) Lyman-α emission that is the major energy source in this atmosphere region during moderate and low solar activity. Lyman-α is created in the solar upper chromosphere and transition region and demarcates the EUV from the FUV spectral regions. It is formed primarily in solar active regions, plage, and network; the photons, arriving at Earth, are absorbed in the mesosphere and lower thermosphere, where they dissociate nitric oxide (NO) and participate in water (H_2O) chemistry. Lyman-α was regularly observed by the SOLSTICE instrument on the UARS and SORCE satellites as well as by the SEE instrument on TIMED (Woods et al., 2000). Starting with the GOES-16 satellite, Lyman-α is also measured.

Since these two solar emissions are competing drivers to the mesosphere and lower thermosphere, a composite solar index of the Xb_{10} and Lyman-α was developed. It does not contain a flare component and is weighted to represent mostly Xb_{10} during solar maximum and to represent mostly Lyman-α during moderate and low solar activity. A normalized F_{81}, F_{81norm}, consisting of the 81-day centered smoothed $F_{10.7}$ divided by its mean value for the common time frame of 1 January 1991 through 16 February 2008 is used as the weighting function and multiplied with the Xb_{10} and Lyman-α (Lya) expressed as ratios to their solar maximum values. The resulting index is called $Y_{10.7}$ and equations (5.5), (5.6), and (5.7) describe this index as reported in sfu.

$$Y_{10.7} = F_{81norm}X_{10} + \left[\left(1 - F_{81norm}\right)L_{10}\right] \tag{5.5}$$

$$L_{10} = -88.3926 + \left(3.35891\times10^{-10}\,Lya\right) + \left(2.40481\times10^{-22}\,Lya2\right) \tag{5.6}$$

$$X_{10} = -42.5991 + \left(0.533669\right)Xb_{10} \tag{5.7}$$

X_{10} has a minimum threshold value of 40.

This daily index was tested with multiday lags, and the 5-day lag was found to have the strongest correlation signal in the satellite drag density residuals after modeled density variations due to the other solar indices were removed. The 81-day running center–smoothed values, Y_{81}, are also used with the 5-day lag. The 5-day lag is consistent with the average molecular oxygen and molecular nitrogen thermal conduction timescales in the lower thermosphere above 85 km, although eddy and turbulent conduction may also play a role. Figure 5.13 shows the $Y_{10.7}$ index during solar cycle 23 from 1 January 1997 to 1 January 2009.

Figure 5.14 provides a comparison between all four indices for the time frame of 1 January 1997 to 1 January 2009. The 81-day smoothed values are used to highlight the trends of each proxy or index. Figure 5.14 shows that each index or proxy captures a different type of solar energy variability. The composite of these four indices and proxies is the most complete representation to date, for empirical thermospheric density models, of the solar XUV, EUV, Lyman- α, and FUV energy that is deposited at optical depths ranging from 85 km to above 200 km. It is these energies that heat the thermosphere and affect satellite orbits. By expressing the proxies and indices in common units, their contribution to the daily density variability in the JB2008 exospheric temperature can be determined. The $F_{10.7}$ contribution to daily exospheric temperature variability is 9.8%, $S_{10.7}$ is 74.1%, $M_{10.7}$ is 10.3%, and $Y_{10.7}$ is 5.8%.

Table 5.1 provides the mapping of each index, including the data used to derive them, from its solar irradiance source(s) to the thermospheric optical depth(s) in which it is effective. The table includes the ISO 21348 spectral category, ISO 21348 spectral subcategory, wavelength range in units of nm, solar source temperature region, solar source feature, altitude region of terrestrial atmosphere absorption at unit optical depth in units of km, and terrestrial atmosphere thermal region of energy absorption. The indices and proxies that are marked with an asterisk (*) have been selected for use in the JB2008 empirical thermospheric density model and specifically in its exospheric temperature equation. Table 5.2 summarizes the characteristics of the daily solar indices including their observing facility, the instrument(s) used for observing the index/proxy, the nominal observation time frame, the measurement cadence, latency, and an assessment of the observational availability.

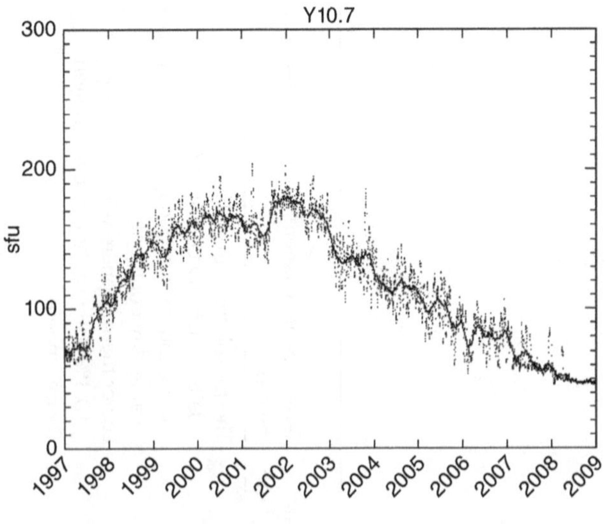

Figure 5.13 $Y_{10.7}$ daily and 81-day smoothed values from 1 January 1997 to 1 January 2009.

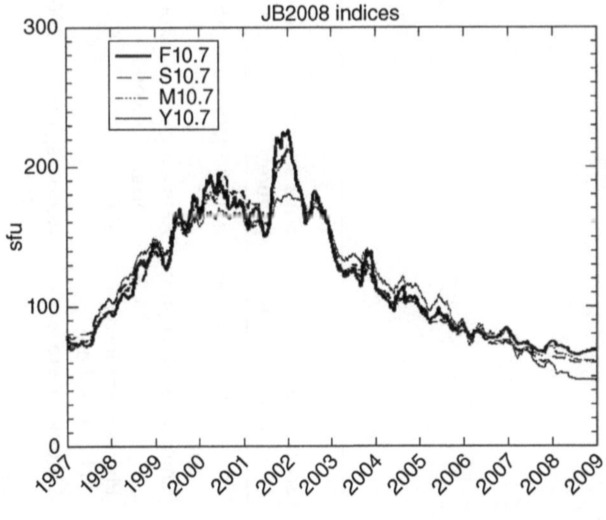

Figure 5.14 $F_{10.7}$, $S_{10.7}$, $M_{10.7}$, and $Y_{10.7}$ daily and 81-day smoothed values from 1 January 1997 to 1 January 2009.

Table 5.1 Solar indices related to atmospheric heating.

Index	IS 21348 Spectral category	IS 21348 Spectral subcategory	Wavelength range (nm)	Solar source temperature region[1]	Solar source feature[1]	Atmosphere absorption (unit optical depth, km)[2]	Terrestrial atmosphere absorption (thermal region)[2]
*$F_{10.7}$	Radio	Radio	10.7E7	Transition region, cool corona	Active region	90–500	Thermosphere with 1-day lag; 9.8% daily variability contribution
*$S_{10.7}$	UV	EUV	26–34	Chromosphere, corona	Active region, plage, network	200–300	Thermosphere with 1-day lag; 74.1% daily variability contribution
*$M_{10.7}$	UV	FUV	160	Photosphere-lower chromosphere	SRC	95–110	Lower thermosphere with 2-day lag; 10.3% daily variability contribution
$MgII_{cwr}$	UV	MUV	280[3]	Chromosphere	Active region, plage, network	200–300	Thermosphere
*$Y_{10.7}$	X-rays and UV	X-rays+H Lyman-α	0.1–0.8, 121	Chromosphere, transition region, hot corona	Active region, plage, network	85–100	Mesopause-lower thermosphere with 5-day lag; 5.8% daily variability contribution
H Lya	UV	H Lyman-α	121	Transition region, chromosphere	Active region, plage, network	85–100	Mesopause-lower thermosphere
X_{b10}	X-rays	X-rays	0.1–0.8	Hot corona	Active region background	85–100	Mesopause-lower thermosphere

* Index or proxy is used in the JB2008 model exospheric temperature equation.

[1] Vernazza et al., 1976; Vernazza et al., 1981.

[2] Banks & Kockarts (1973).

[3] The *h* and *k* lines at the band center are chromospheric and are referenced to blackbody continuum wings at edges of bandpass.

Table 5.2 Characteristics of daily JB2008 solar indices.

Index or proxy	Observing facility	Instrument	Observation time frame	Measurement cadence	Measurement latency	Operational availability
$F_{10.7}$	Penticton ground observatory	Radio telescope	1947–2009	3 times/day	Up to 24 hours	yes
$S_{10.7}$	SOHO, GOES	SEM, EUVS	1996–2009	15-second	Up to 24 hours	(a)
$M_{10.7}$	NOAA-16,17,18, SORCE, ERS-2	SBUV, SOLSTICE, GOME	1991–2009	2 times/day	Up to 24 hours	yes
$Y_{10.7}$	GOES-12, UARS, SORCE, TIMED	XRS, SOLSTICE (2), SEE	1991–2009	1-minute, 16 times/day	Up to 10 minutes, up to 48 hours	(b)

(a) SOHO/SEM is a NASA research instrument but provides daily irradiances on an operational cadence; GOES 13 EUVS B channel makes measurements in the same bandpass as SOHO SEM.
(b) GOES XRS is a NOAA operational instrument, whereas TIMED/SEE and SORCE/SOLSTICE are NASA research instruments providing daily irradiances on an operational measurement cadence.
(c) UARS/SOLSTICE stopped in 2005; SORCE/SOLSTICE intends to provide data for several years.

5.6.7. Geomagnetic Indices

The geomagnetic indices used in JB2008 incorporate existing as well as expanded empirical modeling formulations compared to previous Jacchia and MSIS-type models. A two-index formulation now captures low, unsettled, and substorm/storm-related geomagnetic activity. When the ap value is 40 and below (unsettled to quiet activity), the ap value is used. Above that threshold, the code assumes that there is a storm/substorm in progress and the Disturbance Storm Time (Dst) index is used.

The Dst index is an indicator of the strength of the storm-time ring current in the inner magnetosphere. During the main phase of geomagnetic storms, the ring current becomes highly energized and produces a southward-directed magnetic field perturbation at low latitudes on the Earth's surface. This is opposite to the normal northward-directed main field. The quick-look Dst index is calculated hourly and released through the World Data Center in Kyoto, Japan (http://swdcwww.kugi.kyoto-u.ac.jp/dst_realtime/index.html) using measurements from four off-equatorial magnetic observatories.

Dst is an "absolute" index and is reported in units of nT; magnetic observatory data are required for its calculation. Magnetic observatories are specially designed and carefully operated facilities that provide stable-baseline magnetometer data over long periods of time. Typically, an observatory supports the operation of fluxgate, proton-precession, and declination-inclination (theodolite) magnetometer systems. Sensors are situated on stable piers, and along with their acquisition electronics, they are operated in temperature-controlled conditions. Therefore, measurements from the various sensor systems can be combined to produce data that are extremely accurate. Traditionally, four stations have been used for the Dst calculation: Hermanus (HER), South Africa; Kakioka (KAK), Japan; Honolulu (HON), Hawaii; and San Juan (SJG), Puerto Rico. These stations have been continuously operated by national agencies for decades, and the data that they produce has high quality. These four stations also provide good longitudinal coverage that is needed to make a global estimation of the average ring-current intensity.

Most magnetic storms begin with sharp decreases (southward-directed negative values) in Dst, called the storm sudden commencement, in response to increased solar wind pressure. Following a southward turning of the interplanetary magnetic field, Dst decreases as ring current energy increases during the storm's main phase. During the recovery phase, the ring current energy decreases and Dst increases until the storm's end, when the magnetic field perturbation has ceased. In time series traces, Dst shows a transition from early to late recovery phases that is characterized by significant changes in the rate of change (slope). This occurs as the distribution of the ring current becomes symmetric in local time. It should be noted that a significant fraction of magnetic storms manifest more complex structuring, including multiple main and partial recovery phases.

Use of Dst as an index for the energy deposited in the thermosphere during magnetic storms is more accurate than the use of the ap index. This is because Dst has higher temporal resolution with an ability to segregate storm phases with their corresponding magnitudes. On the other hand, the 3-hour ap index is an indicator of general magnetic activity over the Earth and responds primarily to currents flowing in the ionosphere and only secondarily to magnetospheric variations. The ap index is derived from measurements by observatories at high latitudes that can be blind to energy input during large storms, and it can underestimate the effects of storms on the thermosphere.

During storm periods, the thermosphere acts as a driven-but-dissipative system where the changes in exospheric temperature change due to storm effects can be expressed by a differential equation as a function of Dst. To determine the effects on exospheric temperature, and thereby the thermospheric density distribution at any time in a storm, it is necessary to integrate the differential equation, taking into account the change in exospheric temperature while starting at the storm commencement and proceeding throughout the entire storm period. An algorithm for determining the storm events was developed (Bowman, Tobiska, Marcos, Huang, et al., 2008) that locates the temporal start, minimum, recovery slope change, and final end of the storm as reflected in the Dst index. The algorithm can incorporate the changes due to multiple or complex storms. The ap, Dst, and change in exospheric temperature values used to drive JB2008 thermospheric densities that are a result of geomagnetic activity are updated hourly and provided at the JB2008 website accessible through the JB2008 menu link at http://spacewx.com.

5.7. CONCLUSIONS

This chapter describes remaining issues in describing and predicting upper atmosphere drag on LEO spacecraft. A significant issue is providing a physical description of the LEO drag environment, and how it changes due to external influences, independently of the physical description for gas-surface interactions that determine the sensitivity of the spacecraft to a change in the upper atmosphere. The chief drivers to the atmosphere, i.e. solar and geomagnetic energy sources, are the largest factors in predicting upper atmosphere change. It has been a common approach to use LEO spacecraft orbital behavior as a means to extract space environment properties, particularly gas mass density and winds, and then reapply this description of the space environment to predict future spacecraft behavior using spacecraft orbit propagators. This approach leads to mutual coupling between spacecraft response and atmosphere change that results in convoluted uncertainties that can be challenging to quantify. Unraveling LEO spacecraft sensitivities from space environment changes requires independent gas property measurements of the space environment (particularly temperature and composition), improved physical models of the LEO drag environment to predict the response under varying external solar and geomagnetic influences, improvements in the specification of these external influences, and an independent means to measure the processes involved in gas-surface interactions.

The chapter addressed these particular issues by segmenting the LEO drag environment into a lower register of 100–500 km altitude and an upper register of 500–1000 km. The lower register represents a region where most thermosphere measurements have been made, atmospheric drag is the dominant nongravitational force with high accommodation, the thermosphere gas transitions to predominantly atomic oxygen, the solar and geomagnetic drivers are depositing their energy, and the thermosphere gas is well represented by using continuum gas dynamics. Several remaining issues for the thermosphere gas in the lower register were identified and discussed, including the need to define the turbopause height with location, season, year, and species, describe the 3-D thermosphere circulation globally beyond climatology, observe the thermosphere temperature and composition distribution with height, and describe the solar and magnetospheric energy deposition. Remaining issues in gas-surface interactions for the lower register include modeling of drag coefficients and accounting for complex geometries, identifying errors in energy accommodation and scattering properties, and translating forces and torques to extract mass density and winds.

The upper register represents a region where the most populous LEO objects reside, atmospheric drag and radiation pressure compete, gas-surface interactions are more specular and less accommodating, the thermosphere gas is transitioning to an exosphere where particle motions take on ballistic trajectories, the gas composition transitions to predominantly helium and hydrogen, plasma densities become important, and much fewer measurements of the gas properties are available. Several remaining issues for the thermosphere gas in the upper register were identified and discussed, including the need to specify the lower register of the thermosphere gas as it serves as a necessary lower boundary condition for the upper register; describe oxygen, helium, and hydrogen behavior; simulate thermosphere-exosphere transitions in gas behavior; and understand exosphere and plasmasphere behavior. Remaining issues in gas-surface interactions in the upper register include understanding the impacts of plasma on drag, describing adequately the adsorption and accommodation behavior, understanding diffuse-to-specular scatter transitions, and determining drag coefficient dependencies on composition.

Section 5.6 provided a new contribution to the literature by describing the approach used to account for the solar and geomagnetic indices that drive the JB2008 model. This model was the most significant improvement in empirical thermospheric density modeling since the 1960s and formed the basis for the dramatic improvements in the HASDM system in the 2000s.

Today, greater fidelity in space environment specification is required for LEO spacecraft operations, collision avoidance procedures, and space environment forecasting. This chapter described the processes involved and the remaining issues needing resolution in order to improve the description, prediction, and use of the LEO drag environment in a future that is looking to significantly expand LEO spacecraft operations. These remaining issues are challenging as they require many more observations and well-thought-out experiments in order to advance understanding and capability.

ACKNOWLEDGMENTS

The authors would like to acknowledge helpful discussions with John Emmert, Tim Fuller-Rowell, Matt Hejduk, David Vallado, and Eelco Doornbos. This work was partially supported by University of Colorado's Grand Challenge initiative in the Space Weather Technology, Research, and Education Center, NSF CEDAR Award No. AGS 1452309. Section 5.6 was partially supported by SPAWAR contracts N6600118P7815 and N6600119P6237.

REFERENCES

Bailey, G. J., & Sellek, R. (1990). A mathematical model of the earth's plasmasphere and its application in a study of He+ at L=3. *Ann. Geophys.*, *8*, 171.

Banks, P., & Kockarts, G. (1973). *Aeronomy*. New York: Academic Press.

Bird, G. A. (1994). Molecular gas dynamics and the direct simulation of gas flows. Oxford Engineering Science series, No. 42. Oxford: Clarenden Press.

Bittencourt, J. A. (2004). *Fundamentals of plasma physics*, 3rd ed. Springer-Verlag.

Bowman, B. R., & Moe, K. (2005), Drag coefficient variability at 175–500 km from the orbit decay analyses of spheres. AAS/AIAA, Astrodynamics Specialist Conference, American Astronautical Society, Paper AAS 05-257, San Diego, CA.

Bowman, B. R., Tobiska, K. W., Marcos, F. A., Huang, C. Y., Lin, C., & Burke, W. F. (2008). A new empirical thermospheric density model JB2008 using new solar and geomagnetic indices. AIAA/AAS, Astrodynamics Specialist Conference and Exhibit 18–21 August 2008, Honolulu, Hawaii (AIAA 2008-6438).

Bowman, B. R., Tobiska, W. K., Marcos, F. A., & Valladares, C. (2008). The JB2006 empirical thermospheric density model. *J. Atm. Sol-Terr. Phys.*, *70*, 774–793.

Bruinsma, S. (2015). The DTM-2013 thermosphere model. *J. Space Weather Space Clim.*, *5*, A1. https://doi.org/10.1051/swsc/2015001

Bruinsma, S. L., & Forbes, J. M. (2010). Anomalous behavior of the thermosphere during solar minimum observed by CHAMP and GRACE. *J. Geophys. Res.*, *115*, A11323. doi:10.1029/2010JA015605

Bussy-Virat, C. D., Ridley, A. J., & Getchius, J. W. (2018). Effects of uncertainties in the atmospheric density on the probability of collision between space objects. *Space Weather*, *16*, 519–537. https://doi.org/10.1029/2017SW001705

Capon, C. J., Brown, M., & Boyce, R. R. (2018). Direct and indirect charged aerodynamic mechanisms in the ionosphere. *Adv. Space Res.*, *62*(5), 1090–1101.

Capon, C. J., Smith, B., Brown, M., Abay, R., & Boyce, R.R. (2019). Effect of ionospheric drag on atmospheric density estimation and orbit prediction. *Adv. Space Res.*, *63*, f2495–2505. DOI: 10.1016/j.asr.2019.01.013

Cebula, R. P., & DeLand, M.T. (1998). Comparisons of the NOAA-11 SBUV/2, UARS SOL- STICE, and UARS SUSIM Mg II solar activity proxy indexes. *Solar Phys.*, *177*, 117– 132.

Cook, G. E. (1965). Satellite drag coefficients. *Planet. Space Sci.*, *13*, 929–946.

COSPAR (1972). International Reference Atmosphere. Compiled by the COSPAR Working Group IV. Amsterdam: North-Holland Publishing.

Covington, A. E. (1948). Solar noise observations on 10.7 centimeters. *Proc. of the I.R.E.*, *36*, 454.

Craven, P. D., Comfort, R. H., Richards, P. G., & Grebowsky, J. M. (1995). Comparisons of modeled N+, O+, H+, and He+ in the midlatitude ionosphere with mean densities and temperatures from Atmosphere Explorer. *J. Geophys. Res.*, *100*(A1), 257– 268. doi:10.1029/94JA02306

DeLand, M. T., & Cebula R. P. (1994). Comparisons of the Mg II index products from the NOAA-9 and NOAA-11 SBUV/2 instruments. *Solar Phys.*, *152*, 61–68.

Doornbos, E. (2011). Thermospheric density and wind determination from satellite dynamics (Ph.D. thesis). Delft University of Technology. ISBN: 978-90-9026051-8.

Doornbos, E., van den IJssel, J., Luehr, H., Foerster, M., Koppenwallner, G., Bruinsma, S., et al. (2010). Neutral density and crosswind determination from arbitrarily oriented multiaxis accelerometers on satellites. *J. Spacecraft Rockets*, *47*, 580–589. https://doi.org/10.2514/1.48114

Elrod, M. K., Bougher, S., Bell, J., Mahaffy, P. R., Benna, M., Stone, S., et al. (2017). He bulge revealed: He and CO2 diurnal and seasonal variations in the upper atmosphere of Mars as detected by MAVEN NGIMS. *J. Geophys. Res. Space Physics*, *122*. doi:10.1002/2016JA023482

Emmert, J. T., Picone, J. M., & Meier, R. R. (2008). Thermospheric global average density trends, 1967–2007, derived from orbits of 5000 near-Earth objects. *Geophys. Res. Lett.* *5*, 35. https://doi.org/10.1029/2007GL032809

Emmert, J. T. (2015). Thermospheric mass density: A review. *Advances in Space Research*, *56*(5), 773–824.

FCC (2018). Mitigation of orbital debris in the new space age (IB Docket No. 18-313).

Fuller-Rowell, T. J. (1998). The “thermospheric spoon”: A mechanism for the semiannual density variation. *J. Geophys. Res.*, *103*(A3), 3951– 3956. doi:10.1029/97JA03335

Fuller-Rowell, T. J., Rees, D., Quegan, S., Moffett, R. J., Codrescu, M. V., & Millward, G. H. (1996). A coupled thermosphere-ionosphere model (CTIM). In R. W. Schunk (Ed.), *Handbook of ionospheric models* (pp. 217–238). Logan, Utah: Utah State Univ.

Fuller-Rowell, T., Rees D., Tinsley B., Rishbeth H., Rodger A., & Quegan S. (1990). Modelling the response of the thermosphere and ionosphere to geomagnetic storms: Effects of a mid-latitude heat source. *Advances in Space Research*, *10*(6), 215–223.

Gaposchkin, E. M., & Coster, A. J. (1988). Analysis of satellite drag. *The Lincoln Laboratory Journal*, *1*, 203–224.

González, S. A., Sulzer, M. P., Nicolls, M. J., & Kerr, R. B. (2004). Solar cycle variability of nighttime topside helium ion concentrations over Arecibo. *J. Geophys. Res.*, *109*, A07302. doi:10.1029/2003JA010100

He, C. Y., Yang, B., Carter, E., Kerra, S., Wu, F., Deleflie, H., et al. (2018). Review and comparison of empirical thermospheric mass density models. *Progress in Aerospace Sciences*, *103*, 31–51.

Heath, D. F., & Schlesinger, B. M. (1986). The Mg 280-nm doublet as a monitor of changes in solar ultraviolet irradiance. *J. Geophys. Res.*, *91*, 8672–8682.

Heelis, R. A., Hanson, W. B., & Bailey, G. J. (1990). Distributions of He+ at middle and equatorial latitudes during solar maximum. *J. Geophys. Res.*, *95*, 10313.

Hejduk, M. D., & Snow, D. E. (2018). The effect of neutral density estimation errors on satellite conjunction of the upper atmosphere. *Space Weather*. https://doi.org/10.1029/2017SW001720

Hinteregger, H. E., Fukui, K., & Gilson, B. R. (1981). Observational, reference and model data on solar EUV, from measurements on AE-E. *Geophys. Res. Lett.*, *8*, 1147.

Hoots, F. R., & Roehrich, R. L. (1980). Spacetrack Report #3: Models for Propagation of the NORAD Element Sets. U.S. Air Force Aerospace Defense Command, Colorado Springs, CO.

International Standards Organization (2013). IS 14222. Space environment (natural and artificial) – Earth upper atmosphere. Geneva.

International Standards Organization (2007). IS 21348. Space environment (natural and artificial) – Process for determining solar irradiances. Geneva.

Jacchia, L. G. (1965). Static diffusion models of the upper atmosphere with empirical temperature profiles. *Smithsonian Contr. Astrophys.*, *8*, 215. Smithsonian Institution Astrophysical Observatory.

Jacchia, L. G. (1971). Revised static models of the atmosphere and exosphere with empirical temperature profiles. *Smithsonian Astrophys. Special Report 332*, Smithsonian Institution Astrophysical Observatory.

Jacchia, L. G. (1977). Thermospheric temperature, density, and composition: New models. *Special Report 375*. Smithsonian Institution Astrophysical Observatory.

Judge, D. L., Ogawa, H. S., McMullin, D. R., Gangopadhyay, P., & Pap, J. M. (2001). The SOHO CELIAS/SEM EUV Database from SC23 Minimum to the Present. *Adv. Space Res.*, *29*(12), 1963.

King-Hele, D. G., & Quinn, E. (1965). The variation of upper-atmosphere density between sunspot maximum (1957-8) and minimum (1964). *J. Atmos. Terr. Phys.*, *27*, 197–209.

Klinkrad, H. (2017). Large satellite constellations and related challenges for space debris mitigation. *The Journal of Space Safety Engineering*, *4*, 59–60. http://dx.doi.org/10.1016/j.jsse.2017.06.002

Lie-Svendsen, O., Rees, M. H., & Stamnes, K. (1992). Helium escape from the Earth's atmosphere: The charge exchange mechanism revisited. *Planet. Space Sci.*, *40*, 1639.

Liu, X., Thayer, J. P., Burns, A., Wang, W., & Sutton, E. (2014). Altitude variations in the thermosphere mass density response to geomagnetic activity during the recent solar minimum. *J. Geophys. Res. Space Physics*, *119*. doi:10.1002/2013JA019453

March, G., Doornbos, E. N., & Visser, P. N. (2019). High-fidelity geometry models for improving the consistency of CHAMP, GRACE, GOCE and Swarm thermospheric density data sets. *Adv. Space Res.*, *63*(1), 213–238.

Marcos, F. A., Bowman, B. R., & Sheehan, R. E. (2006). Accuracy of Earth's thermospheric neutral density models, AIAA 2006-6167, AIAA/AAS Astrodynamics Specialist Conference, Keystone, CO, August.

Marcos, F. A., Kendra, M. J., Griffin, J. M., Bass, J. N., Larson, D. R., & Liu, J. J. (1998). Precision low Earth orbit determination using atmospheric density calibration. *J. Astronaut. Sci.*, *46*(4), 395–409.

Mayaud, P. N. (1980). Derivation, meaning, and use of geomagnetic indices. *Geophysical Monograph 22*. Washington, DC: American Geophysical Union.

Mehta, P. M., & Linares, R. (2017). A methodology for reduced order modeling and calibration of the upper atmosphere. *Space Weather*, *15*(10), 1270–1287.

Mehta, P. M., Walker, A., Lawrence, E., Linares, R., Higdon, D., & Koller, J. (2014). Modeling satellite drag coefficients with response surfaces. *Adv. Space Res.*, *54*(8), 1590–1607. doi:10.1016/j.asr.2014.06.033

Mehta, P. M., Walker, A. C., Sutton, E. K., & Godinez, H. C. (2017). New density estimates derived using accelerometers on board the CHAMP and GRACE satellites. *Space Weather*, *15*, 558–576. doi:10.1002/2016SW001562

Moe, K. (1966). Absolute atmospheric densities determined from the spin and orbital decays of Explorer VI. *Planet. Space Sci.*, *14*, 1065–1075.

Moe, K., & Bowman, B. R. (2005). The effects of surface composition and treatment on drag coefficients of spherical satellites (AAS05-258). AAS/AIAA Astrodynamics Conference. South Lake Tahoe, California, USA.

Moe, K., & Moe M. M. (2005). Gas-surface interactions and satellite drag coefficients. *Planet. Space Sci.*, *53*, 793–801. doi:10.1016/j.pss.2005.03.005

Moe, K., Moe, M. M., & Rice, C. J. (2004). Simultaneous analysis of multiinstrument satellite measurements of atmospheric density. *J. Spacecraft Rockets*, *41*(5), 849–853.

Moe, K., & Nebergall D. (1969). Variation of geomagnetic disturbance with latitude. *J. Geo- phys. Res.*, *74*, 1305–1307.

Moe, M. M., Wallace, S. D., & Moe, K. (1995). Recommended drag coefficients for aeronomic satellites. The upper mesosphere and lower thermosphere: A review of experiment and theory. In R. M. Johnson & T. L. Killeen (Eds.), *Geophysical Monograph, Vol. 87* (pp. 349–445), Washington, DC: American Geophysical Union.

Paetzold, H.-K. (1960). COSPAR Space Research, Nice.

Pardini, C., Anselmo, L., Moe, K., & Moe, M. M. (2010). Drag and energy accommodation coefficients during sunspot maximum, *45*, (5), 638–650. https://doi.org/10.1016/j.asr.2009.08.034

Pardini, C., Tobiska, W. K., & Anselmo, L. (2006). Analysis of the orbital decay of spherical satellites using different solar flux proxies and atmospheric density models. *Advances in Space Research*, *37*(2), 392–400. https://doi.org/10.1016/j.asr.2004.10.009

Picone, J. M. Hedin, A. E., Drob, D. P., & Aikin, A. C. (2002). NRLMSISE-00 empirical model of the atmosphere: Statistical comparisons and scientific issues. *J. Geophys. Res., Space Phys.*, https://doi.org/10.1029/2002JA009430 107(A12):SIA 15–1–SIA 15–16.

Pilinski, M. D., & Argrow, B. M. (2013). Aerodynamic analysis based on challenging minisatellite payload satellite lift-to-drag measurements. *Journal of Spacecraft and Rockets*, *50*, 1162–1170. doi:10.2514/1.A32394

Pilinski, M. D., Argrow, B. M., & Palo, S. E. (2010). Semi-empirical model for satellite energy-accommodation coefficients. *Journal of Spacecraft and Rockets*, *47*, 951–956. https://doi.org/10.2514/1.49330

Pilinski, M. D., Argrow, B. M., & Palo, S. E. (2011). Drag coefficients of satellites with concave geometries: Comparing models and observations. *Journal of Spacecraft and Rockets*, *48*, 312–325. https://doi.org/10.2514/1.50915

Pilinski, M. D., Argrow, B. M., & Palo, S. E. (2013). A semi-empirical satellite accommodation model for spherical and randomly tumbling objects. *Journal of Spacecraft and Rockets*. doi: 10.2514/1.A32348

Prieto, D. M., Graziano, B. P., & Roberts, P.C.E. (2014). Spacecraft drag modelling. *Prog. Aerosp. Sci.*, *64*, 56–65. doi:10.1016/j.paerosci.2013.09.001

Prölss, G. (1976). On explaining the negative phase of ionospheric storms. *Planetary and Space Science*, *24*(6), 607–609.

Radtke, J., Kebschull, C., & Stoll, E. (2017). Interactions of the space debris environment with mega constellations: Using the example of the OneWeb constellation. *Acta Astronautica*, *131*, 55–68.

Ren T., Miao, J., & Liu, S. (2018). Atmospheric density determination using high-accuracy satellite GPS data. *Science China Technological Sciences*, *61*, 204. doi: 10.1007/s11431-016-9096-6

Richmond, A. D., Ridley, E. C., & Roble, R. G. (1992). A thermosphere/ionosphere general circulation model with coupled electrodynamics. *Geophysical Research Letters*, *19*(6), 601. https://doi.org/10.1029/92GL00401–604

Ridley, A., Deng, Y., & Toth, G. (2006). The global ionosphere-thermosphere model. *Journal of Atmospheric and Solar-Terrestrial Physics*, *68*(8), 839–864.

Rishbeth, H., Gordon, R., Rees, D., & Fuller-Rowell, T. (1985). Modelling of thermospheric composition changes caused by a severe magnetic storm. *Planetary and Space Science*, *33*(11), 1283–1301.

Roble, R. G., Ridley, E. C., Richmond, A. D., & Dickinson, R. E. (1988). A coupled thermosphere/ionosphere general circulation model. *Geophysical Research Letters*, *15*(12), 1325. https://doi.org/10.1029/GL015i012p01325–1328

Saltsburg, H., Smith Jr. J. M., & Rogers M. (Eds.) (1967). *Fundamentals of gas-surface interactions*. New York: Academic Press.

Schmidtke, G. (1976). *Geophys. Res. Lett.*, *3*, 573.

Schunk, R., & Nagy, A. (2009). *Ionospheres: Physics, plasma physics, and chemistry*. Cambridge Atmospheric and Space Science Series, 2nd ed.

Sentman, L. (1961a). Comparison of the exact and approximate methods for predicting free molecule aerodynamic coefficients. *ARS J.*, *1*, 1576–1579.

Sentman, L. (1961b). *Free molecule flow theory and its application to the determination of aerodynamic forces (Tech. Rep. TR LMSC-448514)*. Lockheed Missile and Space Co., Sunnyvale, CA.

Storz, M. F., Bowman, B. R., Branson, M.J.I., Casali, S. J., & Tobiska, W. K. (2005). High accuracy satellite drag model (HASDM). *Adv. Space Res.*, *36*, 2497–2505. doi:10.1016/j.asr.2004.02.020

Sutton, E. K., Nerem, R. S., & Forbes, J. M. (2007), Density and winds in the thermosphere deduced from accelerometer data. *Journal of Spacecraft and Rockets*, *44*(6), 1210–1219.

Sutton, E. K. (2008). Effects of solar disturbances on the thermosphere densities and winds from CHAMP and GRACE satellite accelerometer data (Ph.D. thesis). University of Colorado at Boulder.

Sutton, E. K. (2009). Normalized force coefficients for satellites with elongated shapes. *J. Spacecraft Rockets*, *46*(1), 112–116. https://doi. org/10.2514/1.40940

Sutton, E. K., Thayer, J. P., Wang, W., Solomon, S. C., Liu, X., & Foster, B. T. (2015). A self-consistent model of helium in the thermosphere. *J. Geophys. Res. Space Physics*, *120*, doi:10.1002/2015JA021223

Tapping, K. F., & DeTracey, B. (1990). The origin of the 10.7 cm flux. *Solar Phys.*, *127*(2), 321–332.

Thayer, J. P., Liu, X., Lei, J., Pilinski, M., & Burns, A. G. (2012). The impact of helium on thermosphere mass density response to geomagnetic activity during the recent solar minimum. *J. Geophys. Res.*, *117*, A07315. doi:10.1029/2012JA017832

Tobiska, W. K. (1988). A solar extreme ultraviolet flux model (Ph.D. thesis). Department of Aerospace Engineering, University of Colorado.

Tobiska, W. K., & Bouwer S. D. (2005). Solar flare evolution model for operational users. 2005 Ionospheric Effects Symposium, ed. J.M. Goodman. JMG Associates, 76.

Tobiska, W. K., Bouwer, S. D., Bowman, B. R. (2008). The development of new solar indices for use in thermospheric density modeling. *J. Atm. Solar-Terr. Phys.*, *70*, 803–819.

Turansky, C. P., & Argrow, B. M. (2014). Rigid-body dynamics in free-molecular and transition flow. *Journal of Spacecraft and Rockets*, *51*, 239–252.

Vallado, D., Crawford, P., Hujsak, R., & Kelso, T. S. (2006). Revisiting Spacetrack Report #3, AIAA/AAS Astrodynamics Specialist Conference and Exhibit, Keystone, Colorado, 21–24 August, 2006.

van den IJssel, J., & Visser, P. (2010). Performance of GPS-based accelerometer: a simulation experiment. *Adv. Space Res.*, *45*(1), 2010.

Vernazza, J. E., Avrett, E. H., & Loeser, R. (1976). Structure of the solar chromosphere. II—The underlying photosphere and temperature-minimum region. *Astrophysical Journal Supplement Series*, *30*, 1–60.

Vernazza, J. E., Avrett, E. H., & Loeser, R. (1981). Structure of the solar chromosphere, III—Models of the EUV brightness components of the quiet-sun. *Astrophysical Journal Supplement Series*, *45*, 635–725.

Viereck, R., Puga, L., McMullin, D., Judge, D., Weber, M., & Tobiska W. K. (2001). The Mg II index: A proxy for solar EUV. *Geophys. Res. Lett.*, *28*(7), 1342.

Walker A., Mehta, P., & Koller, J. (2014). Drag coefficient model using the Cercignani–Lampis–Lord Gas–Surface Interaction Model. *Journal of Spacecraft and Rockets*, *51*, 1544–1563. https://doi.org/10.2514/1.A32677

Woods, T. N., Bailey, S., Eparvier, F., Lawrence, G., Lean, J., McClintok, B., et al. (2000). TIMED Solar EUV Experiment. *Phys. Chem. Earth*, *25*, 393–396.

6

Solar Radio Burst Effects on Radio- and Radar-Based Systems

Dale E. Gary[1] and Timothy S. Bastian[2]

ABSTRACT

Radio emission from solar flares can attain such high flux density that the Sun becomes the dominant source of broadband radio noise in the terrestrial environment. The effects of this radio noise on wireless communication and navigation systems can take many forms, depending on the design and operation of the affected system. These effects can be of special concern for regional or global systems, since the effects can occur simultaneously over the entire sunlit hemisphere of Earth. This chapter reviews the origin of solar radio bursts, the threat they pose based on statistics of the flux-density distribution of such events vs. frequency, and some of the effects that have been documented in the literature. The chapter concludes with a discussion of the potential impacts on current and future technology and how these impacts can be mitigated (1) through improved radio monitoring of the Sun in both circular polarizations to supply meaningful real-time warnings, (2) through improved scientific understanding of the solar phenomena underlying the radio bursts, and (3) through improved system design that takes account of solar radio noise.

6.1. INTRODUCTION

Although the Sun continuously produces radio noise at some level due to normal thermal processes, those emissions went undetected during the early days of radio astronomy. In his excellent history of early radio astronomy, Sullivan (2009) describes several attempts to detect radio emission from the Sun as early as 1894 by Oliver Lodge, with more concerted efforts by Johannes Wilsing and Julius Scheiner in Potsdam in 1896 and Charles Nordmann at Paris Observatory in 1901. Although these efforts failed, Sullivan (2009) quotes Nordmann (1902): "The surface of the sun must emit Hertzian waves . . . in the vicinity of spots and faculae and at the time of maximum in [the cycle of] solar activity." Despite these early efforts and surprisingly accurate expectations for solar radio emission, more than four decades (and three solar cycles) passed before J. S. Hey (1942), in a secret report during World War II, related the accidental discovery of widespread solar "jamming" of British radar installations operating at 55–85 MHz on 27 and 28 February 1942 (Hey, 1946; Sullivan, 2009). Thus, the very first detection of the Sun as a radio source was due to the direct and damaging effect of solar radio noise on the technological systems of the day. Only 4 months later, George Southworth, doing wartime radar research at Bell Labs, made the first detection of the quiet Sun at 3.2 cm wavelength, fully described in two internal "memoranda-for-file" (Southworth, 1942a, 1942b). Unfortunately, wartime censors caused publication of the discovery to be delayed until 1945 (Southworth, 1945), so that the

[1] *Center for Solar-Terrestrial Research, New Jersey Institute of Technology, Newark, New Jersey, USA*

[2] *National Radio Astronomy Observatory, Charlottesville, Virginia, USA*

Space Physics and Aeronomy Collection Volume 5: Space Weather Effects and Applications, Geophysical Monograph 262, First Edition.
Edited by Anthea J. Coster, Philip J. Erickson, and Louis J. Lanzerotti.

DOI: 10.1002/9781119815600.ch6

report actually appeared 5 months after the first published radio detection of the Sun (Reber, 1944) at 160 MHz, completely independently, by the remarkable amateur and radio astronomy pioneer Grote Reber.

Koskinen et al. (2017) list the three pathways by which the Sun influences the Earth's space environment: through direct electromagnetic radiation, propagation and interaction of magnetized plasma, and via high-energy particles. By these pathways, the Sun produces many indirect effects on communications and navigation systems, such as electrical disturbances on long conductors during magnetic substorms, and ionospheric effects that interfere with the reception of shortwave radio communications at lower radio frequencies, and cause scintillation that degrades Global Positioning System (GPS) navigation at higher radio frequencies.

However, this chapter focuses on the first pathway, specifically on effects due to solar radio noise, which acts as a source of direct interference to radio-based (wireless) communication and navigation systems. Despite the fact that the first discovery of radio emission from the Sun was a case of such direct interference, the potential impact of solar radio noise has been underappreciated, mainly because its worst effects arise only at times of extreme events, which occur only rarely yet may have global consequences.

The problem of solar radio noise is compounded by the fact that it often goes unrecognized in its milder forms as slight degradation in the performance of cellular telephone and wireless data systems (Bala et al., 2002; Lanzerotti et al., 1999), GPS receivers (Cerruti et al., 2006; Klobuchar, 1999), and aeronautical radio-navigation (Marqué et al., 2018), and only by careful investigation have these effects come to light. Another problem is that an extreme radio noise event need not be particularly obvious in other measures of solar activity, such as soft X-ray flux, coronal mass ejection mass and speed, or solar energetic particle production. In fact, the processes by which the Sun produces extreme solar radio flux density are not fully established, as we discuss in section 6.2. Coupled with the fact that the effects are transmitted at the speed of light and occur instantaneously without warning, this makes any kind of prediction or forecasting presently beyond reach.

It is only when extreme events do occur that the true scale of impacts on radar systems (Hey, 1946; Knipp et al., 2016), aeronautical systems (Marqué et al., 2018), and GPS location and navigation services (Carrano et al., 2009; Cerruti et al., 2008; Muhammad et al., 2015; Sreeja et al., 2013, 2014) become apparent. The events described by these authors occurred in solar cycles 17, 20, 23, and 24, but it is likely that similar events occurred at other times and went unrecognized or unreported. Hence, it is of great interest to examine the historical record of solar radio burst reports (Bala et al., 2002; Giersch et al., 2017; Nita et al., 2002) to determine the size distribution of solar radio burst events, and from that to determine the likelihood of occurrence of extreme events. The current state of knowledge and its limitations are described in section 6.3.

To appreciate the range and extent of effects of solar radio bursts on wireless systems, we summarize in section 6.4 several of the more consequential impacts that have been reported. Most such impacts are shrouded in secrecy, due to either national security or commercial interests, and hence are hard to study quantitatively. The exception is the effects on the Global Navigation and Satellite System (GNSS), the data for which are openly available from networks of receivers used for scientific research and Space Weather applications. The GNSS receivers thus serve as the "canary in the coal mine" to alert us to the potential global impacts that can occur due to extreme solar radio events, as society becomes ever more reliant on such technological systems.

We conclude in section 6.5 by summarizing the threat from solar radio noise, given this increasing reliance. For example, GPS navigation plays a role in developing technologies such as driverless cars and drone deliveries. New uses for GPS include ever more precise positional information, even in everyday consumer products such as smartphones. We also discuss how the threat might be mitigated through better monitoring and real-time warning of events, as well as system design considerations to make wireless systems more resilient to solar radio noise.

6.2. CAUSE AND GENERAL CHARACTERISTICS OF EXTREME SOLAR RADIO NOISE

Although extreme solar radio noise could in principle occur at any radio frequency range, in practice we are mainly interested in the frequencies used by GNSS and cellular telephone systems, and by aeronautical radar and distance-measuring equipment, which are overwhelmingly in the frequency band referred to as L band (1–2 GHz) or lower. While the expressions below are valid for any radio frequency, in the examples we will focus on L band.

6.2.1. Origin of the Radio Emission and Types of Emission

In general, solar flares produce emission by a variety of processes involving either the independent motion of high-energy electrons to produce incoherent emission, or collective motions of large numbers of electrons to produce coherent emission. The strength of solar radio bursts is typically given in solar flux units (sfu; 1 sfu = 10^{-22} W m^{-2} Hz^{-1}), i.e. as a flux spectral density, or flux density S_ν. Quite generally, the flux density is proportional to the product of source brightness, measured in units of brightness temperature, T_b, and source area, related by

$$S_\nu = \frac{2k\nu^2}{c^2}\int_\Omega T_b d\Omega \approx 12 T_7 \nu_{\rm GHz}^2 L_8^2 \text{ sfu}, \quad (6.1)$$

where k is Boltzmann's constant, c is the speed of light, ν is the frequency, and the integral is taken over the angular area (solid angle Ω) of the source. The approximation is for T_b in units of 10^7 K, frequency in GHz, and a length scale of 10^8 m (the size of a large solar active region). Obviously, the most extreme events will be characterized by being simultaneously both extremely bright (high T_b) and large in angular size.

Incoherent processes are limited to a maximum $T_b \approx 10^{10}$ K (Dulk, 1985), which limits the flux density to about 12000 sfu at 1 GHz (27000 sfu at 1.5 GHz), even when the source covers an entire active region. Klobuchar (1999) analyzed the susceptibility of GPS receivers and concluded that a solar radio burst of 20,000 sfu in right-hand circular polarization (RCP) would produce only a factor of 2 (3 dB) increase in noise in the receiver. From this, we can conclude that incoherent emission processes are not the cause of extreme events at L band.

Certain types of coherent processes, in contrast, can produce far brighter emission, with $T_b \approx 10^{15}$ K (Benz, 1986), but generally with restricted source area because special conditions are required for the emission that are generally not met over a large area. In addition, coherent processes tend to be of short duration. The most common form of coherent emission is the type III burst, which is due to a short-duration (of order 1 s) energetic electron beam that excites Langmuir waves oscillating at the local plasma frequency. Individual beams are accelerated in highly localized regions, limiting their flux density and duration, but sometimes a great number can be accelerated close in time, to produce a spatially broader and longer-lasting source of emission. Still, such emission rarely exceeds 20,000 sfu at decimeter wavelengths, or durations longer than a few minutes.

A more likely candidate can be found among the plethora of burst types collectively called type IV emission. Type IV emission is associated with solar flares and is characterized by a fluctuating broadband continuum superposed with various spectral fine structures, such as those shown in Figure 6.1 (see also Isliker & Benz, 1994). The continuum is caused by high-energy electrons accelerated by the associated flare, while the fine structures are due to interaction with various wave modes excited by them. Recently, radio imaging with high frequency and time resolution (Chen et al., 2015) has revealed that different spectral features generally come from spatially distinct locations within the affected active region, and while the overall spatial extent of the emission can be of order 10^8 m, the size occupied by a single type of fine structure is smaller. Only some of these spectral fine structures typically attain the high $T_b \gg 10^{10}$ K required for an extreme event. Examples of such spectral fine structures are shown in Figure 6.1. Notable are the so-called "spike bursts" (Figure 6.1b), known to have been the cause of the most extreme radio flux density ever reported at decimeter wavelengths, the $>10^6$ sfu event of 6 December 2006 (Cliver et al., 2011; Gary, 2008), which is discussed further in section 6.4.3.

Given their role in producing the most extreme decimeter-wavelength event yet recorded, it is worthwhile to look further at the characteristics and origin of such bursts. As may be apparent in Figure 6.1b, spike bursts are bright, short-duration (~10 ms at L band), narrowband ($\Delta\nu/\nu < 1\%$) bursts that are generally accepted to be a form of electron-cyclotron maser emission (ECME; see reviews by Fleishman & Melnikov, 1998; Melrose, 2017; Treumann, 2006). This mechanism refers to a linear plasma instability that pumps free energy available in a nonthermal electron distribution (an anisotropy in momentum space) into free-space electromagnetic modes. ECME is well established as the mechanism for coherent emissions from planetary magnetospheres: e.g. terrestrial auroral kilometric radiation (AKR) and some Jovian decametric bursts (Melrose, 2017). In its original form, a loss-cone electron distribution function was proposed as the source of free energy to drive AKR emission (Wu & Lee, 1979), an idea that was subsequently adopted by Melrose & Dulk (1982) to account for solar spike bursts. In the case of AKR, however, in situ observations have established that loss-cone distributions play no role; rather, "horseshoe" or "shell" distributions resulting from electrons accelerated in electric fields aligned with the magnetic field produce ECME (Ergun et al., 2000; Treumann, 2006, and references therein). While ECME is implicated in the production of solar spike bursts and, more broadly, type IV continua and "zebra bursts" (Figure 6.1a; Chen et al., 2011; Winglee & Dulk, 1986), the details of the plasma conditions in which it arises from the Sun are not well established, although plausible adaptations to the theory have been proposed (Fleishman & Melnikov, 1998), including consideration of magnetosphere-like conditions giving rise to horseshoe distributions (Ergun et al., 2000; Melrose & Wheatland, 2016).

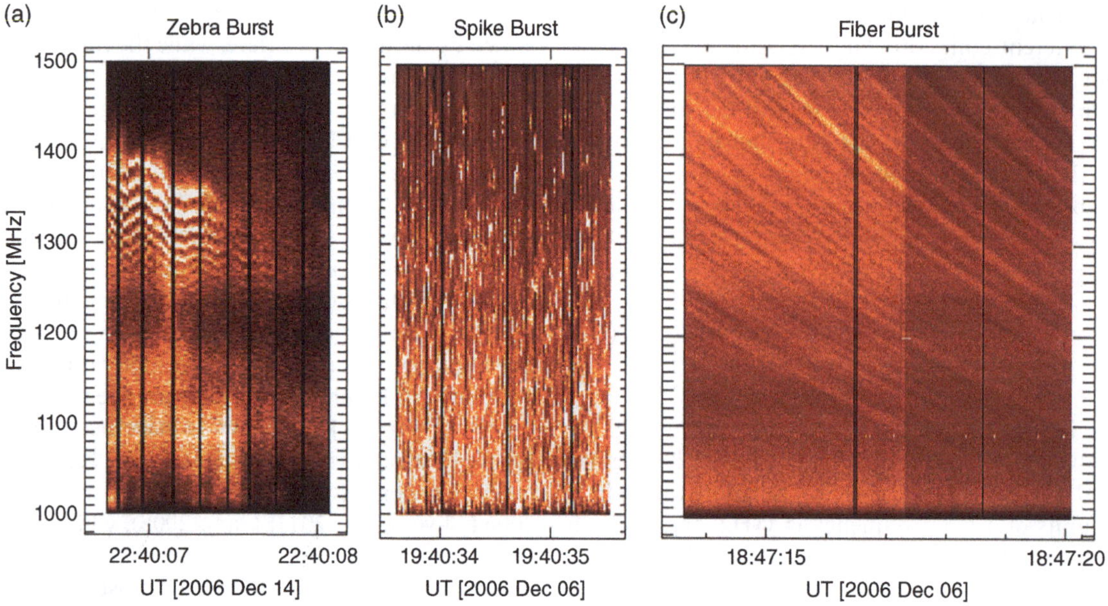

Figure 6.1 Examples of solar type IV spectral fine structures observed at L band (1–1.5 GHz) during December 2006. (a) A dynamic spectrogram of 1 s duration, showing a Zebra burst (see Chen et al., 2011). (b) 2 s of data showing spike bursts, due to electron-cyclotron maser emission. This coherent emission lasted nearly 1 h and was the cause of the record-setting flux density (1 million sfu) of the SOL20061206 burst, which resulted in widespread GPS outages. (c) 6 s of data showing fiber bursts, thought to be due to whistler waves propagating along closed magnetic loops and interacting with Langmuir waves to produce the emission. (Source: Chen et al., 2015. Reproduced with permission of AAAS.)

These studies have shown that ECME may produce intense type IV continuum and spectral fine structures under a wide range of plasma conditions, and the ECME mechanism may therefore be quite common at decimeter wavelengths. In the case of spike bursts in L band, the radiation is emitted near the local electron cyclotron frequency $\omega_B = eB/\gamma mc$, and the mechanism also requires $(\omega_p/\omega_B)^2 \ll 1$, where ω_p is the local electron plasma frequency. At L band, these conditions require a region of high magnetic field strength (300–500 G) and low electron density ($<10^{10}\,\mathrm{cm}^{-3}$). Additionally, to escape the plasma, the radiation must somehow avoid being resonantly absorbed in overlying, weaker-field regions where the magnetic field strength matches the second and third harmonics of ω_B. This may be mitigated if emission from density cavities plays a role (Ergun et al., 2000; Treumann, 2006).

The occurrence of *extreme events* due to spike burst emission is apparently rare (see section 6.3). Most spike bursts are seen to have only modest numbers (perhaps hundreds or thousands) of individual spikes, whereas the 6 December 2006 event was estimated to have more than 10 million over 10s of minutes (Gary, 2008). While individual spikes come from a tiny source region of order 200 km (deduced from their narrow bandwidth; Fleishman & Melnikov, 1998), the vast number of spikes in the 6 December 2006 event suggests the nearly simultaneous emission of spikes over a larger (but still small) region of the corona. We can use equation (6.1) to estimate the size needed to attain $S_\nu = 10^6$ sfu at 1.2 GHz, for spikes of $T_b = 10^{15}$ K, which gives a linear size of 2400 km. This implies that of order 100 bursts would have to occur simultaneously at a given frequency in a region 3–4 arcsec across, assuming unit filling factor, although a larger region with lower filling factor is likely. To cover a band of several hundred MHz, as in the 6 December 2006 event, another factor of 100 is needed. Once conditions exist for such extreme events, they can persist for many days, as evidenced by the fact that the solar active region responsible for the 6 December 2006 event produced two other unusually strong events on 13 and 14 December (Cliver et al., 2011; Gary, 2008).

6.2.2. Frequency Distribution and Polarization Characteristics of Radio Noise

To assess the threat of solar radio noise on specific navigation and communication systems, it is important to characterize the frequency distribution and polarization of solar radio bursts. The frequency distribution is important because only radio emission at the specific operating frequency of a given technology is relevant, while polarization is important for some technologies, notably GNSS, which uniformly operates in RCP. Solar bursts at GNSS frequencies but polarized in the left-hand circular sense will have little or no effect on GNSS operation.

The ECME mechanism, as with most other coherent emission mechanisms, can produce highly (up to 100%) circularly polarized emission, with the sense of polarization determined by the direction of the line-of-sight component of magnetic field in the source region. Since either direction may occur with equal probability in solar active regions, it follows that both senses of circular polarization are equally likely. The polarization of the December 2006 events was 100% RCP, and hence they were maximally effective at interfering with GNSS. It should be noted that linear polarization does not occur for solar bursts seen at Earth, because any linear polarization produced at the source is highly Faraday rotated during propagation in the solar corona, and hence only circular polarization can survive. Since circular polarization can be considered a superposition of two linearly polarized waves of equal amplitude, a 100% circularly polarized solar burst will be 50% effective at interfering with a technological system having linear polarization sensitivity.

Although the ECME mechanism produces very narrow-band spikes, such spikes are produced at a range of frequencies depending on the range of magnetic field strengths in the region over which the ECM instability is operable. Hence, the effective band over which extreme events can produce adverse effects can be quite broad. The 6 December 2006 event was observed to cause extreme flux densities from 1000 to 1800 MHz, and likely extended to unobserved frequencies well below 1000 MHz. Like most such events, however, the extreme flux densities abated by the next higher observed frequency, $\nu = 2400$ MHz, which presumably reflects an upper limit to the magnetic field strength ($B \approx \nu/2.8 \times 10^6 \sim 850$ G) in the emitting region. Nita et al. (2004) studied a uniform set of 412 bursts in the range 1–18 GHz observed with the Owens Valley Solar Array during two solar maximum years, 2001–2002, and found that the probability of occurrence of the highest flux density events increases markedly for the decimetric frequency range, 1–2.6 GHz, falling abruptly at 2.8 GHz and above. This is entirely consistent with the tendency of high-T_b coherent emission processes to extend to lower frequencies, with a cutoff at around 2.6 GHz, while the lower-T_b incoherent processes dominate at higher frequencies.

Thus, solar radio bursts pose the greatest risk for technologies operating at L band, which unfortunately is exactly the band where the most common navigation and communication systems currently operate. The threat comes from the rare occurrence of extremely bright coherent emission in sufficient numbers (or equivalently covering a sufficient angular area) to raise the flux density above the noise threshold for the affected system. So far, such bursts have been seen up to a flux density of 10^6 sfu, but since a relatively small source area is implied relative to the available area of an active region, there is nothing to prevent events of far higher flux density in future events. In contrast, emission above about 2.6 GHz is overwhelmingly due to much lower T_b incoherent processes. Although these, too, can achieve high flux densities of order 3×10^4 sfu, this only happens when the source area grows to cover much of the active region, and hence there is little room for such bursts to gain more than a factor of a few higher flux densities, at least for normal active region sizes.

6.3. OCCURRENCE STATISTICS OF SOLAR RADIO BURSTS

6.3.1. Expected Flux Density Threshold for Effects

The flux density threshold above which a technological system might begin to be affected depends greatly on the system and how it is used. For GNSS signals, as already previously mentioned, Klobuchar et al. (1999) evaluated the then-available systems to determine a noise threshold of 40,000 sfu for an unpolarized burst, or equivalently 20,000 sfu in RCP. That is the flux density required to equal the noise floor of a circa 1999 GPS receiver, and this expectation was largely borne out by actual observations (Cerruti et al., 2006). The determining figure of merit, called the noise figure, was assumed by Klobuchar et al. to be 3.8 dB, which led to a receiver temperature estimated to be 413 K. However, technology has improved greatly since then, and current smartphone receivers have a noise figure closer to 1.5 dB. This increased sensitivity lowers the receiver temperature to 145 K, and following the calculation of Klobuchar et al., that results in a roughly factor of 2 lower noise floor equivalent to 9300 sfu for modern smartphone receivers.

A similar threshold estimate was done for cellular telephone systems by Bala et al. (2002); however, in that case it is not the individual handset that is the limitation but rather interference on tower-to-tower relays. The antennas on such relay towers point at the local horizon in the direction of neighboring towers, and so east- and west-facing antennas can suffer interference from solar radio bursts near local sunrise and sunset, respectively, when the Sun is in the antenna beam (field of view), but not generally at other times of day. The Bala et al. threshold estimates were 960 sfu at 1 GHz, rising to 6000 sfu at 2.4 GHz.

Thresholds for other systems, such as radar receivers or wifi data links, are harder to characterize in general, due to lack of standardization, but are likely to be in the same range of order 10,000 sfu. It should be recognized that these thresholds are valid when the Sun is within the directional beam of the receiver. GPS receivers are open to the entire visible sky because they must track multiple satellites, whereas cell tower antennas, radars, and wifi links generally have a narrower cone of susceptibility. Also, these thresholds are for a doubling of the noise floor, which may lead to a detectable effect, but not necessarily one that causes system failure. Rather, system failures become likely only when the solar burst flux density exceeds the threshold by factors of order 100.

The remaining question is what are the occurrence statistics of solar radio bursts, as a function of frequency, for a given flux density threshold? And to assess the threat of system failure, how often can such bursts exceed the threshold by 2 orders of magnitude? To answer these questions, we must consult the historical record.

6.3.2. Data Sources and Limitations

The observation of solar radio bursts has a long history, starting with the initial discovery by Hey (1942) mentioned in section 6.1. However, routine monitoring that includes microwaves only began in 1960, using a worldwide network of radio observing sites, each with its own instrumental characteristics and frequencies. Bala et al. (2002) and Nita et al. (2002) used a database of 40 years (1960–1999) of solar radio burst reports compiled by the National Geophysical Data Center of the National Oceanic and Atmospheric Administration (NOAA) to investigate this historical record. Beginning in 1966, a uniform network operated by the US Air Force, called the Radio Solar Telescope Network (RSTN), and its predecessors provided the longest-running dataset at a select set of eight radio frequencies. Giersch et al. (2017) performed a similar study of this dataset, extended to the year 2010. The results show that the occurrence rate of solar bursts at a given frequency is distributed as a power law in flux density, with slightly different power law indexes for different frequencies and different phases of the 11-year solar activity cycle.

Both Nita et al. (2004) and Giersch et al. (2017) describe some limitations of the dataset, including missing bursts and saturation of the radiometers above certain fixed flux density levels (10^5 sfu at L band), which unfortunately limit the usefulness of the dataset for counting the most extreme events. Another shortcoming is the lack of polarization information (the network measures only total intensity), which as previously mentioned is important for assessing the threat to GNSS. Monitoring with the Nobeyama Radio Polarimeters (NoRP) was developed in various stages beginning in 1978 (Nakajima et al., 1985; Torii et al., 1979), and although it operates at a single longitude, it has the advantages that it saturates only at a higher flux density (not well determined) and can measure circular polarization.

In the next subsection we summarize these results, but it should be kept in mind that the results are based on an incomplete historical record. Missing bursts will affect the occurrence rate distribution. If we assume that the missing bursts are proportionally distributed in flux density, then it will not affect the shape of the distribution but will shift the distribution downward to lower rates. Nita et al. (2002) defined a "geographic" correction factor of order 1.5–2 to crudely account for missing bursts. Additionally, radiometer saturation will remove events from the high flux density tail of the distribution, which in principle can be recovered by extrapolating the distribution to higher flux densities. However, not measuring the extreme events in the tail has led to an underappreciation for how large the events can get, and there is always a danger that occurrence rate of the largest events is underestimated from a simple extrapolation of the distribution.

6.3.3. Flux Density Distribution and Dependence on Frequency

When the number of observed events in the NOAA database are plotted against their peak flux density, the number per flux density bin $\frac{dN}{dS}$ is found to closely approximate a power law distribution, $\frac{dN}{dS} \propto S_\nu^\lambda$, where S_ν is the flux density given in equation (6.1), and λ is the power law index. Of interest for this discussion is the cumulative distribution that gives the number of events above a given flux density,

$$N(S) = \int_{S_0}^{\infty} \frac{dN}{dS} dS = -\frac{1}{\lambda+1} \frac{dN}{dS}\Big|_{S_0} S_\nu^{\lambda+1}. \quad (6.2)$$

Figure 6.2 shows an example of the solar radio burst cumulative distribution (events per day) from Nita et al. (2002), summing bursts at all frequencies above 2 GHz, separately for solar maximum (Figure 6.2a) and solar minimum (Figure 6.2b). The power law fit (solid line) is excellent except at flux densities below about 20 sfu, where signal-to-noise limits affect the reported numbers, and at high flux densities where saturation no doubt affects the measurements.

Nita et al. (2002) give tables of $N(S_0 > 1$ sfu) and λ for a variety of frequency ranges and observing periods (cycle 21, 22, and 23, and solar maximum vs. solar minimum), from which one can estimate the expected number of bursts per day above any given flux density. These results are largely replicated by Giersch et al. (2017) from

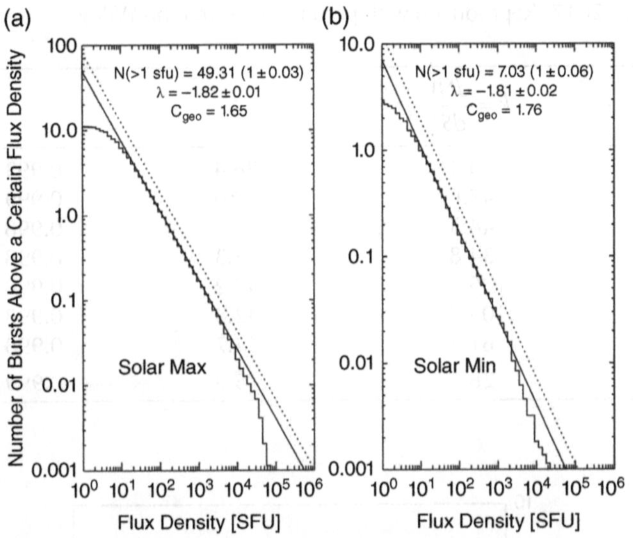

Figure 6.2 Cumulative number of bursts per day at frequencies above 2 GHz (a) during solar maximum years and (b) during solar minimum. The histograms show the actual, observed distribution. The solid line shows the best power-law fit to the observed distribution, and the dotted line shows the fit after application of the geographical correction factor C_{geo}, which is 1.76 during solar maximum, and 1.65 during solar minimum epochs. (Source: Nita et al., 2002. Reproduced with permission of IOP Science.)

their study of a subset of RSTN events, extended to include cycle 24.

However, the NOAA database, dominated by the RSTN measurements that make up the bulk of the burst reports, has only crude frequency coverage (the eight RSTN frequencies are listed in Table 6.1). As already noted in section 6.2.2, the Nita et al. (2004) paper based on higher-frequency-resolution data from the Owens Valley Solar Array (OVSA; 40 frequencies in the range 1–18 GHz) demonstrates that there is a qualitative difference in character of bursts in the decimetric range (specifically 1–2.6 GHz), where the prevalence of coherent emission processes increases the probability of extreme events. This can also be seen in the results of Giersch et al. (2017) in the form of flatter power law indexes (around −1.8) below 2695 MHz, as shown in Table 6.1, adapted from their Table 7.

Although the power law indexes in Table 6.1 may seem similar, the difference $\Delta\lambda \sim 0.2$ in power law indexes at 1415 MHz and 2695 MHz result in more than an order of magnitude difference in occurrence rate when extrapolated to 10^6 sfu.

6.3.4. Extreme Events

The distributions of Figure 6.2 do not end abruptly at 10^5 sfu, as expected for saturation in the form of truncation, but rather trail off starting at flux densities as low as 20,000 sfu. In addition, there are *no* events in the database above the 10^5 sfu saturation level. However, the NoRP database does include several bursts above 10^5 sfu, and likewise Gary (2008) reports that OVSA observed two events in December 2006 that exceed 10^5 sfu (one of which reached the record 10^6 sfu). Additionally, Gary (2008) notes that these two bursts were in the RSTN database, but with reported flux densities more than 2 orders of magnitude lower.

These spot checks from other observatory sites demonstrate that the NOAA/RSTN database is not reliable for extreme events and suggests that the dearth of large events in Figure 6.2 is an artifact of the database. This is particularly problematic for assessing the threat of solar radio bursts on technological systems. Another problem with the database is the relatively large gaps in available frequency coverage (Table 6.1). Spectrometer records show that the flux density can vary greatly and abruptly with frequency, so for example, a relatively minor burst at the RSTN 1415 MHz frequency may be extremely strong at the nearby L1 (1575 MHz) or L2 (1228 MHz) frequencies. This was indeed the case for the event described by Marqué et al. (2018), which we detail in the next section.

Taking the extrapolation to extreme events from Figure 6.2 at face value, we would estimate that events of order 10^6 sfu should occur once every 1000 days during solar maximum years, which is to say about twice during the 4 maximum years of a solar cycle. However, events tend to cluster in time, like the events of December 2006, with multiple bursts being produced by a single, particularly productive active region. Given the small-number statistics and the obvious gaps in the historical record, we must conclude that we do not currently have reliable statistics on the rate of occurrence of extreme events, or a reliable limit on how extreme they can get. However, the fact that extreme events do occur and do have important deleterious effects on technological systems is amply shown by a number of high-profile examples that we discuss in the next section.

6.4. NOTABLE SOLAR RADIO BURST IMPACTS REPORTED IN THE LITERATURE

6.4.1. The May 1967 Great Storm and the Cold War

This story, related in detail by Knipp et al. (2016), is notable not only for the direct interference of a solar radio burst on the US Department of Defense Ballistic Missile Early Warning System (BMEWS) radars, but also for the response of the Strategic Air Command and North American Air Defense Command (NORAD) to the disturbance, which was interpreted initially as jamming. As stated by Knipp et al., "Cold War military

Table 6.1 Burst PDF regression parameters. (Source: Giersch et al., 2017. Reproduced with permission from John Wiley & Sons.)

Frequency (MHz)	λ	σ_λ	$k = \frac{dN}{dS}\Big\vert_{1\,sfu}$	σ_k	r
245	−2.032	0.076	119.4	86.4	0.993
410	−1.843	0.022	47.6	8.0	0.999
610	−1.875	0.037	44.5	17.6	0.998
1415	−1.815	0.041	34.8	12.3	0.998
2695	−2.011	0.054	106.6	47.2	0.998
4995	−2.010	0.049	105.7	42.4	0.998
8800	−1.913	0.049	61.2	24.7	0.998
15400	−1.779	0.023	28.2	5.3	0.999

commanders viewed full scale jamming of surveillance radars as a potential act of war." Contributing to this conclusion was the fact that the Sun was visible to all three BMEWS radars, situated in Alaska, Greenland, and England, with the Thule, Greenland, radar being particularly well aligned with the Sun.

Of several strong radio bursts that occurred in May 1967, the strongest occurred on May 23 and reached peak flux densities of > 20000 sfu at 8800 MHz, 85100 sfu at 1.415 GHz, and 3.7×10^5 sfu at 606 MHz. Although no monitoring was done at the 440 MHz operating frequency of BMEWS, it is highly likely that extreme flux density was also present on that frequency. As related by Citrone (1995), the Air Weather Service "notified NORAD in real time of the event and [its potential] impacts," but "outside agencies made uninformed decisions without considering the drastic impacts the event imparted to NORAD's early warning system."

Exacerbating the danger of scrambling the Strategic Air Command bombers in response to a perceived threat, as Knipp et al. (2016) point out, high-frequency–ultra-high-frequency (HF-UHF) communications were simultaneously impacted by solar energetic particle deposition (polar cap absorption) and the ionospheric disturbance due to X-rays from a series of flares during the period. They state that "during the politically tense days of late May 1967 a full out aircraft launch by western forces could have been very provocative and, just as importantly, difficult to recall in the greatly challenged HF-UHF radio environment." Fortunately, Col C. K. Anderson and his NORAD solar forecasting staff were able to convince the military command that the jamming was due to the at-the-time unprecedented effects of interference by the solar radio burst.

6.4.2. A Reported Impact on a Cellular Telephone System

Cellular telephone companies are highly guarded about performance data to protect their commercial interests.

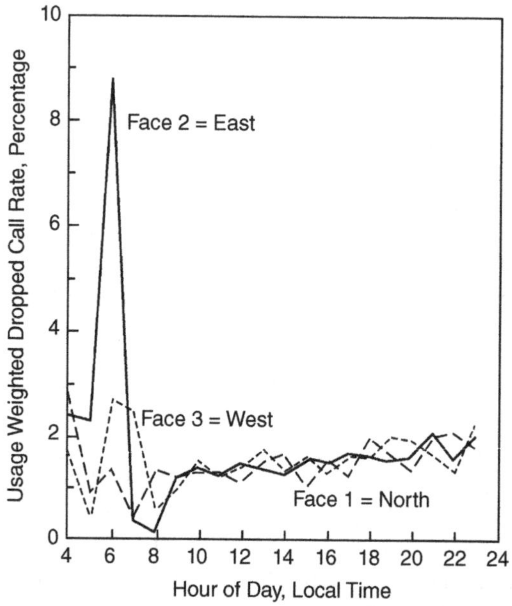

Figure 6.3 The percentages (on an hourly basis) of dropped calls averaged over all base stations in a cellular system within a major US state as a function of local time during one day in 1996. There was solar flare activity during early local morning as the Sun was rising at the cell sites. The east-facing antennas clearly suffered very high dropped rates just after sunrise in comparison to the other two antenna directions shown. (Source: Lanzerotti et al., 1999. Reproduced with permission of John Wiley & Sons.)

To our knowledge, the only published report of a direct impact on a cellular telephone system that can be plausibly associated with a solar flare is that of Lanzerotti et al. (1999). That paper reports an incident where an east-facing antenna of a cellular base station experienced an enhanced number of dropped calls (Figure 6.3) relative to antennas facing other directions, at a time when the Sun was flaring near local sunrise. Attempts have been made to extend the study of this phenomenon by seeking additional data from cell phone companies, but to date these have been unsuccessful.

Although this incident is anecdotal and lacks specificity compared to the much-better-established case for GNSS impacts, it is plausible that cellular base stations can be impacted by solar radio bursts during times near sunrise and sunset when the Sun is in the beam of a base station antenna (Bala et al., 2002). Cell phones often operate near the edges of coverage areas, where the signal-to-noise ratio is about 10. A doubling of the noise level due to a solar radio burst can cause the bit error rate to rise by 2 orders of magnitude (Proakis & Manolakis, 2006). Although bit error correction in modern smartphones can mitigate the impact on dropped calls, it is not unreasonable that burst flux densities as low as 1000 sfu can have a significant impact (Gary et al., 2005), while flux densities of order 10^6 sfu would unavoidably cause outages. When one considers the widespread prevalence of cellular telephone networks, the fact that it is always sunrise or sunset somewhere in the world suggests that widespread regional outages must have occurred before now and gone unreported. This is perhaps because only a fraction of customers (those near the edges of coverage areas) would be impacted.

6.4.3. The 6 December 2006 Solar Radio Burst

As already mentioned, this event is the largest reported so far at decimeter wavelengths, reaching $>10^6$ sfu (Gary, 2008). This event was documented by Cerruti et al. (2008) to cause widespread effects on the Global GPS Network semi-codeless dual-frequency (L1 and L2) receivers operated by NASA's Jet Propulsion Laboratory, which uses them to run a real-time differential timing service, with corrections sent to users to improve the GPS signals of their own receivers. The solar radio burst was equally strong at both L1 and L2 frequencies and was nearly 100% polarized in the same RCP sense as used by the GPS satellites. The radio burst affected all receivers of the global network on the sunlit hemisphere of Earth, with a large fraction of them affected to the extent that they could track fewer than four satellites, and hence failed to attain a full positioning solution. The outage was instantaneous and simultaneous over the entire network and lasted for more than 10 minutes. Cerruti et al. (2008) also found that the radio burst had similar impacts and loss of positioning capability on the International GNSS Service and Continuously Operating Reference System networks.

Cerruti et al. (2008) also examined the effect of the burst on the Wide-Area Augmentation System (WAAS) network, which is an accurate and safe navigation system designed by the US Federal Aviation Administration and the Department of Transportation for civil air navigation. WAAS is certified in safety of life applications and is used to provide en route and guided approach navigation to within 200 feet above a runway. In 2011, the equivalent European Geostationary Navigation Overlay Service was certified (see e.g. Grzegorzewski et al., 2012), and similar services are being adopted elsewhere. Cerruti et al. (2008) found that although the radio burst was detected by the WAAS system, its effects were less intense than on other operational systems because of its robust system design. Although WAAS receivers did experience anomalous loss of lock with GPS satellites, it maintained contact with enough satellites to provide basic navigation services. Still, Cerruti et al. caution that still larger solar radio burst flux densities are possible, and under more extreme conditions the WAAS and other robust receivers still face the prospect of widespread failure.

We remarked earlier that extreme events tend to cluster in time due to the appearance of particularly active sunspot regions on the Sun, and the December 2006 region is no exception (Gary, 2008). Strong solar radio bursts occurred on 13 December, with Afraimovich et al. (2009) discussing its affects to the Asia-Pacific region, and on 14 December again affecting the Americas and discussed by Carrano et al. (2009). Both were strongly RCP polarized, in common with the 6 December event, suggesting that the same general active region structure was maintained over 7–8 days. One interesting aspect of the 14 December event documented by Carrano et al. (2009) is that the burst was much stronger (10^5 sfu) at the L1 frequency than at the L2 frequency ($< 10^4$ sfu), and accordingly the GPS receiver signal fade was much stronger at L1.

6.4.4. Solar Radio Burst Effects on Air Traffic Control Radar

Marqué et al. (2018) have recently published a fascinating account of a solar radio burst on 4 November 2015 that strongly disturbed secondary air traffic control radar in Sweden and caused milder disturbances elsewhere in Europe. While primary radars provide information about azimuth and distance of aircraft, secondary radars send coded queries to transponders, which return additional information such as aircraft identification and altitude (older mode A/C radars), and for newer mode-S radars, some additional information. Both operate at 1030 MHz for uplink and 1090 MHz for downlink.

As related by Marqué et al. (2018), during the incident secondary A/C radar in Belgium gave false echoes representing non-existent aircraft on azimuths in the direction of the Sun, but these were automatically filtered and did not cause traffic interruption. In the southern part of Sweden, however, the secondary radar at several sites could not display proper information, which led the authorities to reduce air traffic for safety reasons. The disruptions first occurred between 14:19 and 14:34 UT, and again from 14:47 to 14:50 UT. Other air traffic–related incidents also seem to have occurred in Norway and Greenland.

The associated solar flare was not exceptionally large, being a modest M3.7 in soft X-rays, and seems unremarkable in radio bands monitored by RSTN. The flux density at 1415 MHz frequency was reported as 5800 sfu. However, three solar spectrometers in Europe, the Humain Solar Radio Spectrograph (HSRS) in Belgium, the Observations Radio pour FEDOME et l'Étude des Éruptions Solaires (ORFEES) in France, and a Callisto spectrometer in Switzerland, were operating at or near the secondary radar frequencies. Although separately calibrated, all three spectrometers agree that the solar radio burst exceeded about 10^5 sfu at 1060 MHz (the mean of the radar uplink and downlink frequencies) during two periods, as shown in Figure 6.4. These two time periods agree remarkably well with the times of disruption of the air surveillance radars. This highlights the importance of monitoring not only a few frequencies, as is currently done by the RSTN network; rather, a dense frequency coverage is needed, as provided by the spectrometers. It also shows that soft X-ray class or other nonradio classifications is not a reliable guide to the potential of a given flare to disrupt wireless systems.

Marqué et al. (2018) examine the effect of a 10^5 sfu solar burst in the beam of a radar relative to a conservative estimate of the typical interference threshold and find that the radio interference would have been about 34 dB (a factor of 2500) above the threshold. It is therefore not surprising that the radars fail when pointed at the flaring Sun. However, disruptions to air traffic control need not have occurred if the air traffic controllers and the systems they use had been aware of the solar radio environment in real time.

Marqué et al. discuss the mechanism responsible for the strong solar radio emission, noting that the burst is a typical type IV burst with several well-known types of fine structure such as that shown in Figure 6.1. With the 250 ms time resolution of the HSRS, though, it was not possible to resolve individual spikes, and therefore the emission could not definitively be associated with the ECME mechanism. The authors point out that the timing of the abrupt onset of flux density enhancement at 14:15 UT in Figure 6.4 corresponds to expansion of a flare ribbon through the umbra of one of the sunspots in the region, which could lend support to the ECME mechanism, but such a link could not be confirmed.

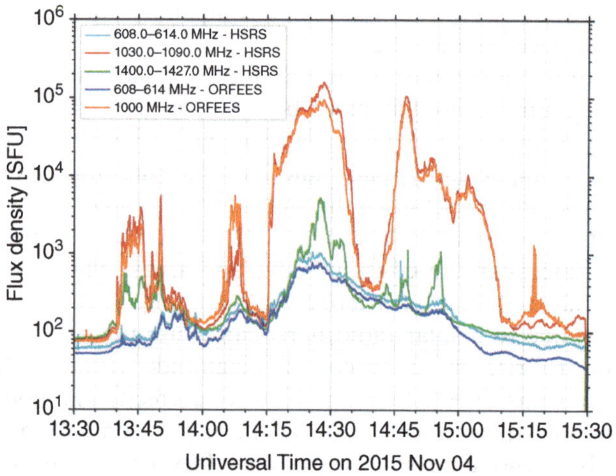

Figure 6.4 Radio light curves in sfu of the 4 November 2015 event from the HSRS and ORFEES spectrometers. (Source: Marqué, https://www.swsc-journal.org/articles/swsc/abs/2018/01/swsc170090/swsc170090.html. Licensed Under CC BY SA 4.0)

6.5. CONCLUSIONS

The four example events of the previous section clearly demonstrate that a broad range of modern-day technological systems based on wireless transmission and reception are threatened by strong radio emission from the Sun. Reports of other less severe events appear in the literature, and no doubt many similar events have gone unreported, either not recognized as being caused by the Sun or suppressed for commercial or security reasons. The operational threat to navigational and communication systems due to solar radio bursts is qualitatively different from other potential threats because of several factors: (1) they occur suddenly and without warning, and sometimes out of proportion with associated emission in other monitored wavelengths; (2) they simultaneously affect large geographic regions, making their effects systemic; and (3) when bursts of sufficiently large flux density occur, they have the potential to cause not just degradation, but complete operational failure, especially in the case of GNSS.

The extent to which outages affect society depends on the uses of the threatened systems. Use of GNSS for safety of life situations is ever increasing. In addition to their role in navigation and landing of aircraft (Cerruti, 2008), plans are underway to use GNSS in autonomous (driverless) automobiles (although aided by alternative signals, e.g. Morales, Royson & Kassas, 2016) and vital tasks of emergency services (Hapgood, 2017). Hapgood notes that even when systems such as driverless cars are designed to be resilient to loss of GPS, it is not clear that the case of sudden and simultaneous loss of GPS (potentially for every such vehicle in an entire nation) is being considered in design and testing.

The existence of the threat of radio frequency interference (RFI) from solar radio bursts is by now widely recognized, and several areas have been suggested where more effort is needed in response. Given the sudden onset of the RFI, and the lack of proportionality of extreme radio emission to other forms of solar activity, as well as the highly variable flux density in frequency, time, and sense of polarization, it is essential that dual polarization radio monitoring of the Sun with high frequency and time resolution be available 24/7, covering the entire frequency range of interest. No such system exists at the present time. Since terrestrial weather has little effect on microwave to meter-wave frequencies, ground-based monitoring at 3–4 sites around the globe are sufficient to meet the need. The monitoring systems must, however, be capable of reasonably accurate flux density measurements over a wide range of flux densities without suffering saturation. The ability to locate the radio emission relative to the solar disk (e.g. Dougherty et al., 2000) is not needed for this monitoring purpose. However, high-resolution radio imaging observations with similarly high frequency and time resolution are needed to improve our understanding of the solar phenomena that give rise to extreme events, which ultimately may lead to an ability to forecast their occurrence.

From the system design perspective, certain steps can be taken to minimize the risk, such as (1) the ability to quickly change operating frequency or polarization to avoid RFI in near real time, or redundant systems that operate at more than one well-separated frequency; (2) designing systems to operate in frequency ranges less prone to interference (e.g. above about 2.8 GHz); (3) designing systems with increased dynamic range so that they can operate through most solar bursts; and (4) designing systems with directional antennas to avoid the Sun. How practical these steps might be for a given system depends on the size, cost, and complexity of the system, and on the consequences of a failure. It is equally important to design and test the system to cope gracefully and safely with widespread failure. This can be greatly helped by the availability of a robust, real-time monitoring system of the type mentioned in the previous paragraph.

In conclusion, this chapter has presented what is known about how extreme solar radio bursts are generated in the Sun, how often they occur, and what their potential impacts are on a wide variety of wireless communication and navigation systems. The currently available database of solar radio events has some limitations that compromise our knowledge of how often such extreme events occur, and how strong they can be. However, documented cases of actual, serious impacts on modern-day technological systems show that such bursts represent a real threat. In today's increasingly interconnected and automated world, it is more important than ever that design engineers be cognizant of the threat, and that systems be designed and tested to cope with future global effects.

ACKNOWLEDGMENTS

DG acknowledges support for this work through NSF grant AST-1910354 and NASA grant 80NSSC18K1128 to New Jersey Institute of Technology. The National Radio Astronomy Observatory is a facility of the National Science Foundation operated under cooperative agreement by Associated Universities, Inc.

REFERENCES

Afraimovich, E. L., Demyanov, V. V., & Smolkov, G. Y. (2009). The total failures of GPS functioning caused by the powerful solar radio burst on December 13, 2006. *Earth, Planets, and Space*, *61*, 637–641. doi:10.1186/BF03352940

Bala, B., Lanzerotti, L. J., Gary, D. E., & Thomson, D. J. (2002). Noise in wireless systems produced by solar radio bursts. *Radio Science*, *37*, 1018. doi:3010.1029/2001RS002481

Benz, A. O. (1986). Millisecond radio spikes. *Solar Physics*, *104*, 99–110. doi:10.1007/BF00159950

Carrano, C. S., Bridgwood, C. T., & Groves, K. M. (2009). Impacts of the December 2006 solar radio bursts on the performance of GPS. *Radio Science*, *44*, RS0A25. doi:10.1029/2008RS004071

Cerruti, A. P., Kintner, P. M., Gary, D. E., Lanzerotti, L. J., de Paula, E. R., & Vo, H. B. (2006). Observed solar radio burst effects on GPS/Wide Area Augmentation System carrier-to-noise ratio. *Space Weather*, *4*, S10006. doi:3910.1029/2006SW000254

Cerruti, A. P., Kintner, P. M., Gary, D. E., Mannucci, A. J., Meyer, R. F., Doherty, P., & Coster, A. J. (2008). Effect of intense December 2006 solar radio bursts on GPS receivers. *Space Weather*, *6*, S10D07. doi:4310.1029/2007SW000375

Chen, B., Bastian, T. S., Gary, D. E., Jing, J. (2011). Spatially and spectrally resolved observations of a zebra pattern in a solar decimetric radio burst. *Astrophysical Journal*, *736*, 64. doi: 10.1088/0004-637X/736/1/64

Chen, B., Bastian, T. S., Shen, C., Gary, D. E., Krucker, S., & Glesener, L. (2015). Particle acceleration by a solar flare termination shock. *Science*, *350*, 1238–1242. doi:10.1126/science.aac8467

Citrone, P. J. (1995). Paper session 1-B—USAF space weather support. In *Proceedings: People and technology—the case for space*. The Space Congress. (Paper 20, http://commons.erau.edu/space-congress-proceedings/proceedings-1995-32nd/april-25-1995/20)

Cliver, E. W., White, S. M., & Balasubramaniam, K. S. (2011). The solar decimetric spike burst of 2006 December 6: Possible evidence for field-aligned potential drops in post-eruption loops. *Astrophysical Journal*, *743*, 145. doi: 10.1088/0004-637X/743/2/145

Dougherty, B. L., Freely, W. B., Zirin, H., Gary, D. E., & Hurford, G. J. (2000). Solar radio burst locator. In proc. High Energy Solar Physics Workshop - Anticipating Hessi, *Astronomical Society of the Pacific Conference Series*, *205*, 367.

Dulk, G. A. (1985). Radio emission from the sun and stars. *Annual Reviews of Astronomy and Astrophysics*, *23*, 169–224. doi:1056.1146/annurev.aa.23.090185.001125

Ergun, R. E., Carlson, C. W., McFadden, J. P., Delory, G. T., Strangeway, R. J., & Pritchett, P. L. (2000). Electron-cyclotron maser driven by charged-particle acceleration from magnetic field-aligned electric fields. *Astrophysical Journal, 538*, 456–466. doi: 10.1086/309094

Fleishman, G. D., & Melnikov, V. F. (1998). Reviews of topical problems: Millisecond solar radio spikes. *Physics Uspekhi, 41*, 1157–1189. doi:10.1070/PU1998v041n12ABEH000510

Gary, D. E. (2008). Cause and extent of the extreme radio flux density reached by the solar flare of 2006 December 06. In J. M. Goodman (Ed.), *2008 ionospheric effects symposium*. JMG Associates Ltd., Sheridan Books (accession No. PB2008-112709). arXiv:1901.09262 (2019).

Gary, D. E., Lanzerotti, L. J, Nita, G. M., & Thomson, D. J. (2005). Effects of solar radio bursts on wireless systems. In I. Daglis (Ed.), *Effects of space weather on technology infrastructure* (vol. *176*, pp. 203–213). Netherlands: Kluwer Acad.

Giersch, O. D., Kennewell, J., & Lynch, M. (2017). Solar radio burst statistics and implications for space weather effects. *Space Weather, 15*, 1511–1522. doi:10.1002/2017SW001658

Grzegorzewski, M., Swiatek, A., Oszczak, S., Ciecko, A., & Cwiklak, J. (2012). Study of EGNOS safety of life service during the period of solar maximum activity. *Artificial Satellites, 47*, 137–145. doi:10.2478/v10018-012-0019-5

Hapgood, M. (2017). Satellite navigation—Amazing technology but insidious risk: Why everyone needs to understand space weather. *Space Weather, 15*, 545–548. doi: 10.1002/2017SW001638

Hey, J. S. (1942). Solar radiations in the 4–6 metre radio wavelength band on 27th and 18th February 1942. *Army Operations Research Group Report*, 275.

Hey, J. S. (1946). Solar radiations in the 4–6 metre radio wavelength band. *Nature, 157*, 47–48. doi:10.1038/157047b0

Isliker, H., & Benz, A. O. (1994). Catalogue of 1–3 GHz solar flare radio emission. *Astronomy and Astrophysics Supplement Series, 104*, 145–160.

Klobuchar, J. A., Kunches, J. M., & VanDierendonck, A. J. (1999). Eye on the ionosphere: Potential solar radio burst effects on GPS signal to noise. *GPS Solutions, 3*, 69–71. doi:10.1007/PL00012794

Knipp, D. J., Ramsay, A. C., Beard, E. D., Boright, A. L., Cade, W. B., Hewins, I. M., et al. (2016). The May 1967 great storm and radio disruption event: Extreme space weather and extraordinary responses. *Space Weather, 14*, 614–633. doi:10.1002/2016SW001423

Koskinen, H. E. J., Baker, D. N., Balogh, A., Gombosi, T., Veronig, A., & von Steiger, R. (2017). Achievements and challenges in the science of space weather. *Space Science Reviews, 212*, 1137–1157. doi:8710.1007/s11214-017-0390-4

Lanzerotti, L. J., Thomson, D. J., & Maclennan, C. G. (1999). Engineering issues in space weather. In M. A. Stuchly (Ed.), *Modern radio science* (pp. 25–50). New York: Wiley/IEEE.

Marqué, C., Klein, K.-L., Monstein, C., Opgenoorth, H., Pulkkinen, A., Buchert, S., et al. (2018). Solar radio emission as a disturbance of aeronautical radionavigation. *Journal of Space Weather and Space Climate, 8*(27), A42. doi:10.1051/swsc/2018029

Melrose, D. B. (2017). Coherent emission mechanisms in astrophysical plasmas. *Reviews of Modern Plasma Physics, 1*, 5. doi:10.1007/s41614-017-0007-0

Melrose, D. B., & Dulk, G. A. (1982). Electron-cyclotron masers as the source of certain solar and stellar radio bursts. *Astrophysical Journal, 259*, 844–858. doi:10.1086/160219

Melrose, D. B., & Wheatland, M. S. (2016). Is cyclotron maser emission in solar flares driven by a horseshoe distribution? *Solar Physics, 291*, 3637–3658. doi:10.1007/s11207-016-1006-y

Morales, J., Roysdon, P., & Kassas, Z. (*2016*) Signals of opportunity aided inertial navigation. In *Proc. ION GNSS Conf.*, Sept. 2016, 1492–1501.

Muhammad, B., Alberti, V., Marassi, A., Cianca, E., & Messerotti, M. (2015). Performance assessment of GPS receivers during the September 24, 2011 solar radio burst event. *Journal of Space Weather and Space Climate, 5*, A32. doi:10.1051/swsc/2015034

Nakajima, H., Sekiguchi, H., Sawa, M., Kai, K., & Kawashima, S. (1985). Theradiometer and polarimeters at 80, 35, and 17 GHz for solar observations at Nobeyama. *Astronomical Society of Japan, 37*, 163–170.

Nita, G. M., Gary, D. E., Lanzerotti, L. J., & Thomson, D. J. (2002). The peak flux distribution of solar radio bursts. *Astrophysical Journal, 570*, 423–438. doi:10.1086/339577

Nita, G. M., Gary, D. E., & Lee, J. (2004). Statistical study of two years of solar flare radio spectra obtained with the Owens Valley solar array. *Astrophysical Journal, 605*, 528–545. doi:10.1086/382219

Nordmann, C. (1902). Recherche sur le rôle des ondes hertziennes en astrophysique. *Revue Générale des Sciences, 13*, 379–388.

Proakis, J. G., & Manolakis, D. G. (2006). *Digital signal processing: Principles, algorithms, and applications*. Pearson; 4 edition.

Reber, G. (1944). Cosmic static. *Astrophysical Journal, 100*, 279.

Southworth, G. C. (1942a). Evidence for microwave radiation in the Sun's spectrum, 14 July 1942 memo-for-file MM-42-160-73.

Southworth, G. C. (1942b). Evidence for microwave components in the Sun's spectrum, 20 November 1942 memo-for-file MM-42-160-140.

Southworth, G. C. (1945). Microwave radiation from the Sun. *J. Franklin Inst., 239*, 285.

Sreeja, V., Aquino, M., & de Jong, K. (2013). Impact of the 24 September 2011 solar radio burst on the performance of GNSS receivers. *Space Weather, 11*, 306–312. doi: 10.1002/swe.20057

Sreeja, V., Aquino, M., de Jong, K. & Visser, H. (2014). Effect of the 24 September 2011 solar radio burst on precise point positioning service. *Space Weather, 12*, 143–147. doi:10.1002/2013SW001011

Sullivan III, W. T. (2009). *Cosmic noise: A history of early radio astronomy*. Cambridge University Press.

Torii, C., Tsukiji, Y., Kobayashi, S., Yoshimi, N., Tanaka, H., & Enome, S. (1979). Full-automatic radiopolarimeters for solar patrol at microwave frequencies. *Proceedings of the Research Institute of Atmospherics, Nagoya University, 26*, 129–132.

Treumann, R. (2006). The electron-cyclotron maser for astrophysical application. *Astronomy and Astrophysics Review, 13*, 229–315. doi: 10.1007/s00159-006-0001-y

Winglee, R. M., & Dulk, G. A. (1986). The electron-cyclotron maser instability as a source of plasma radiation. *Astrophysical Journal, 307*, 808–819. doi: 10.1086/164467

Wu, C. S., & Lee, L. C. (1979). A theory of the terrestrial kilometric radiation. *Astrophysical Journal, 230*, 621. doi: 10.1086/157120

7

Space Weather Influences on HF, UHF, and VHF Radio Propagation

Gary S. Bust[1], William Liles[2], and Cathryn Mitchell[3]

ABSTRACT

This Chapter considers space weather effects of the ionosphere on the propagation of radio frequencies classed by the International Telecommunications Union as HF (3-30 MHz), VHF (30-300 MHz) and UHF (300-3000 MHz). The ionosphere and its variation is important for these radio frequencies. The ionosphere is the part of the upper atmosphere above 70-80 km altitude, where the density of the neutral atmosphere is low and the ionisation is significant enough to affect radio propagation. Even though the amount of ionisation is still below 1% of the neutrals, it is sufficient to exert an important effect on radio signals. In this Chapter the ionosphere is introduced, then a description of the theory of radio propagation at HF, VHF and UHF is given. The effects of solar storms and space weather on the ionosphere and hence on radio propagation are outlined. Finally, anomalous propagation conditions and outstanding problems and questions are also discussed.

7.1. INTRODUCTION

Space weather predominately affects the propagation of radio frequencies classed by the International Telecommunications Union as HF (3–30 MHz), VHF (30–300 MHz) and UHF (300–3000 MHz). Most of these effects are imparted through the ionosphere, and therefore its variations are important for these radio frequencies. Ionization of electron density exerts an important effect on radio signals, even though a very small fraction of neutral densities are ionized (below 1%). The ionosphere impacts radio propagation at upper atmosphere altitudes extending from 70–80 km to the beginning of the plasmasphere at ~ 1000 km.

[1] *Johns Hopkins University Applied Physics Laboratory, Laurel, Maryland, USA*

[2] *HamSCI Community, Virginia, USA*

[3] *Department of Electronic and Electrical Engineering, University of Bath, Bath, UK*

In this chapter the ionosphere is introduced, then a description of the theory of radio propagation at HF, VHF, and UHF is given. The effects of solar storms and space weather on the ionosphere and hence on radio propagation are outlined. Finally, anomalous propagation conditions and outstanding problems and questions are also discussed.

7.2. THE IONOSPHERE

The ionosphere is ionized mainly by solar extreme ultraviolet radiating onto the tenuous outer layers of our planet's atmosphere. The ionosphere has a vertical profile of free electron density, that is the ionization, with regions or layers named as D, E, F1, and F2 (Silberstein, 1959). The D region is the lowest altitude layer at around 70–90 km, the E region at 90–150 km, and the F layer at 150 to over 400 km, which is divided into the lower (F1) and the upper (F2) layer (Standards on wave propagation, 1950; IEEE, 2018). Above the F layer is the topside ionosphere that has a gradual decrease in electron density up to the next region, the plasmasphere.

Space Physics and Aeronomy Collection Volume 5: Space Weather Effects and Applications, Geophysical Monograph 262, First Edition.
Edited by Anthea J. Coster, Philip J. Erickson, and Louis J. Lanzerotti.

DOI: 10.1002/9781119815600.ch7

The global scale distribution of the ionosphere can be broadly considered as four regions that are categorized by their geomagnetic latitude. The low-latitude ionosphere is the region around the geomagnetic equator. Approximately 15 degrees of latitude either side of the geomagnetic equator are the peaks of the equatorial anomaly where the electron density is at its highest (Appleton, 1947, 1954). Moving poleward from this region is the midlatitude ionosphere, characterized by less dramatic gradients and sometimes populated with waves known as traveling ionospheric disturbances (Munro, 1950) and a nightside large-scale depletion known as the main trough. At higher latitudes the ionosphere is characterized by the auroral oval with particle precipitation ionization that can extend into the F region, forming structures known as boundary blobs (Rino et al., 1983; Vickrey et al., 1980; Weber et al., 1984) and also into the E region, producing enhancements known as auroral E (CCIR, 1986). Further poleward still, the ionosphere is influenced by the solar wind and in particular the large-scale polar electric field. This latter field in the form of ExB drift can drive ionization from the mid-latitudes into the polar cap, resulting in convecting patches of enhanced electron density (Weber et al., 1984).

In addition to the geographical variation, the ionosphere varies with time of day, time of year, and the solar activity. The ionosphere usually has a higher electron density in the daytime than at night because it is illuminated by the Sun in the day, generally producing a higher density when the solar zenith angle is higher. Broadly speaking, the ionosphere is also higher in density at solar maximum than at solar minimum and is higher in density in January than in July. The main reason for the higher density at solar maximum is the increase in the solar radiation and hence photo-ionization. Part of the explanation for the higher overall ionospheric density in January is the orbital ellipticity of the Earth around the Sun: the Earth is closer to the Sun in January, causing a rise in neutral temperature and therefore an increase in neutral atmospheric constituents (the ionospheric source) at a given altitude. There are also important seasonal variations, and of note is the winter anomaly – the ionospheric density at mid-latitudes is higher in the winter than in summer. An overview of the ionosphere is given in Hargreaves (1992).

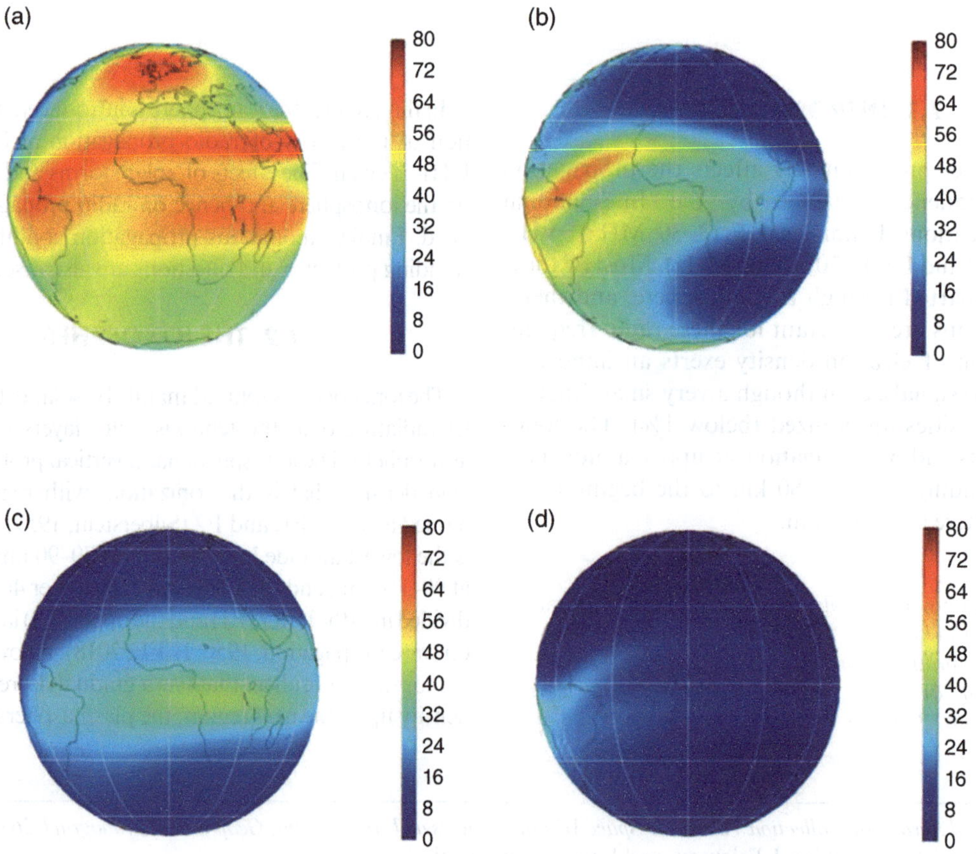

Figure 7.1 TEC from the International Reference Ionosphere that illustrates different characteristic integrated ionospheric density values during (a) solar maximum day, (b) solar maximum night, (c) solar minimum day, and (d) solar minimum night. (Source: Bilitza, 2014.)

The ionosphere can be characterized by a number of parameters, but a particularly important one is a measurement of the integrated electron density, known as the ionospheric total electron content (TEC). Figure 7.1 shows a global view of the vertical TEC produced using a statistical ionospheric model known as the International Reference Ionosphere (Bilitza, 2014). This model uses a proxy for the solar radiance, given by a sunspot number or by the 10.7 cm radio flux from the Sun. These are used as model drivers because the EUV spectrum that is incident on the ionosphere does not penetrate the atmosphere and hence is not measured on the Earth's surface. Figure 7.1 illustrates the different values of TEC at solar maximum and minimum and at day and night (in January). The images reveal the global importance of the high values of TEC in the equatorial anomaly and the variability between day and night and across the solar cycle. It is also apparent that the ionosphere has significant geographic variations. For example, the enhancement in ionospheric density over the northern hemisphere in winter (the winter anomaly) is captured even in this statistical model. However, short-term variability in the ionosphere is not captured in this type of climatological model, and actual measurements reveal significant variability and deviations from a model such as this over a number of time scales. These ionospheric changes on time scales of seconds to minutes are very important factors when interpreting the effects on radio systems. The physical processes that cause some of the short-term variations are discussed in the later sections of this chapter.

7.3. RADIO PROPAGATION THROUGH AN IONIZED MEDIUM

The theory of electromagnetic propagation is based upon Maxwell's equations, which are described in various basic texts on electromagnetics. These equations describe the full set of electric and magnetic interactions and propagation. However, except for fairly simple cases, they are difficult to solve exactly, even with state-of-the-art numerical methods. For many applications and frequencies, an approximation known as ray theory can be used instead of the full set of equations (Budden, 1961). With ray theory, back scatter and forward scatter are neglected, and the interest is primarily in following the characteristics of the electromagnetic wave as it propagates through a smooth, slowly varying medium.

Ray theory has been implemented many times in numerical ray-trace algorithms, which trace out the path of the primary energy of the electromagnetic wave as it propagates through a complicated medium such as the ionosphere. Ray-trace algorithms compute the path of the propagation, the time of flight, the phase change, and other characteristics of the propagation. In order to use ray-trace methods, it is necessary to have a full description of the phase refractive index everywhere along the path of the wave. The formula for the phase refractive index is given by the equation created and named by Mary Taylor (1933), the Appleton-Hartree equation:

$$\mu^2 = 1 - \frac{X}{1 - jZ - \left[\frac{Y_T^2}{2(1-X-jZ)}\right] \pm \left[\frac{Y_T^4}{4(1-X-jZ)^2} + Y_L^2\right]^{\frac{1}{2}}}, \tag{7.1}$$

where $X = \omega_N^2 / \omega^2$, $Y = \omega_B/\omega$, $Y_L = \omega_L/\omega$, $Y_T = \omega_T/\omega$, $Z = \nu/\omega$.

The equation relates the refractive index of the medium to the frequency of a particular radio transmission through it, ω, where ω_N is the angular plasma frequency, ω_B represents the electron gyrofrequency, and ω_L and ω_T are its longitudinal and transverse components such that $\omega_L = \omega_B \cos\theta$ and $\omega_T = \omega_B \sin\theta$, where θ is the angle between the geomagnetic field and the direction of propagation. The Z term is dissipative as it contains ν, the electron collision frequency.

The plasma frequency is given by

$$\omega_N = \sqrt{\frac{Ne^2}{\varepsilon_0 m}}, \tag{7.2}$$

where N is the electron density, e the electron charge, ε_0 the permittivity of free space, and m the particle mass (electron mass in this case). Cold electrons are assumed.

The Appleton-Hartree equation can be used most easily by looking at specific cases. If absorption of the radio wave is negligible, we can set $Z = 0$, and if we choose to ignore the magnetic field of the Earth, then $Y = 0$. This allows us to use the following simplified formula for the phase refractive index:

$$\mu^2 = 1 - X = 1 - \frac{\omega_N^2}{\omega^2} = 1 - k\frac{N}{\omega^2}. \tag{7.3}$$

This is a useful form of the Appleton-Hartree equation because it allows a calculation of the trajectory of a ray path in accordance with the electron density along the path, which relates to the plasma frequency through the equation above. For cases of higher radio frequency where the ray bending can be neglected, the phase advance and the group delay of the signal are of primary interest.

Binomial expansion then allows an approximation for the phase refractive index as:

$$\mu = 1 - k\frac{N}{2\omega^2}, \tag{7.4}$$

where the constant k is given by $k = \frac{e^2}{4\pi\varepsilon_0 m} = 80.6\ \text{m}^3\ \text{Hz}^2$.

Therefore,

$$\mu = 1 - 40.3\frac{N}{\omega^2}. \tag{7.5}$$

This resulting approximation is particularly straightforward because, for a fixed radio frequency, the departure of the refractive index from unity changes linearly with changes in the electron density.

The group refractive index, the ratio of the velocity of a wave group in a vacuum to the velocity in a particular medium, can now be easily derived (Davies, 1990) as

$$\mu' = \frac{d}{dw}\left(\omega(\mu)\right) = 1 + 40.3\frac{N}{\omega^2}. \tag{7.6}$$

For UHF and VHF signals, transionospheric group delay and phase advance are easily treated with these approximate formulae. The group delay or the phase advance are proportional to the integral of the electron density along the path for a given frequency.

7.3.1. Magnetic Field Effects on Propagation

The expressions in the previous section describe transionospheric propagation for the VHF/UHF case, where the propagation frequency is much larger than the largest plasma frequency in the ionosphere, and both the magnetic field and collisions may be ignored. These are useful approximations when the measurements of interest are time-delay or phase along the path, since the magnetic field effect is very small. However, if the desired measurement is polarization (Faraday rotation), then the magnetic field must be included. In this case, for HF, VHF, and UHF propagation, the refractive index includes both electron density and magnetic field effects and can depend on both space and time.

Inclusion of the magnetic field term in the Appleton-Hartree equation leads to solutions where there are two different refractive indexes, due to the ± term in the denominator, which leads to two independent states of polarization that then propagate separately through the ionosphere. The modes are known in the literature as the "ordinary" or O-mode and the "extra-ordinary" or X-mode. In the northern hemisphere for a radio wave propagating along the field line toward the direction of the magnetic north pole, the O-mode is left-hand circularly polarized because by definition it rotates in the opposite direction to the electrons. This arises from the electron gyration, which is the circular motion of an electron in the plane perpendicular to the magnetic field and rotates along the direction given by the right hand toward the direction of the magnetic north pole, i.e. in the direction of the magnetic field. Conversely, the X-mode is right-hand polarized for a radio wave propagating along the field line toward the direction of the magnetic north pole.

For VHF and UHF frequencies, polarization should be taken into account when designing certain antennas and systems, and the Faraday rotation, which arises from the combination of the O and X modes, is also important in this case.

For transionospheric propagation, the O and X are circularly polarized with an opposite sense of rotation. The amount of Faraday rotation (FR) is obtained by measuring both polarization rotation states from a reference (such as the x-axis) and taking the sum of them, divided by two. The resulting Faraday rotation measure produces values proportional to the integral of the product of electron density and the component of the magnetic field aligned along the propagation path. Knowledge of the component along the magnetic field (B_p) from other means, e.g. a reference model, then gives a measurement proportional to the line integral of the electron density or TEC.

$$FR = K_{FR}\frac{\int B_p N ds}{f^2}, \tag{7.7}$$

where the integration constant K_{FR} is $2.97\text{x}10^{-2}$. This is explained further in Hargreaves (1992).

Because HF propagation can refract from the ionosphere and return to Earth (sometimes repeating this process for multiple hops), it is useful for long-range communications. For a signal transmitted from the ground, the refractive index is predominantly decreasing (as the electron density increases) and thus refracts the path of the signal towards the Earth. Depending on the propagation frequency, launch elevation angle, and the ionosphere, the propagation path may reach a maximum altitude and refract back down to the Earth. Figure 7.2 shows an example of propagation for a range of different launch elevation angles through a nominal example ionosphere. For high enough elevation angles, the propagation penetrates the ionosphere. For lower elevations, the signal is refracted back to Earth. Note that in the example given, no rays return to Earth between the transmitters and ~2200 km away. This is known as the skip zone. The skip zone is a region of ground distance from the transmitter where the signal cannot be received. It depends upon the frequency of propagation and particular spatial distribution characteristics of the ionospheric electron density and the antenna configuration.

In general, radio propagation paths can be quite complicated for HF. The actual path of a particular signal depends on the 3D ionospheric electron density at any one instant of time and is influenced by gradients, both vertically and horizontally. The signal path will also vary for different geographical regions, time of day, local dynamics, and geomagnetic conditions (magnetic storms).

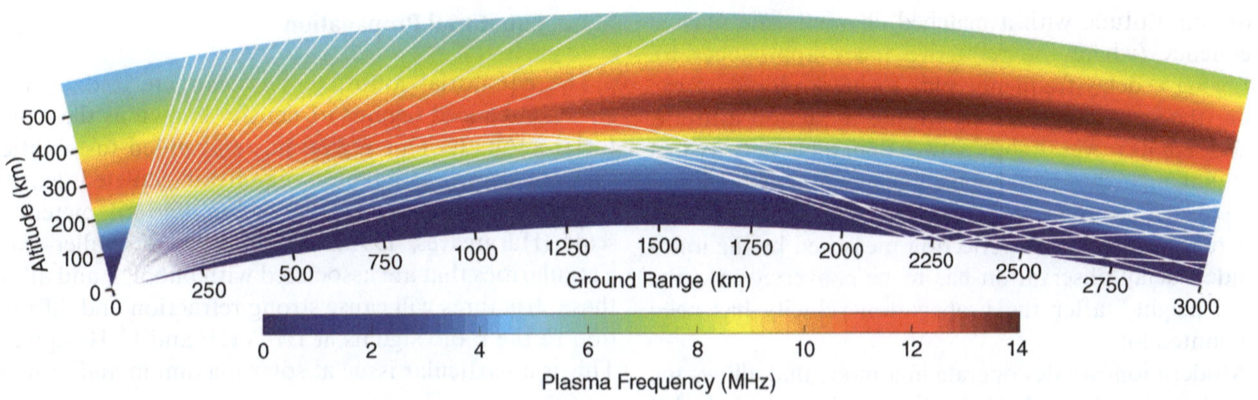

Figure 7.2 Example image showing the ionospheric plasma frequency (color) with the propagation of a radio signal and a fixed frequency above the peak plasma frequency. Note that the higher-elevation rays pass through the ionosphere, but the lower elevation rays are refracted back to Earth. The distance from the transmitter to the closest return of the signal to the ground is called the skip zone.

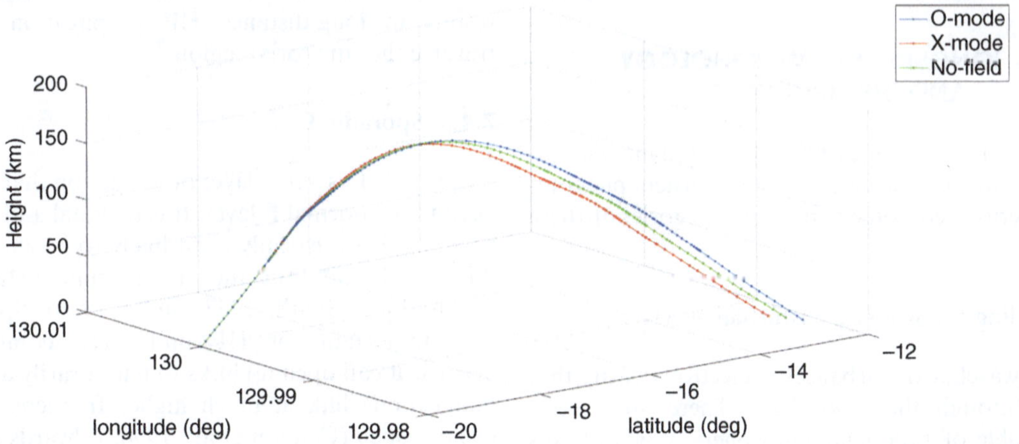

Figure 7.3 Three ionospheric propagation paths, for the same frequency and launch angles, for ordinary (O – blue) propagation, extra-ordinary (X – red) propagation, and the no magnetic field case (green).

For HF propagation, the effect of the magnetic field on the two modes of polarization can be significant. In particular, HF propagation through the ionosphere with a magnetic field is more complicated than for VHF/UHF transionospheric propagation due to the changing angle of magnetic field with respect to the direction of propagation (i.e. k vector). The O and X mode propagate separately and independently, taking different paths through the ionosphere with different times of flight, and arriving at different angles at the receiver. Figure 7.3 shows three propagation paths, for the same frequency and launch angles, for ordinary (O – blue) propagation, extra-ordinary (X – red) propagation, and the no magnetic field case (green). The O and X refract from different altitudes and land at different locations on the Earth. An interesting effect arises from the presence of a magnetic field and the resulting two polarization states. Depending on frequency, the ionosphere, and time of day, an HF communication link may be open for O but not X or vice-versa.

Note that up to now, the collision term in the Appleton-Hartree equation has been ignored. Collisions for HF propagation lead to absorption of the energy of the propagated wave. Collisions are most important in the D and E region of the ionosphere due to the high number of neutral atoms and molecules relative to ions and electrons, and for long distance signal communication links, knowledge of the amount of absorption and loss of energy can be very important.

7.3.2. Vertical HF Sounding

The ionosonde is an instrument developed to probe the vertical structure of the ionosphere. The ionosonde is a radar and in fact was the basis of all modern radar systems. It uses the principle that a sweep of HF frequencies will be individually transmitted and reflected from the ionosphere. Each particular transmission will reflect from an altitude with a matched characteristic plasma frequency (where the refractive index approaches 0), allowing a delay-frequency plot to be produced. Away from the immediate vicinity of the refraction altitude, radio signals travel slightly slower than their velocity in a vacuum, c, as they traverse the lower region of the ionosphere before their reflection. For this reason, the "virtual height" of the reflection measured by the ionosonde's delay observation has to be converted into the "real height," after the propagation velocity has been accounted for.

Modern ionosondes operate in a mode that allows the O and the X polarized propagation modes to be sounded and recorded independently. The longest continuous ionospheric record from ionosondes comes from the UK, where the ionosonde at Slough began operation in 1931 and, after several instrument revisions, is still operational today from Chilton in the UK.

7.4. IONOSPHERIC MORPHOLOGY AND DISTURBANCES

The ionosphere contains a number of dynamic features, both in intrinsic structure and in transient phenomena. We describe here some important categories of these features.

7.4.1. Traveling Ionospheric Disturbances (TIDs)

TIDs are wavelike disturbances in electron density that propagate through the ionosphere. There are several sources capable of producing ionospheric waves of this nature, including large-scale atmospheric gravity waves propagating from the polar region towards the equator, medium-scale gravity waves generated by local sources in the lower atmosphere, electrodynamic waves, and magnetospheric waves. It is thought that one primary source of TIDs is atmospheric gravity waves produced by such processes as winds blowing over orographic features that eventually propagate and break up in the mesosphere and lower thermosphere (Hines, 1960).

TIDs, driven by gravity waves in the neutral atmosphere, produce significant (typically up to 10%) variations in the bottom side ionospheric electron density, and therefore their presence can severely impact HF propagation. In particular, though TIDs produce a small perturbation in ionospheric density of just a few percent, they contain strong gradients that refract the HF signal. This time-variable refraction can cause deviations in the path of tens of degrees azimuth and elevation over time scales of minutes. In addition to variations in the angles of arrival, TIDs produce large variations in the group delay and Doppler along the path. They are also important for their effects on transionospheric VHF/UHF navigation signals.

7.4.2. Equatorial Propagation

The equatorial ionosphere is characterized by two strong enhanced regions of density known as the equatorial anomaly. This region has embedded instabilities that are known as bubbles: regions of depleted density that may extend for a couple of hundred kilometers in scale (Hargreaves, 1992). There are also smaller-scale irregularities that are associated with bubbles, and all of these structures will cause strong refraction and diffraction to the radio signals at HF/VHF and UHF signals. This is a particular issue at solar maximum and around sunset.

Ground-to-ground HF signals are known to cross the equatorial ionosphere in a phenomenon known as transequatorial propagation. This is a well-known phenomenon and is considered to be an ionospheric mode that is enabled by the refraction from the two anomaly peaks. It results in long-distance HF propagation paths that traverse the equatorial region.

7.4.3. Sporadic E

Sporadic E is a thin layer of ionization that is enhanced beyond the normal E layer. It is regional and highly variable, and is statistically most likely to be encountered at midlatitudes and in the summertime (Matsushita & Smith, 1975; Smith, 1957; Smith 1978). Sporadic E is very important for HF and VHF communications because it can open up links to temporarily allow a communications link at much higher frequencies than are usually open (Cannon et al., 1996; Edwards et al., 1984; Miya et al., 1978). For example, sporadic E can allow communication links at 50 MHz on the amateur bands (Owen, 1987; Pierce, 1938; Pocock, 1988; Wilson 1970, 1971).

7.5. SOLAR STORM EFFECTS

Solar activity varies throughout the 11-year solar cycle, but there is significant short-term variation as well. These processes affect the coupled Earth-Sun system, and therefore ionospheric propagation is also affected.

Perhaps the most significant solar event is a coronal mass ejection (CME; Baker & Martyn, 1953; Birkeland, 1908, 1913; Chapman & Bartels, 1949; Störmer, 1907). A CME is a release of radiation and magnetic fields from the sun that can result in enhanced magnetic fields and particles reaching the environment around the Earth via the solar wind. CMEs are often accompanied by solar flares, which are sudden releases of electromagnetic radiation across the electromagnetic spectrum from radio waves to gamma rays (Mitra, 1974). Solar flares can also occur away from CMEs.

There are several effects on the ionosphere from these events. The electromagnetic radiation will reach the Earth typically around 8 minutes after leaving the Sun and, depending on the exact profile of the energy with wavelength, will cause the ionosphere on the dayside to increase in density through photo-ionization of different constituent atoms and molecules. This may result in an overall increase in electron density, but the important point to note for radio propagation is the relative increase at different altitudes or the change to the vertical electron density profile. This is important because ultimately the nature of the resulting effect, whether beneficial or detrimental to propagation, depends on where the changes have occurred, as is discussed further below.

Following the electromagnetic radiation from the Sun, fast energetic solar particles may arrive at the Earth within a few more minutes and penetrate into the upper atmosphere, causing further ionization at D-region altitudes. Energetic protons released by a CME can cause an increase in the ionization, especially in the high-latitude polar regions. The increase in free electrons at low altitudes can enhance radio wave absorption, leading to polar cap absorption events (Bailey, 1957, 1959; Chapman & Little, 1957; Reid & Collins, 1959; Reid & Leinbach, 1959).

After the initial arrival of electromagnetic radiation, a cloud of plasma and magnetic fields that are carried in the solar wind encounter the Earth's magnetic field in the bow shock region. The direction of the CME and the exact configuration of the interplanetary magnetic field will determine the coupling of the fields. The shock wave of traveling mass disrupts the Earth's magnetosphere, compressing it on the dayside and extending the nightside magnetic tail. When the magnetosphere reconnects on the nightside, it releases energy in the form of a substorm. The entire event is called a geomagnetic storm, and the effect on the ionosphere is called an ionospheric storm. It is often characterized in severity by the amount of disturbance to magnetometers on a global scale. The ionospheric storm typically involves an initial enhancement in the electron density for the first few hours, followed by a collapse and a slow restoration over several days.

During a large solar storm, there are several effects on radio signals. At UHF and VHF frequencies, irregularities in electron density may cause more severe and geographically extensive scintillation at mid- and high latitudes. The scintillation at low latitudes may be enhanced in some regions and depressed in others, depending on the exact changes in ionospheric dynamics for a given storm. There are also important changes in TEC that result in changes to the group delay and in phase advances that affect navigation signals. The overall pattern of storm effects will first have a positive phase where the TEC is enhanced in selected spatial regions, then a negative phase where it is depressed. This will cause a change to the ionospheric delay of the signal (and will of course affect the signal phase in the opposite way). Typically, initially the delay will increase and the phase advance for signals crossing the regions experiencing the positive storm effects, and then later on the opposite will occur. These effects may or may not be coincident with scintillation and may be accompanied by very strong TEC gradients, that can affect certain applications that reply on statistical ionospheric models that do not replicate such events.

Additionally, solar storms can be very important for HF communications and radar systems because increased ionization at low altitudes in the D region causes HF signals to be absorbed. The exact form of the effects depends on the type and duration of the storm but can last for many days. The effects of solar storms on radio propagation are discussed further in McNamara (1991).

7.6. SPACE WEATHER EFFECTS ON UHF/VHF AND HF USER APPLICATIONS

7.6.1. Ground-Based Applications

Operational use of UHF/VHF and HF along ionospheric paths, further classified here as those radio links which are both transmitted and received on the ground, falls into several categories. All three frequency bands are used for radars for defense and for scientific studies.

Over-the-horizon radars (OHR) operate in the HF band and backscatter from electron density irregularities in the ionosphere in the Bragg condition, where incident radar energy is perpendicular to the irregularity that tends to be elongated along the magnetic field due to electron mobility. The backscatter Doppler values from these systems can inform the user of the velocity of the irregularity field and hence produce useful information about the ionospheric dynamics. Recent work has shown the use of the ground-scatter of such signal to infer information about TIDs (Grocott, 2013).

OHR is also used as a conventional radar to observe moving targets at ranges over the horizon. These radars require expert interpretation and ionospheric knowledge to ensure optimal operation. Comprehensive details of OHR can be found in Fabrizio (2013).

UHF/VHF radars are used for many different defense applications. UHF radars are used as missile warning systems as well as for satellite surveillance and tracking. Space weather has important effects on VHF/UHF radars operating at low and high latitudes because the irregularities in the ionosphere cause radio scintillation. This is particularly important at equatorial regions, where irregular structures in the ionosphere can cause very strong refraction and diffraction, resulting in enhancements and fading of the radar signals with as much as 3-dB increase in the target (and clutter) signal-to-noise ratio (Knepp & Houpis, 1991). Effects on delay and Doppler characteristics are also important for transionospheric radars, particularly for lower frequencies. UHF and VHF radars are also used as incoherent scatter radars, first proposed by Gordon (1958), to undertake comprehensive scientific investigations on the ionosphere.

All three frequency bands discussed in this chapter are used for communications. The most important space weather effects on communications are on HF. HF communications have been used for beyond line-of-sight communications since the early 20th century. Before the advent of satellite communications, this was an important means of radio communication that was used extensively by the defense sectors and by amateur radio enthusiasts. HF remains an important band today for communications, but for many applications, higher frequencies with their associated larger bandwidth and greater data throughput are preferred, even if they require a satellite link (cf. next section). HF communication planning does require knowledge of the ionosphere and of current and future space weather conditions, which can have adverse effects ranging from requiring a change in frequency to the complete inability to make an HF link. Many modern systems use automatic link establishment to find the best link in real time.

A recent and exciting development in the field of radio communications has been the proliferation of software-based radio systems. These low-cost devices have opened up new opportunities for distributed networks or transmitters and receivers, usually at HF but sometimes up to VHF and down to lower frequencies too. Two of these, the RBN (reversebeacon.net) and the Weak Signal Propagation Reporting Network (WSPRNet, wsprnet.org), operate continuously and contain several years of communication link records. They are operated on a volunteer basis by members of the amateur radio community.

These networks are already showing their utility as wide-area ionospheric sensors that can help to provide new perspective on the ionosphere over regions that would traditionally have been data sparse. For example, Frissell et al. (2014) demonstrate that the data can be used as an ionospheric sounding instrument. Frissell et al. (2019) use the data to study the response to solar variations during a critical period in September 2017 when HF radio was needed for emergency operation in the Caribbean region and found daytime radio blackouts with recovery periods of minutes to hours. Further, the overall suppression of effective HF propagation bands extended globally for 2 weeks.

7.6.2. Satellite/Transionospheric Applications

HF signals have been transmitted from satellites since the first artificial satellite in Earth orbit, Sputnik (in 1957). Later in 1962, the first topside sounder was launched in the form of the Canadian satellite Alouette 1 (Thomas, 1963). A topside sounder is an ionosonde-like instrument that sounds the ionosphere from above the peak density in the F region. Modern satellite-based HF receivers include the enhanced Polar Outflow Probe (e-POP) Radio Receiver Instrument (RRI) that has reported on amateur radio signals across the US (Perry et al., 2018).

Prior to GPS becoming operational, the United States Navy used a navigational system known as Transit, or the Navy Navigational Satellite System. The USSR had a similar system known as CICADA. Transit satellites were in polar orbit at approximately 1100 km altitude and transmitted dual frequency coherent beacons on 150 and 400 MHz. The Doppler was measured by ground receivers, over the period of a pass of the satellite (~20 minutes), and the observations were used to estimate the position of the ground receiver. The ionosphere produces a significant error on the accuracy of the Doppler positioning. As the satellite moves across the sky, the change in TEC with time produces a change in phase with time, or Doppler. The use of two coherent frequencies allows for the correction of the ionospheric error term. After GPS became operational, Transit ceased operational service as a navigational system in 1996.

UHF/VHF signals are used for satellite communications. The ionosphere can disrupt the communications, particularly for wideband signals in regions of scintillation.

UHF/VHF signals are also used for satellite orbit determination. DORIS is a French ground-satellite system that was primarily designed for the precise orbit computation required for observing the oceans by altimetry. DORIS consists of approximately 60 ground stations distributed over the globe, transmitting coherent beacon signals on two frequencies of 2036.25 MHz and 401.25 MHz. These ground transmitters are currently being received by six low Earth orbiting satellites, including JASON-2, JASON-3, and Sentinel-3. DORIS uses the Doppler technique to obtain precise orbit determination. The ionosphere produces an error on the Doppler that must be eliminated in order to achieve the precision required. The dual coherent beacon frequency transmission allows for the elimination of the ionospheric error term.

The DORIS system has been used for several other applications in addition to altimetry, such as determination of the Earth's gravitational field, the Earth's rotation, precise positioning, navigation and time tagging.

DORIS measurements have also been used to study the ionosphere. The dual frequency observations can be combined to eliminate the satellite motion and produce an observation proportional to the TEC along the path from ground transmitter to the satellite.

Signals in the UHF band are sometimes used for satellite-based altimetry. This altimeter is a radar mounted in a satellite that makes measurements of the radar return time, and with knowledge of the satellite altitude is able to infer altitude of the reflector. An example mission is the JASON-2 mission to monitor the oceans. The use of UHF signals in a synthetic aperture radar mission is currently planned to measure the biomass of the Earth. This project, titled Biomass, is planned for launch in 2022.

7.7. SUMMARY AND OUTSTANDING QUESTIONS

This chapter has provided an overview of the key aspects of HF, VHF, and UHF propagation that are influenced by the ionosphere and space weather. It also provides examples of the scientific use of those frequency bands and gives some practical applications.

Finally, we now consider some outstanding questions.

The theory of radio propagation through the ionosphere at these frequencies is indeed complicated but apparently is complete. Whether there are outstanding questions about the theory can ultimately only be determined through experiment and comparison back to theory. This is a very difficult topic to address, since the ionosphere is a natural laboratory that changes continuously and has no perfect measurement system to characterize it at a given moment. Thus, the testing of radio propagation theory through the ionosphere has no classical control or truth ionosphere to compare against, and the experimenter may be unsure about the exact result that the theory would predict for a given measurement at a given moment. However, there is no current experimental result we are aware of that makes us doubt Appleton-Hartree magnetoionic propagation theory, because we always have the possibility of real instantaneous ionospheric variability to account for in our interpretation.

Even when we look at scintillation and diffraction theory (not covered here), again the theory has not been shown to be lacking. A pertinent question to ask: if we were to know the time-dependent 3D electron density *and* magnetic field *and* neutral density *and* composition in sufficient detail, would the propagation theory then allow us to replicate a particular measurement? Again, we have no evidence to doubt the propagation theory, even at the level of diffraction.

Having said this, the real challenge is having a capability to nowcast the propagation environment. This would involve nowcasting the ionosphere, the magnetic field, and the neutral density and composition in sufficient detail to adequately predict the propagation characteristics of the affected radio signals.

Forecasting the ionosphere is of great interest for radio system planning for communications. It is also of importance on a long-term basis of months to years ahead for satellite missions. Forecasting the ionosphere is very difficult and relies on short-term extrapolation for minutes to hours ahead and statistical models for longer time scales. There has been some recent research on the use of full physics ionosphere models for forecasting, and the use of such approaches is under evaluation.

There are a number of outstanding questions about the physics of the ionosphere that are manifestly not solved because they limit our ability to model and to forecast (e.g. Rishbeth, 2000; Rishbeth, 2007). One of the key unknowns is still the variable influence of the lower atmosphere on the ionosphere. Anything that limits our ability to nowcast and forecast the ionosphere has the potential to disrupt HF communications planning.

The question of sporadic E morphology and drivers still results today in a lack of scientific consensus about the underlying cause and remains elusive to investigation. Moreover, the applicability of ray theory to the investigation of certain aspects of the ionosphere, for example sporadic E, is sometimes unclear and warrants more investigation. Essentially, the sharpness of the gradient encountered is the fundamental consideration when choosing ray theory.

The effects of solar storms are challenging to forecast ahead. Nevertheless, the pattern of events and the progression of effects on radio systems have been catalogued in a way that does allow us to anticipate the range of effects to be a multi-day disruption to radio communication from changes in the delay, Doppler, absorption, and scintillation environment. The big challenge is to specify where and when the disruption will occur to the HF/VHF/UHF links within the overall storm period. This requires an accurate nowcast and forecast ability that does not currently exist.

ACKNOWLEDGMENTS

The results published in this chapter were obtained using the HF propagation toolbox, PHaRLAP, created by Dr Manuel Cervera, Defence Science and Technology Group, Australia (manuel.cervera@dst.defence.gov.au). This toolbox is available by request from its author. The authors of this chapter are grateful for the use of the International Reference Ionosphere Model (2016).

REFERENCES

Appleton, E.V. (1947). Geomagnetic control of F2-layer ionization. *Science*, *106*, 17.

Appleton, E.V. (1954). The anomalous equatorial belt in the F2 layer. *J. Atmos. Terr. Phys.*, *5*, 348.

Bailey, D. E. (1957). Disturbances of the lower ionosphere at VHF. *J. Geophys. Research*, *62*, 431.

Bailey, D. K. (1959). Abnormal ionization in the ionosphere associated with cosmic ray flux enhancements. *Proc. IRE*, *47*, 255.

Baker, W. G., & Martyn (1953). Electric currents in the ionosphere. *D.F., Phil. Trans. of Roy, Soc.*, *A246*, 281.

Bilitza, D., Altadill, D., Zhang, Y., Merten, C., Truhlik, V., Richards, P., et al. (2014). The International Reference Ionosphere 2012 – A model of international collaboration. *J. Space Weather Space Clim.*, *4*, A07.

Birkeland, K. (1908). *Norwegian Aurora Polaris Expedition 1902–1903*, vol. *1*, part 1, pp 39–315. H. Aschehoug and Co, Christiania.

Birkeland, K. (1913). *Norwegian Aurora Polaris Expedition 1902–1903*, vol. *1*, part 2, pp. 319–551. H. Aschehoug and Co, Christiania.

Budden, K. G. (1961). *Radio waves in the ionosphere.* Cambridge University Press.

Cannon, P. S., Weitzen, J. A., Ostergaard, J., & Rasmussen, J. E. (1996). Relative impact of meteor scatter and other long-distance high-latitude propagation modes on VHF communication systems. *Radio Sci.*, *31*(5), 1129–1140. doi:10.1029/96RS01208

CCIR (1986). Recommendations and reports of the CCIR. *VI Propagation in ionized media*. Geneva: Int. Telecom. Union.

Chapman, S., & Bartels, J. (1949). *Geomagnetism*, vols. *1* and 2. New York: Oxford Univ. Press.

Chapman, S., & Little (1957). The non-deviative absorption of high-frequency radio waves in auroral latitudes. *C.G., J. Atmospheric and Terrest. Phys.*, *10*, 20.

Davies, K. (1990). *Ionospheric radio*. IET.

Edwards, K. J., Kersley, L., & Shrubsole, L. F. (1984). Sporadic E propagation at frequencies around 70 MHz. *Radio Electr. Eng.*, *54*, 231–237.

Fabrizio, G. (2013). *High frequency over-the-horizon radar: fundamental principles, signal processing, and practical applications*. McGraw Hill.

Frissell, N. A., Miller, E. S., Kaeppler, S. R., Ceglia, F., Pascoe, D., Sinanis, N., et al. (2014). Ionospheric sounding using real-time amateur radio reporting networks. *Space Weather*, *12*, 651–656. https://doi.org/10.1002/2014SW001132

Frissell, N. A. et al. (2019). High frequency communications response to solar activity in september 2017 as observed by amateur radio networks. *Space Weather*, *17* doi:10.1029/2018SW002008.

Gordon, W. E. (1958). Incoherent scattering of radio waves by free electrons with applications to space exploration by radar. *Proc. IRE*, *46*, 1824.

Grocott, A., Hosokawa, K., Ishida, T., Lester, M., Milan, S. E., Freeman, M. P., et al. (2013). Characteristics of medium-scale traveling ionospheric disturbances observed near the Antarctic Peninsula by HF radar. *J. Geophys. Res. Space Physics*, *118*, 5830–5841. doi:10.1002/jgra.50515.

Hargreaves, J. K. (1992). *The solar-terrestrial environment: An introduction to geospace—The science of the terrestrial upper atmosphere, ionosphere, and magnetosphere*. Cambridge University Press.

Hines, C.O. (1960). Internal gravity waves at ionospheric heights. *Can. J. Phys.*, *38*, 1441–1481.

IEEE Std 211TM-2018 (Revision of IEEE Std 211-1997). IEEE Standard Definitions of Terms for Radio Wave Propagation.

Knepp, D. L., & Houpis, H.L.F. (1991). ALTAIR VHF/UHF observations of multipath and backscatter enhancement. *IEEE Transactions on Antennas and Propagation*, *9*(4).

Matsushita, S., & Smith, L. G. (Eds.) (1975). Special issue on recent advances on the physics and chemistry of the E region. *Radio Sci.*, *10*, 229.

McNamara, L. F. (1991). *The ionosphere: Communications, surveillance, and direction finding*. Malabar, FL: L Krieger.

Mitra, A. P. (1974). *Ionospheric effects of solar flare*, vol. *46*, ISBN: 978-94-010-2233-0. Norwell, MA: D. Reidel.

Miya, K., Shimizu, K., & Kojima, T. (1978). Oblique-incidence sporadic-*E* propagation and its ionospheric attenuation. *Radio Sci.*, *13*(3), 559–570. doi:10.1029/RS013i003p00559

Munro, G. H. (1950). Traveling disturbances in the ionosphere. *Proc. Roy. Soc.* (London), *202*, 208–223.

Owen, M. R. (1987). Midlatitude Es at 220.12 MHz. *WEX*, *4*, 9.

Perry, G. W., James, H. G., Gillies, R. G., Howarth, A., Hussey, G. C., McWilliams, K. A., et al. (2017). First results of HF radio science with e-POP RRI and SuperDARN. *Radio Science*, *52*, 78–93.

Pierce, J. A. (1938). Interpreting 1938's 56 Megacycle DX. *QST*, *23-24*, 72.

Pocock, E. (1988). Sporadic-E propagation at VHF: A review of progress and prospects. *QST*, 33–39.

Reid, G. C., & Collins, C. (1959). Observations of abnormal VHF radio wave absorption at medium and high latitudes. *J. Atmospheric and Terrest. Phys.*, *14*, 63.

Reid, G. C., & Leinbach, H. (1959). Low-energy cosmic ray events associated with solar flares, *J. Geophys. Research*, *64*, 1801.

Rino, C. L., Livingston, R. C., Tsunoda, R. T., Robinson, R. M., Vickrey, J. F., Senior, C., et al. (1983). Recent studies of the structure and morphology of auroral zone Fregion irregularities. *Radio Sci.*, *18*, 1167.

Rishbeth, H. (2000). The equatorial F-layer: Progress and puzzles. *Annales Geophysicae*, *18*, 730.

Rishbeth, H. (2007). *Thermospheric targets*. *EOS*, June.

Silberstein, R. (1959). The origin of the current nomenclature for the ionospheric layers. *J. Atmos. Terr. Phys.*, *1*, 382.

Smith, E. K. (1957). Worldwide occurrence of sporadic-E. *NBS Circ.*, *582*. Washington, DC: US Government Printing Office.

Smith, E. K. (1978). Temperate zone sporadic-*E* maps (f_0E_s > 7 MHz), *Radio Sci.*, *13*(3), 571–575. doi:10.1029/RS013i003p00571

Standards on wave-propagation (1950). Definition of terms, 1950, *Proc. IRE*, *38*, 1264–1268.

Störmer, C. (1907). Sur les trajectoires des corpuscules électrisés. *Arch. Sci. Phys. et nat.*, *24*, 1.

Taylor, M. (1933). The Appleton-Hartree formula and dispersion curves for the propagation of electromagnetic waves through an ionized medium in the presence of an external magnetic field. Part 1: Curves for zero absorption. *Proceedings of the Physical Society*, *45*(2), 245.

Taylor, M. (1934). The Appleton-Hartree formula and dispersion curves for the propagation of electromagnetic waves through an ionized medium in the presence of an external magnetic field. Part 2: Curves with collisional friction. *Proceedings of the Physical Society*, *46*(3), 408.

Thomas, J. O. (1963). Canadian satellite: The Topside Sounder Alouette, 139, no. 3551, 229.

Vickrey, J. F., Rino, C. L., & Potemra, R. A. (1980). Chatanika/TRIAD observations of unstable ionization enhancements in the auroral F region. *Geophys. Res. Lett.*, *7*, 789.

Weber, E. J., Buchau, J., Moore, J. G., Sharber, J. R., Livingstone, R. C., Winningham, J. D., & Reinisch B. W. (1984). F-layer ionization patches in the polar cap. *J. Geophys. Res.*, *89*, 1683.

Wilson, M. (1970). Midlatitude intense sporadic-E propagation: Part I. *QST* (Dec.), 52–58.

Wilson, M. (1971). Midlatitude intense sporadic-E propagation: Part II. *QST* (Mar.), 54–57.

8

GNSS/GPS Degradation from Space Weather

Anthea J. Coster[1] and Endawoke Yizengaw[2]

ABSTRACT

This chapter provides an introduction to space weather effects on GNSS. These effects, which can at times be significant, are introduced to the signal as it propagates through the atmosphere and include: significant range errors due to large gradients in the ionospheric total electron content; loss of lock and signal degradation due to presence of small scale density fluctuations or irregularities; and loss of lock of the signal due to the increase in background L-band radiation during certain solar radio bursts. The significant space weather effects observed in each of the different ionospheric regions (low, mid, and high-latitudes) are identified. GNSS techniques and methodologies to monitor these effects are described. Finally, the significance of space weather effects in various GNSS applications including aviation, marine, emergency response, and autonomous vehicle control will be discussed.

8.1. INTRODUCTION

The umbrella term for all satellite navigation systems that provide autonomous geospatial positioning with global coverage is Global Navigation Satellite System (GNSS). The Global Positioning System (GPS; Parkinson, 1980), one of the main GNSS constellations, was begun in the early 1970s and expanded on concepts developed in earlier navigation systems such as Transit. Transit, the US Navy Navigation Satellite System (NNSS), flew satellites that broadcast signals at 150 and 400 MHz (Guier & Weiffenbach, 1959, 1997). From the set of Doppler shifts and the satellites' locations, positions could be derived to within 200 meters. The Transit system ceased navigation service in 1996 after GPS was fully operational, as GPS offered several improvements, including continuous operation and higher accuracy.

There are now a number of global navigation systems besides GPS, including GLONASS (Coster et al., 1999), Galileo (Mayer et al., 2010), Beidou (Betz et al., 2019), and other regional systems. The basic mode of operation of all GNSS satellites is to continuously transmit navigation signals through the atmosphere on two or more radio frequencies. Radio waves, either transmitted through the ionosphere for satellite communication and navigation or reflected off the ionosphere for HF and radar applications, are all impacted by space weather–driven ionospheric dynamics (Booker & Wells, 1938; Doherty et al., 2004). In this manner, all GNSS systems are thus susceptible to degradation due to space weather (Coster & Komjathy, 2008).

The standard definition of space weather refers to conditions on the Sun and in the magnetosphere, ionosphere, and thermosphere that can affect performance and reliability of space-based and ground-based technological systems. Space weather effects are introduced to GNSS signals due to their propagation through atmospheric regions, specifically through the ionosphere and the troposphere. In these regions, the index of refraction is not equal to 1, and because of this, a range delay and phase advance are introduced. In addition, because the index of refraction changes as a function of altitude, the propagation path is also bent, adding a correction to the elevation of the signal. In this introduction, we will briefly describe

[1] *Haystack Observatory, Massachusetts Institute of Technology, Westford, Massachusetts, USA*

[2] *Space Science Application Laboratory, The Aerospace Corporation, El Segundo, California, USA*

Space Physics and Aeronomy Collection Volume 5: Space Weather Effects and Applications, Geophysical Monograph 262, First Edition.
Edited by Anthea J. Coster, Philip J. Erickson, and Louis J. Lanzerotti.

DOI: 10.1002/9781119815600.ch8

how propagation through the ionosphere and the troposphere affects the GNSS signals. This will provide the basis for our discussion of space weather impacts on GNSS throughout the rest of this chapter.

8.1.1. The Ionosphere

The ionosphere is defined to be the region of free electrons and ions in the atmosphere that begins around 100 km and extends beyond 1000 km with a small but measurable number of free cold electrons above the ionosphere called the plasmasphere. The electrons and ions are generated primarily by solar EUV radiation and are strongly influenced by the earth's magnetic field. At night, the ionosphere diminishes but does not completely decay due to a combination of factors.

In the ionosphere, the index of refraction is a function of both the electron density and frequency, and radio signals do not propagate at the speed of light but at a slower speed due to the group velocity. Thus, ionospheric range correction (or error) is due to this additional path delay and remains one of the largest error correction terms within the GPS error budget which is directly proportional to the total electron content (TEC). Mathematically, the index of refraction, the ionospheric group delay, time delay (Δt), can be approximated as

$$\Delta t = \frac{40.3}{cf^2} \int N\,dl,$$

where f is the frequency in Hz, c is the speed of light in m/s, N is the number of electrons/m^3, and the integral quantity is the total electron content (TEC) in el/m^2.

On the other hand, the carrier phase information on the received signal on the ground appears to be advanced relative to the received code phase in the presence of the ionosphere. The ionospheric carrier phase advance (in cycles) can be approximated as

$$\Delta \varphi = \frac{40.3}{cf} \int Ndl = \frac{1.34 \times 10^{-7}}{f} \int Ndl.$$

By comparing the delays of two GNSS signals at different frequencies, the ionospheric error can be directly calculated. The average daytime ionospheric zenith delay at L1 frequency is 5–15 meters; however, this delay can be as large as 36 m at the equator during solar maximum. Furthermore, zenith delays must be multiplied by mapping functions to account for the longer path length observed through the ionosphere at lower elevations. For example, at 5° elevation, the total delay is about 3 times higher than the zenith delay due to the TEC difference along the two ray paths. The ionospheric TEC at a specific place and time can vary by 25%–50% about a monthly average, although there are times this effect is much larger (for example, following a large geomagnetic storm).

In general, with respect to GNSS applications, the primary space weather effects are the introduction of range errors and the loss of signal reception. Space weather–driven effects on GNSS signals can be divided into three categories: (1) large gradients in ionospheric electron density can introduce significant range errors and bending; (2) small-scale irregularities in the ionosphere can cause loss of lock on the GNSS signal fluctuations (or scintillation); and (3) solar radio bursts can also introduce notable effects on GNSS signals by raising the background noise level. Detailed information about each category is given in section 8.2.

8.1.2. The Troposphere

The troposphere is the region of the atmosphere from about 0 to 15 km from the Earth's surface. In the troposphere, the index of refraction is a function of pressure, temperature, and humidity. Unlike the ionosphere delay, the tropospheric delay is not dispersive at radio frequencies. For GNSS users, the tropospheric delay is typically derived using models, or estimated after all other error terms are calculated (e.g. Fernandes et al., 2010). A typical tropospheric delay is 2.5 m at zenith. The zenith delay is converted to a line of sight delay through the use of mapping functions, which increases the delay by a factor of approximately 10 at 5° elevation. The tropospheric delay is normally divided up into two principal components: a hydrostatic (dry) term and a wet delay term. The dry term depends principally on the pressure and temperature, and it can be predicted with a high degree of accuracy. The dry term typically accounts for 80%–90% of the total delay due to the troposphere. The wet term, which depends on the humidity and temperature, is the highly variable component of the tropospheric delay. On average, it accounts for only 10%–20% of the total delay.

For the tropospheric research, there has been significant utilization of radio occultation measurements received onboard low Earth orbiting (LEO) satellites. For radio occultation measurements, the GNSS signals are influenced by the atmospheric (tropospheric) refractivity, resulting in both time delay and bending of the signal. The primary observable is the atmospheric excess phase, and it is measured with millimetric accuracy (Wickert et al., 2002). These measurements are the basis for high vertical resolution and precise refractivity and temperature profiles. However, the tropospheric effect on GNSS signals is not as strong as those effects that are directly influenced by changes in conditions on the Sun and in the solar wind. We will not be discussing tropospheric effects in further detail in this chapter.

This chapter begins with a description of space weather effects on GNSS as a function of latitude in section 8.2. In section 8.3, the role GNSS plays in monitoring these space weather effects is discussed, as well as the current and future methodologies used for mitigation. Finally, in section 8.4, the significance of space weather mitigation for various GNSS applications is described.

8.2. GNSS SPACE WEATHER EFFECTS, MONITORING, AND APPLICATIONS

8.2.1. Space Weather Effects

Space weather–driven ionospheric variability has a profound influence on GNSS systems. For example, the range error on GNSS signals due to the large gradients in the ionospheric TEC can cause significant errors in positioning. In the ionosphere, it is the large horizontal gradients that cause significant errors in GNSS positioning. Vertical gradients, which to some extent are always in the ionosphere, are not typically of consequence for GNSS positioning, although they may lead to scintillation effects described later in this section. A large horizontal density gradient is more prevalent in the midlatitude region due to the westward extended storm-enhanced density plume (see Section 2.1.2; Foster, 1993) and in the equatorial region due to the equatorial anomaly (Kelley, 1989). In general, large horizontal ionospheric density gradients make it difficult to correctly predict the ionospheric range correction, which is critically needed for high-precision real-time positioning applications such as autonomous vehicle control, marine positioning, emergency response, and railway control.

In addition, small-scale ionospheric density fluctuations or irregularities affect GNSS signal propagation. These irregularities are generated by complex dynamics and are prevalent in the equatorial and high-latitude regions, and to a lesser extent in the midlatitudes. The predominant ionospheric irregularities at the equator span several degrees of scale-size and unfavorably affect GNSS signal propagations, leading to possible loss of lock on the GNSS signals. When scintillation (the rapid amplitude and phase fluctuations of GNSS signals due to turbulence generated by ionospheric irregularities) occurs, several GNSS applications can be affected, such as the integrity and performance of positioning applications as well as the detection and tracking of aircraft, missiles, satellites, and other targets. Such effects on GNSS applications become more severe during geomagnetic storm periods (e.g. Basu et al., 2001).

Finally, what may be considered as a separate category of space weather effects, is the degradation of GNSS signal reception due to powerful solar radio bursts. Cornell University researchers first demonstrated in 2007 that a solar radio burst in the L-band frequency could have a powerful impact on GNSS systems. When these radio bursts occur, they can raise the background noise level in the L-band channel (and thereby decrease the signal-to-noise ratio of the incoming signal), causing the receivers to lose lock (Cerruti et al., 2006, 2008; Langley, 2007). We recommend readers refer to chapter 6 in this volume for more detailed information about this effect.

In general, the importance of ionospheric density structures has become a growing concern, particularly to new applications that are life-critical, such as GNSS-aided military applications, search and rescue applications, and certain commercial and civilian applications. These applications have been increasing exponentially (Conker et al., 2003), and ionospheric density fluctuations are as much an engineering concern as they are a scientific quest (Kintner et al., 2007; Hlubek et al., 2014). Ionospheric dynamics and its impacts occur at all latitudes: low, middle, and high. Understanding ionospheric dynamics, in different forms, through GNSS measurements is crucial to mitigate the space weather impacts on the technological systems that our society has become increasingly reliant on. In the next section, space weather effects are identified for the different regions.

Low-latitude space weather effects

Typically, ionospheric effects are strongest in the low-latitude region, and these effects may extend to the midlatitude region, especially during intense magnetic storm periods (e.g. Basu et al., 2001; Keskinen et al., 2004). The equatorial ionospheric region is an extremely complex region that hosts numerous plasma instabilities and many unresolved problems. Examples of ionospheric dynamics in the equatorial region include the variability of the equatorial ionospheric anomaly (e.g. Mannucci et al., 2005), the presence of equatorial irregularities (e.g. Basu et al. 2001), the dynamo efficiency (e.g. Maruyama et al., 2005), the equatorial electrojet (e.g. Yizengaw et al., 2014), the enhanced postsunset vertical drift (prereversal enhancement; e.g. Fejer et al., 1999), and the tidal behavior of thermospheric winds and tides (e.g. Immel et al., 2006). As an example, during large geomagnetic storms, the equatorial anomaly region may be strongly enhanced due to penetration electric fields (Mannucci et al., 2005). These storm-time penetration electric fields can generate extremely large increases in the TEC and produce significant scintillation at low latitudes by producing strong vertical drifts that uplift the plasma to higher altitudes where the recombination rate is low. This can be considered one of the most dramatic consequences of magnetosphere-thermosphere-ionosphere coupling (Mannucci et al., 2005). In general, during severe storm periods, GNSS receivers suffer from strong scintillation effects in the equatorial region (from ±10° geomagnetic latitude to

±20° geomagnetic latitude). During severe storm periods, this region even extends to ±30° geomagnetic latitude.

Equatorial scintillations are also produced by irregularities in the F-layer of the equatorial ionosphere following the passage of the evening terminator (e.g. Basu et al., 2001; Groves et al., 1997) and may even extend to post-midnight sectors (e.g. Patra et al., 2009; Yizengaw et al., 2013). In these regions, the most severe scintillations are associated with the crests of the equatorial anomaly, which are centered approximately 15° either side of the magnetic equator (Knight et al., 1999). Since it is coupled to the anomaly, equatorial scintillation tends to be more severe during the years of solar maximum, when the anomaly is at its greatest.

Ionospheric scintillation affects all space-based communication, surveillance, broadcast, and navigation radio systems. As early as the 1970s, Taur (1973) recognized that ionospheric scintillation impacted communication satellites at 4 and 6 GHz. Scintillation effects were noted to be the strongest in the lower latitude regions between ±30° latitude, peaking about an hour after local sunset and showing strong seasonal and diurnal variations. In Taur (1976), this analysis was extended to cover frequencies down to 1.5 GHz (GNSS frequencies).

Scintillation's impact on GNSS receivers is of particular importance, as the effect may range from degradation of positioning, velocity, and timing accuracy to receiver loss-of-lock. Thus, understanding the effect of scintillations on radio signals and mitigating those effects requires an understanding of ionospheric dynamics. Different instruments, such as UHF and high rate (50–100 Hz) GNSS receivers, have been used to monitor ionospheric scintillation activity at different locations and times. Even standard GNSS receivers have been used to estimate scintillation activity through the calculation of the time rate of TEC (ROT) index or ROTI (Pi et al., 1997). This scintillation information has been utilized in modeling and forecasting efforts for quite a while. Nevertheless, the prediction and modeling of scintillation occurrence is still a challenge for the space science community, primarily because the ionospheric irregularities that produce scintillation depend on the presence of plasma instabilities in the ionosphere. How these instabilities develop is extremely sensitive on the strength and nature of the ionospheric processes that act as drivers (Kintner et al., 2007).

Figure 8.1 shows a comparison between the physics-based ionospheric model, PBMOD, (Retterer, 2005, 2010) predicted scintillation S4 values (red curves in all panels) with two independent S4 measurements, UHF (top panels), and GNSS (bottom panel) receivers, during quiet time (on 25 October 2004) at Ancon (77.2°W, 11.8°S) locations. The PBMOD ionospheric model is a system of Physics-Based MODels that describes the 3D time-dependent evolution of the low-latitude ionosphere on

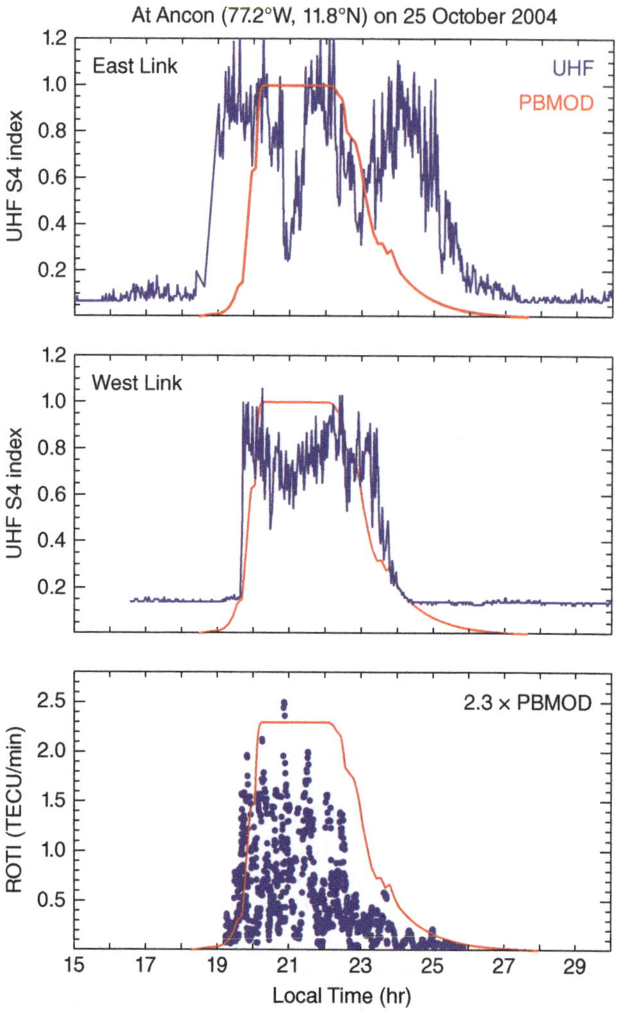

Figure 8.1 The top and middle panels show the comparison between model estimated (red curve) and UHF measured (blue curves) amplitude scintillation (S4-index), and the bottom panel shows model S4 value comparison with ROTI (blue curves) estimated from GNSS at measurements at Ancon.

several different time scales. PBMOD was run using climatological driver models for the times that had no vertical plasma drift data to input as well as other parameters, such as neutral atmospheric densities of N_2, O_2, and O; the neutral temperature from the MSIS model; horizontal neutral wind components from the empirical wind model HWM-90; and others listed in Retterer, 2005, to drive PBMOD. It is noteworthy to mention that scintillation is not a phenomenon particularly associated with geomagnetic activity (at least in the equatorial region) and can occur during magnetically quiet periods. During geomagnetically quiet times, scintillation activity is driven by seeding perturbations from the lower thermosphere that initiate an ionospheric bottom-side ionospheric instability, usually referred to as Rayleigh-Taylor instability. Forcing from the mesosphere and lower thermosphere will not be addressed further in this chapter, although its presence is crucially important and is currently a subject of intense research (e.g. Goncharenko et al., 2010; Yizengaw & Groves, 2018).

The blue curves in the top and middle panels of Figure 8.1 depict the amplitude scintillation index S4 (a measure of the intensity fluctuations in the signal) estimated from UHF receiver measurements, looking to the east (top panel) and west (middle panel) directions. The SCINDA (Scintillation Network Decision Aid) (Groves et al., 1997) UHF receivers located at Ancon, Peru, monitor the strength of radio signal transmitted from two geostationary communication satellites (Fleetsat 7 and Fleetsat 8) operated by the US Navy for UHF radio communications between ships, submarines, airplanes, and ground stations. Thus, when the radio signals traverse through unstable regions of the ionosphere with density irregularities, the UHF receivers detect strong S4 values, as shown in the top and middle panels of Figure 8.1. Comparisons of the model and data curves by eye gives a general sense of the agreement of the model with the data, but for simple quantitative comparisons, specific metric values from the curves need to be extracted. The metrics include the onset time, the duration of scintillation, and some measure of the strength of scintillation, such as its peak value. The model performs fairly well in reproducing these observational metrics, except it fails to reproduce the two different onset times for the east and west links, as shown in the top two panels. This comes about because the two links look in opposite directions, and the onset of the Rayleigh Taylor instability will occur at different times in the different directions. In addition, the links have different elevation angles and are being compared to the S4 calculated from the PBMOD output. PBMOD models the S4, assuming that the signal is vertically incident on the local ionosphere; see Retterer (2010) for a more thorough discussion of this comparison.

The bottom panel in Figure 8.1 shows the comparison between the modeled (red curve) S4 values and the ROTI observations (Pi et al., 1997). ROTI is calculated from the TEC values estimated from the GNSS measurements at Ancon. ROTI is defined based on the standard deviation of ROT over a 5-minute period (i.e. $\mathrm{ROTI} = \sqrt{\langle \mathrm{ROT}^2 \rangle - \langle \mathrm{ROT}^2 \rangle}$) (Pi et al., 1997) and can be derived using output from the standard scientific GNSS receiver operating at a rate between 0.03 and 0.1 Hz. Both the S4 and phase scintillation, $\sigma_{\varphi,}$ indices, typically require specialized high-resolution GNSS receivers to characterize the small-scale (down to hundreds of meters) ionospheric irregularity distribution. ROTI, on the other hand, provides significant information about scale sizes on the size of a few kilometers (plasma irregularity scale-lengths of about 6 km (Basu et al., 1999).

The ROTI measurements agree with the modeled values as well as the UHF-measured S4 values do. This shows that ROTI, may be considered as a third index to indicate scintillation occurrence. Yizengaw and Groves (2018) have clearly demonstrated that ROTI can be used as a proxy for assessing the presence of ionospheric scintillation, especially in regions where no high-rate GNSS scintillation receivers are available. In fact, they showed that ROTI can serve as a proxy for both the S4 scintillation index in the low latitudes and the phase scintillation index, sigma-phi (σ_{φ}), in the high latitudes. Recently, Carrano et al. (2019) provided a quantitative theory for ROTI that provides its dependence on the sampling interval, satellite motion, propagation geometry, and the spectral shape, strength, anisotropy, and drift of the irregularities. They concluded that ROTI may be considered more than just an indicator of the presence of irregularities and can be used as a proxy for S_4.

Midlatitude space weather effects

The midlatitude ionosphere is normally a fairly benign region; however, during geomagnetic storms, it experiences severe ionospheric disturbances, such as the storm-enhanced density (SED) feature. The SED feature was originally recognized in the early 1990s with the Millstone incoherent scatter radar (Foster, 1993). The SED plume is defined as a spatially narrow and distinct region of enhanced density, observed frequently during geomagnetically active periods in the afternoon and premidnight sector at the equatorward edge of the midlatitude trough. The SED plume evolves to convect plasma sunward and poleward toward the high-latitude cusp. It can be detected from the ground using GNSS TEC maps (e.g. Coster et al., 2001, 2003; Foster et al., 2002, 2007; Heelis et al., 2009; Yizengaw et al., 2006) and from space using in-situ density measurements onboard LEO satellites (e.g. Yizengaw et al., 2008). Within the SED plume, the plasma is dominated by O+ ions at ionospheric heights and is carried by the low-latitude edge of the storm time magnetospheric origin subauroral disturbance electric field, termed the subauroral polarization stream (SAPS; e.g. Foster & Burke, 2002; Foster & Vo, 2002). SAPS is a hallmark of a dynamic two-way magnetosphere-ionosphere coupling and presents in the ionosphere as a large poleward-directed electric field that drives a closure current between Region 1 and Region 2 field-aligned current sheets across the low-conductance region located equatorward of the auroral electron precipitation region (e.g. Foster et al., 2005).

The SED plume is responsible for some of the largest space weather effects in the midlatitudes (Coster & Foster, 2007). Large horizontal gradients in the range of 50 TEC units per degree have been observed, and these gradients can persist for several hours, leading to large errors in the measured range and in the differential range. Impacts on the GNSS differential range measurement are particularly important due to their heavy use in high-precision applications such as surveying, navigation of inland waterways, geodesy, off-shore oil drilling, and satellite-based augmentation systems such as the FAA's Wide-Area Augmentation System (WAAS) in North America. Skone and Yousuf (2007) demonstrated how the SED plume affected marine navigation. It has also been shown that the SED plume can significantly affect the availability for the localizer performance with vertical guidance (LPV) service that provides vertical guidance to aircraft to enable descent to 200–250 feet above a runway in the WAAS (e.g. Datta-Barua et al., 2005; Doherty et al., 2004). For example, during the geomagnetic storms on 29 and 30 October 2003, due to the presence of large ionospheric gradients, no LPV service was provided by the WAAS over the United States for 15 hours on the 29th and 11.3 hours on the 30th (Doherty, et al., 2004). LPV approaches are operationally equivalent to the instrument landing systems installed at many runways. The benefit of WAAS LPV over instrument landing systems is that it does not depend on ground-based infrastructure.

Midlatitude ionospheric irregularities are not common (Kintner et al., 2006). Nevertheless, there are three distinct plasma processes that give rise to midlatitude ionospheric irregularities, and all are related to geomagnetic storms. The first is associated with auroral plasma processes, the second with SED gradients, and the third process is observed with large TEC gradients that form during geomagnetically active times in the lower midlatitudes. The first instability is associated with the plasma density gradient at the trough wall, which is at the poleward edge of SED plume, due to opposing density and temperature gradients (Basu et al., 2008). The second type of instability is driven by SAPS convection. When the SAPS convection stream is associated with ion convection velocities that exceed a threshold of 300 m/s, density instabilities are triggered that eventually generate ionospheric plasma irregularities with scale lengths varying from tens of kilometers to hundreds of meters (Basu et al., 2008). Finally, irregularities associated with the SED feature were first measured by GNSS by Ledvina et al. (2002, 2004). The presence of ionospheric irregularities in the vicinity of SED plumes can also be estimated using ROTI.

Figure 8.2 shows a typical example of the presence of ionospheric irregularities in the vicinity of SED plumes. The bottom panel in the figure depicts 2D maps of ROTIs that are binned in 2° × 2° latitude and longitude bins over 30-minute intervals. The figure clearly indicates the spatial distribution of strong plasma irregularities in the midlatitude region. In general, better understanding of

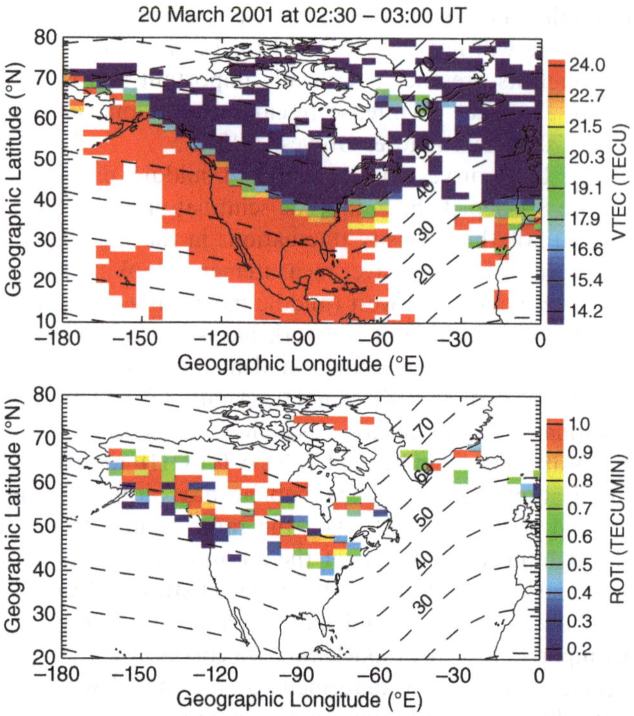

Figure 8.2 The 30-minute average 2D map of a GNSS TEC snapshot that shows stretched SED plume on 20 March 2001 at 0230–0300 UT over the North American sector (top panel). The bottom panel shows the 2D snapshot of ROTI calculated from ground-based GNSS receivers; dashed curved lines represent the geomagnetic latitudes. (Adapted from Yizengaw, 2019.)

the formation of SED plumes and associated plasma irregularities at different longitudes as a function of geomagnetic storm magnitude and other drivers remains an important goal for accurate characterization of the global distribution of midlatitude ionospheric structure and dynamics, including SED plumes.

High latitudes

At high latitudes, the ionospheric weather phenomena of interest include auroral dynamics, polar patches, traveling ionospheric disturbances, and the SED/tongue of ionization features, all readily observed in GNSS electron content measurements. These phenomena produce different dynamic ionospheric features due to the different physical mechanisms involved. Auroral substorms are characterized by increased spatial and temporal decorrelations of ionospheric range delay, which may be difficult to model using a wide-area ionosphere grid approach. However, the effect of ionospheric gradients in the range measurements is smaller than in the midlatitudes and low latitudes, simply because there is less ionosphere. Nevertheless, as in the midlatitude case, an example of GPS scintillation data combined with GPS TEC mapping is shown in Figure 8.3, demonstrating a clear relationship

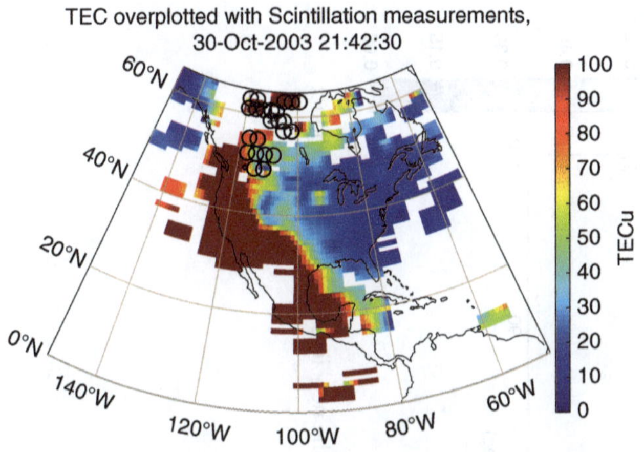

Figure 8.3 GPS scintillation observations (black circles) overlaid onto a GPS mapping of the ionospheric TEC during a strong geomagnetic storm and associated SED plume to study their relationship. (Source: Coster et al., 2005. Reproduced with permission of Anthea Coster.)

between observed high-latitude GNSS scintillations and electron density gradients associated with the SED plume.

Auroral activity is also characterized by the presence of small-scale irregularities in electron density associated with scintillations. These small-scale irregularities are observed in polar cap patches and auroral structures. Polar cap plasma patches are regions or islands of high-density plasma within the high-latitude region. Many researchers have analyzed effects in this region (Doherty et al., 2000; Pullen et al., 1998; Skone & Knudsen, 2000), and others have conducted simulation studies to determine the impact on receiver tracking performance (El-Arini et al., 2003; Ganguly et al., 2003; Morrissey et al., 2004).

High-latitude scintillation occurrence is known to correlate in general with changes in the solar and interplanetary conditions and has a consequential impact on GNSS receiver tracking performance. Large-phase variations (phase scintillations) are present in the auroral region (65°–75° geomagnetic latitude) and polar cap (>75° geomagnetic latitude) regions (Pi et al., 2002). Phase scintillations may also lead to cycle slips and loss of lock for receivers as they track the signal (Datta-Barua et al., 2003). Phase scintillation measures the severity of signal phase and is quantified by σ_φ, which is the standard deviation of high-frequency detrended phase observations over an interval of up to 60 seconds. The value of σ_φ can vary from 0.05 radians for very weak scintillation to 1.0 radians for a strong scintillation case. Auroral scintillations also show a seasonal dependence that is opposite of the observed climatology at low latitudes, being greatest from the autumn equinox through winter to the spring equinox, and a minimum during the summer months (Aarons & Basu, 1994). Indeed, the geomagnetic disturbances that excite auroral and polar cap scintillations tend to suppress the onset of equatorial scintillations during solar maximum periods (Aarons et al., 1983; Davies, 1990; Klobuchar, 1996). Because geomagnetic storm activity is linked to solar activity through solar flares and coronal holes, auroral and polar cap scintillations also show a strong dependence on the 11-year solar cycle, being most intense during solar maximum periods but almost nonexistent during minima.

8.3. SCIENTIFIC APPLICATION OF GNSS

8.3.1. Space Weather Monitoring

To accurately assess the effects on GNSS due to refraction, delay, and disruption, required space weather information, observation of ionospheric and plasmaspheric dynamics as a function of time and locations, is critically important. Because such information is highly crucial to performing real-time predictions of the time and location of ionospheric disturbances, GNSS's ability to perform these observations makes it an important instrument.

Monitoring the horizontal structure of ionosphere with GPS (2D TEC map)

TEC maps are among the most important tools of the space weather community for monitoring changes in the ionosphere during quiet and large space weather events, including monitoring of SED plumes. As was pointed out earlier, with the advent of GNSS TEC maps, the full scale of the SED feature (Coster et al., 2001; Foster et al., 2002) and its connection with the tongue of ionization in the polar regions (Foster et al., 2005), as well as the large enhancements in the equatorial anomaly during geomagnetic storms (Kelley et al., 2004; Thomas et al., 2013; Vlasov et al., 2003), were all clearly identified. An understanding of all these phenomena, and the ability to predict their effects, is required by the navigation and surveying communities to assess the expected impact on GNSS users in near real time.

For example, 2D TEC maps can be used to estimate the TEC horizontal gradients, which produce positioning errors, for operation. Several organizations provide TEC maps, including the NSF CEDAR–funded Madrigal site. The Madrigal site provides TEC data binned into 1° bins of latitude and longitude every 5 minutes, utilizing ~6000 GNSS sites a day (http://cedar.openmadrigal.org/). However, there is a delay of 2 to 3 days in obtaining the CEDAR TEC maps (Rideout & Coster, 2006; Vierinen et al., 2015). This is strictly because the majority of receiver sites do not provide real-time data. Figure 8.4 shows typical 2D maps of TEC obtained on different date and time. The left plot

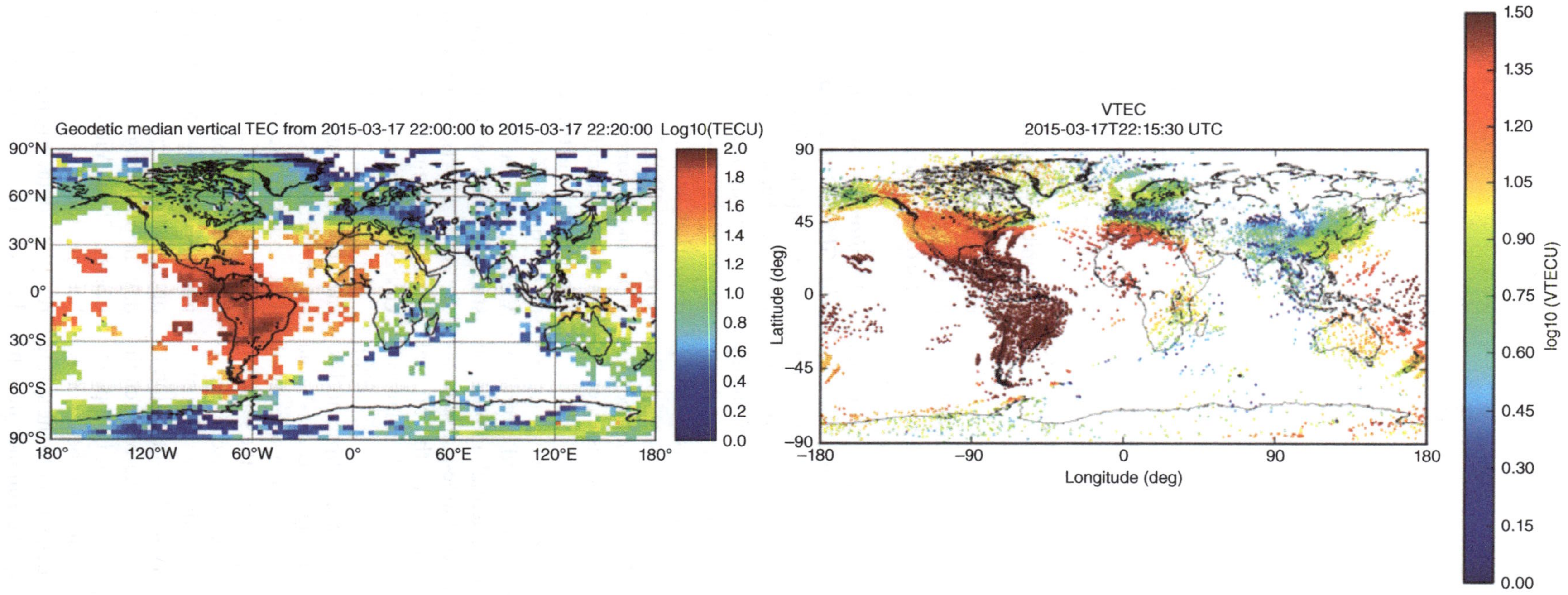

Figure 8.4 Plots available through the Cedar Madrigal web site illustrating the differences in detail in TEC between the new line of sight (LOS) product (on right) and the standard binned TEC product (on left).

shows the TEC data binned every 5 minutes and every 1° latitude and longitude. The right plot shows the new line of site data, which provides a TEC estimate to every satellite in view from every receiver. The data are not binned or averaged, but the data are delimited to 20-second sampling. This data set provides far more detail and enables different types of processing at later stages. The difference in the detail of the observed density structure is clearly indicated in these two figures. Utilizing a subset of far fewer receivers, NOAA's real-time USTEC provides a near real-time assessment of the TEC over the continental United States (CONUS; Anghel et al., 2009). The lack of availability of real-time data streams globally distributed is the major drawback for space weather now-casting (https://www.swpc.noaa.gov/products/us-total-electron-content).

Monitoring the 3D structure of ionosphere and plasmasphere using tomographic reconstruction technique

Operational knowledge of the 3D ionospheric electron density distribution is extremely beneficial for satellite communication and navigation, and for single-frequency GNSS users to achieve optimal positioning accuracy. Radio wave transmission and reception at fixed locations on the ground, such as pulsed HF ionosonde and incoherent scatter radar (ISR; e.g. Foster, 1993), topside sounders onboard satellites (e.g. Reinisch & Huang, 2001), and in situ rocket and satellite observations (e.g. Stankov et al., 2003) are among the different techniques that have been used to study the vertical electron density of the ionosphere. However, these instruments are sparsely located, and it is impossible to image global density structures using these instruments for two main reasons: (1) they are too expensive to place at multiple longitudinal and latitudinal sectors, and (2) they only provide density profiles up to a certain altitude (typically 1000 km and sometimes higher, depending on conditions and instrument).

Ionospheric TEC measurements from the growing number of GNSS receivers have been very effectively used to characterize the wide-scale temporal and spatial structure and dynamics of the ionosphere. However, the GNSS TEC data provide only the line integral of the free electron density along individual lines of sight between the transmitter and receiver. These signals alone offer no information on the 3D profile of the density structures. In order to image the ionospheric dynamics and structure continuously, the tomographic inversion method (based on a linear mathematical inversion technique) can be applied to the multiple GNSS TEC lines of sight that are available at any instant and location. The reconstruction plane is discretized into 2D pixels (Heise et al., 2002; Yizengaw & Moldwin, 2005). Usually, but not necessarily, the grid of these 2D boxes is subdivided equidistantly with height and angular spacing (e.g. Raymund, 1995). By representing as a finite sum of shorter integrals along segments of the ray path length in each pixel, and assuming that the electron density is constant in each pixel (Tsai et al., 2002), the integrated GPS TEC can be inverted into a vertical density distribution. Thus, by assuming that the density is constant in each pixel (Tsai et al., 2002), the TEC along the ray path can then be represented as a finite sum of shorter integrals along segments of the ray path length and is given by

$$STEC_i = \int n(r)dl \approx \sum_{j=1}^{M} n_j d_{ij}$$

where $STEC_i$ is the integrated density along the line-of-sight of ray i, $n(r)$ is the electron density at a point r along the ray path, d_{ij} is the ith ray path length inside pixel j, and n_j corresponds to the average electron density inside pixel j. The algebraic reconstruction technique is used to invert the slant TEC into vertical density profiles.

It is likely that in the next decade, the tomographic reconstruction will gradually replace 2D TEC maps due to the growing number of ground- and space-based GNSS receivers. However, ionospheric tomography has some limitations that include (1) a limited number of ray paths between the transmitters and receivers, (2) ray path intersections with the ionosphere and plasmasphere, in that available ray paths may not be uniformly distributed across regions of interest, (3) limited time to complete data sampling due to stationarity requirements, and (4) GNSS satellite geometry. Typically, tomographic algorithms assume that the data measurement paths and the region of interest are coplanar. In the case of GPS ionospheric tomography, however, the chain of receivers, the ionospheric region of interest, and the spacecraft orbit are not necessarily entirely coplanar. Thus, the TEC ray paths may have passed through regions of the ionosphere far different from the desired imaging plane. If the set of TEC data does not accurately reflect the ionosphere in the plane of interest, reconstructions obtained from the data should be expected to differ from other independent measurements, such as ionosonde and ISR density profiles. Therefore, in order to address such geometry problems, not all GPS ray paths from satellites that were in view to the chain of receivers in the region of interest are used for the inversion process. Instead, data obtained only from a number of suitable satellite passes viewed by the available receivers near the plane of interest are used. Detailed descriptions of the algebraic reconstruction technique algorithm and discussion about the potential and limitations of the technique can be found (e.g. Bust

& Mitchell, 2008; Na et al., 1995; Raymund, 1995; Yizengaw et al., 2012, and references therein).

In order to reveal the detailed structure of the 3-D ionospheric dynamics, especially the topside ionosphere structure that is dominated by the F-region density, the tomographic reconstruction is divided in two steps to obtain the profiles of the ionosphere and the plasmasphere. The bottom side ionospheric density structure and dynamics can be imaged using the GNSS network on the ground (from here onward termed as ground-based tomography in this chapter) and has been successfully verified using independently measured density profiles, such as incoherent scatter radar and ionosondes (e.g. Pryse et al., 1995; Yizengaw et al., 2004), as well as through in-situ density observations (e.g. Yizengaw et al., 2012). The GPS observations on board LEO satellites can be used to obtain the vertical density distribution of the topside ionosphere and plasmasphere by applying tomographic reconstruction on the slant TEC with elevation angles greater than zero (referred as space-based tomography from here onward). The advantage of space-based tomography is its ability to image the density structure of the topside ionosphere and plasmasphere, which cannot be done using ground-based tomography due to the "domination" of F-region plasma (e.g. Yizengaw et al., 2006). The reconstruction plane is discretized into pixels, and for the space-based tomography the grid height is not uniform; rather, it increases as a function of altitude (e.g. Heise et al., 2002; Jakowski et al., 2007; Yizengaw et al., 2006).

The left panels in Figure 8.5 show an example comparison between electron density profiles either tomographically reconstructed (top panel) or measured by the Jicamarca ISR (bottom panel) on 29 October 2008. The vertical white line in the top panel marked local midnight, and the white area in the bottom panel indicates where ISR data were not available. The white curve over plotted on both panels represents the vertical drift velocity separately estimated from the Jicamarca–Piura pair of magnetometers using the technique described in Yizengaw et al. (2012). The white horizontal dashed lines show the zero level of the vertical drift velocity, with a velocity scale indicated on the right side. In general, the two independent measurements show excellent agreement in density magnitude, structure, and peak altitudes, demonstrating the power of the tomographic reconstruction technique for satisfactory rendering of the vertical density distribution in a cost-effective way. Due to current resource limitations,

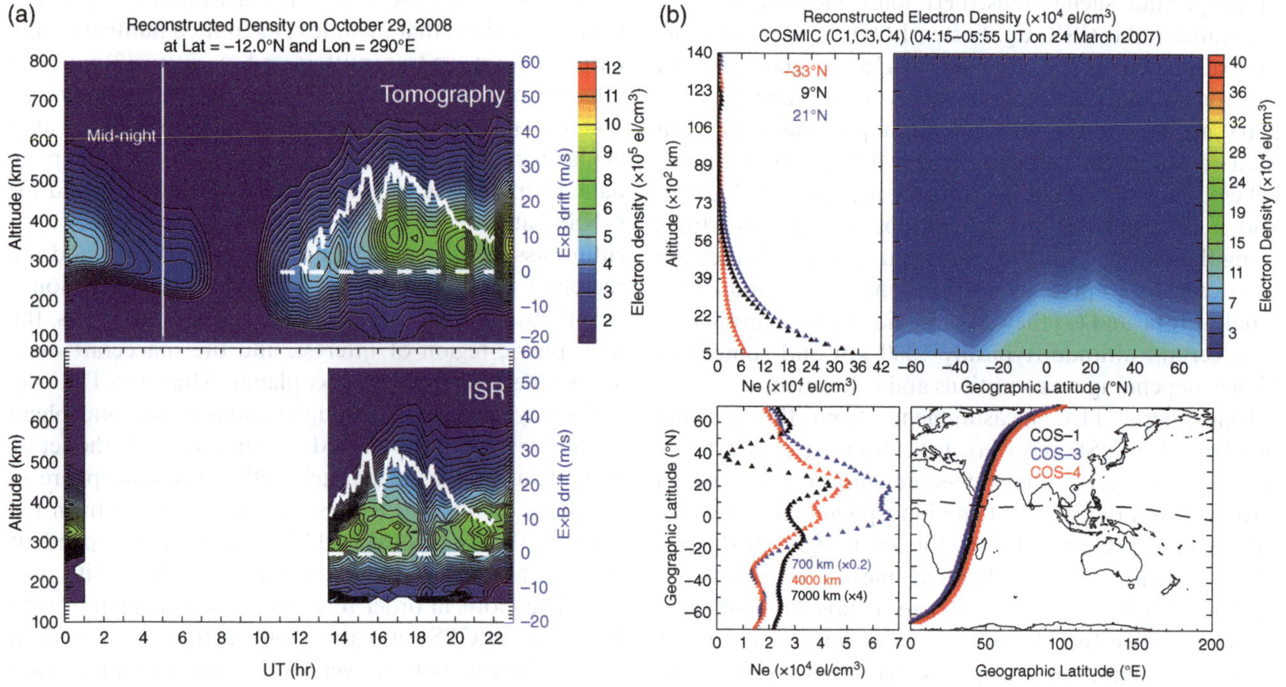

Figure 8.5 The panels on the left side show the vertical electron density distribution obtained by tomographic reconstruction technique (top) and Jicamarca ISR measurement (bottom) at the location of Jicamarca in Lima, Peru (after Yizengaw et al., 2012). The panels on the far right show tomographically reconstructed topside ionosphere and plasmasphere density profiles (top) using TEC data from three tailgating COSMIC (Alizadeh et al., 2011; Rocken et al., 2000) satellites along their ground track trajectories (bottom). The top and bottom panels on the left of this group show the vertical and horizontal density profiles at three different latitudes and altitudes, respectively (after Yizengaw & Carter, 2016).

the Jicamarca ISR does not operate in oblique mode for routine vertical electron density profiles. In these situations, GNSS-based tomographic reconstruction can be performed, yielding reasonable samples of the vertical ionospheric density structure on a synoptic basis at a number of latitudes and longitudes. Tomographically reconstructed vertical density profiles can also be augmented with occultation profiles obtained from the growing number of LEO satellites equipped with a GNSS occultation antenna. The upcoming COSMIC-2 mission (Lee et al., 2013) will have a GNSS occultation sensor on each satellite, which will allow electron density profiles to be extracted not only from GPS but also from the GLONASS signals. This will significantly increase the daily number of occultations available and improve the distribution of these measurements. In the long term, the increasing number of GNSS constellations available, combined with the increasing number LEO satellites with GNSS occultation sensors onboard, promises significant improvement in our ability to image global ionospheric dynamics.

An example of tomographically imaged topside ionosphere and plasmasphere density profiles using data from LEO satellite constellations is shown in Figure 8.5. The right panels in the figure show a typical example of tomographically reconstructed electron density distribution (top right panel) of the topside ionosphere and plasmasphere as a function of altitude and latitude, using topside TEC observations by GPS receivers onboard three tailgating COSMIC (Alizadeh et al., 2011; Rocken et al., 2000) satellites (their ground-track is shown in the bottom right panel). The time span for the three satellites to cover the given latitudes was about 70 minutes, and by merging the data from these satellites, a 70-minute average topside ionosphere and plasmasphere density profile is reconstructed. Once the density is reconstructed, it is possible to extract a density profile at different latitudes (see top left panel) and investigate the latitudinal difference of the in-situ density structure and magnitude at different altitudes (see bottom left panel). This may help us understand the physics behind the horizontal and vertical plasma transport both during dayside and nightside at a very high altitude (topside ionosphere and plasmasphere regions).

In general, such promising space-based tomographic reconstruction techniques are the most valuable means to remotely monitor the topside ionosphere and plasmasphere density structures. The growing number of LEO satellites with GNSS antenna onboard enhances the ray path coverage to accurately capture medium and possibly even small-scale structures of the topside ionosphere and plasmasphere density distributions. These data will contribute to our understanding of space weather–driven impacts on our communication and navigation systems.

Longitudinal dependence of space weather–associated scintillation measurements

Ionospheric scintillations are one of the primary space weather impacts on GNSS systems. The vulnerability of GPS receivers to scintillations, for example, is primarily dictated by the lower GPS frequency L2 ($f2$ = 1.2276 GHz). Small-scale irregularities in electron density can diffract the signal, leading to rapid fluctuations in both signal intensity, known as amplitude scintillation, and phase, known as phase scintillation (Skone & Knudsen, 2000). Amplitude scintillation can be severe enough that the received GPS signal intensity drops below a receiver's clock threshold, forcing the receiver to reacquire the signal. This situation is termed loss of lock. Amplitude scintillation is measured by the S4 index, which is the ratio of the standard deviation of the intensity of the received signal to its mean. S4 can vary from 0.1 for very weak ionospheric scintillation to 0.9 for strong amplitude scintillation in the equatorial region (Secan, 1996).

Despite the fact that much progress has been made in the study of ionospheric scintillations, there are still many gaps remaining in the effort toward understanding the global equatorial ionospheric irregularity/scintillation distributions and its driving mechanisms. The longitudinal and day-to-day variabilities of ionospheric irregularities/scintillations are still challenging issues for the modeling community. For example, irregularities appear to be more prevalent over the African continent and active all year round, even during the postmidnight sector (Yizengaw et al., 2013), compared to other longitudes (e.g. Burke et al., 2004; Hei et al., 2005; Su et al., 2006; Yizengaw et al., 2014; Yizengaw & Groves, 2018).

Figure 8.6 presents the statistical longitudinal variability of ionospheric irregularity distributions (local time/day of the year) estimated in the vicinity of geomagnetic equator using the time rate of TEC (ROT) index or ROTI, described earlier. The different panels represent different longitudes (given at the top of each panel). The geographic locations of GNSS receivers are also indicated on the bottom panel. The color index represents the TEC irregularities in terms of ROTI. In general, the irregularity distributions show several different features at different longitudes. These include following observations: (1) irregularities are weaker in the Pacific sector and stronger in the African and Asian sectors; (2) during the June solstice, there are almost no irregularities detected in the American sector, yet substantial irregularities are observed in the African and Pacific sectors; and (3) during the equinoxes, significant postmidnight bubbles are

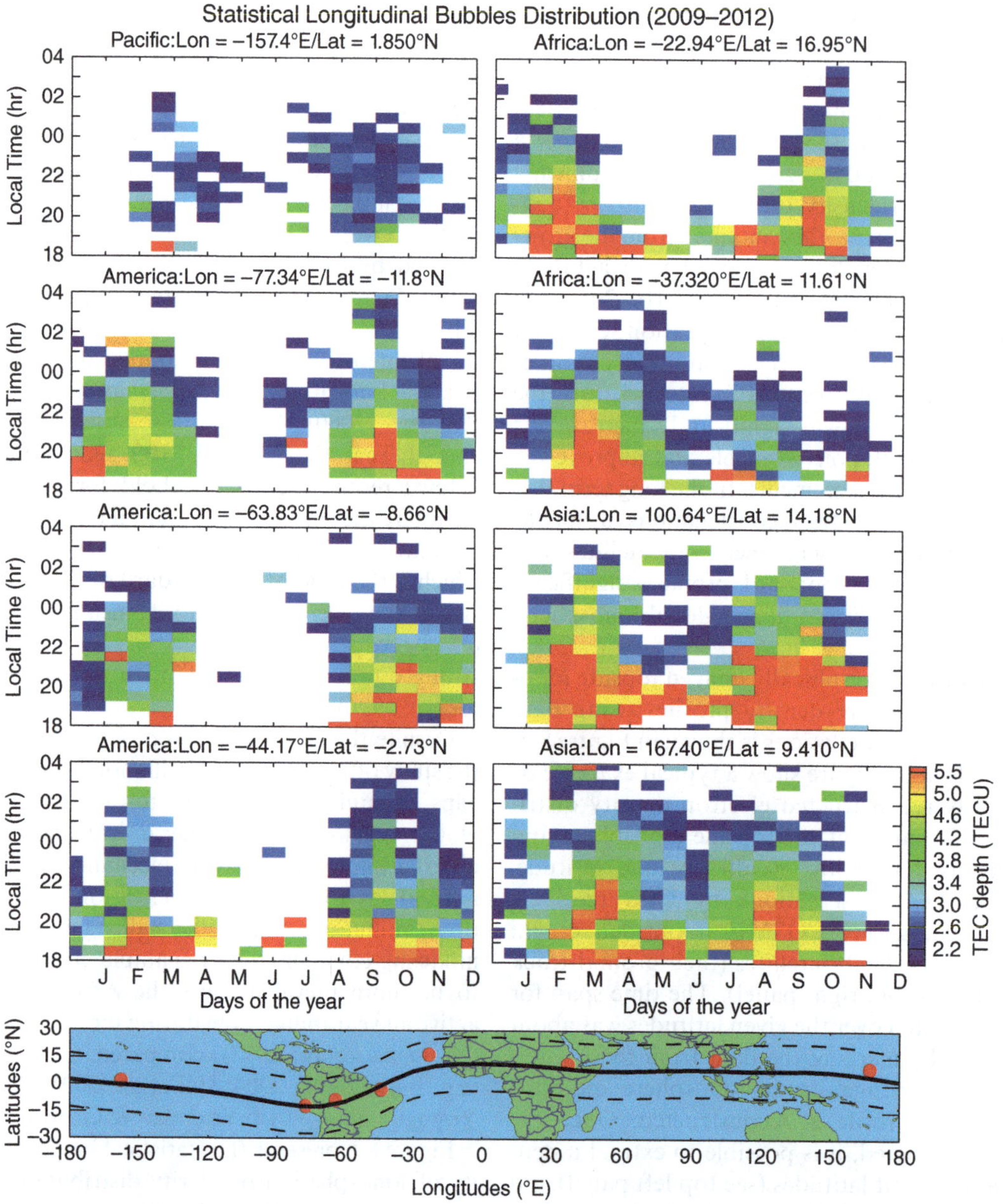

Figure 8.6 The statistical view of GNSS ROTI distribution as a function of local time versus day of the year. Different panels depict different longitudes that are given at the top of each panel and also shown on the map at the bottom panel. (Source: Yizengaw & Groves, 2018. Reproduced with permission of John Wiley & Sons.)

also seen in the Brazilian, African, and Asian sectors. Figure 8.6 clearly illustrates that density irregularities behave differently at different longitudes and seasons. Therefore, understanding the physics of irregularities at different longitudes is essential to model and accurately capture the global irregularity structure and dynamics of the ionosphere. This information is also required to estimate associated effects on GNSS signal propagation. In general, these aggregate observations demonstrate that GNSS signal propagation may differ significantly in different longitudinal sectors.

8.4. GNSS APPLICATIONS

GNSS services are deeply integrated into our society, and our reliance on them has significantly grown. Certain critical applications such as railway control, highway traffic management, emergency response, commercial aviation, and marine navigation require high-precision positioning. Understanding and mitigating the impacts of space weather on these applications is critically important. We discuss here specific air, marine, and land navigation problems with space weather effects. Following this, we will discuss the changing conditions in the Arctic to provide a clear example of the importance of understanding and mitigating space weather and to emphasize the need for a larger observational space weather network.

With respect to space weather impacts on air navigation, we have described the space weather issues the FAA's Wide-Area Augmentation System (WAAS) experienced during the October 2003 solar storms. WAAS is an air navigation aid developed by the FAA to augment GPS, with the goal of improving its accuracy, integrity, and availability. Essentially, WAAS is intended to enable aircraft to rely on GPS for all phases of flight, including precision approaches to any airport within its coverage area. It may be further enhanced with the Local Area Augmentation System, also known by the United Nations International Civil Aviation Organization's preferred term, the Ground-Based Augmentation System, in critical areas. WAAS is based on a network of ground-based GPS reference stations in the North American continent that continuously measure ionospheric slant delays (the line-of-sight range delays between the GPS receiver and the satellite). All of these data are transferred to master stations, where the integrity of the system is assessed, and the measurements are combined to compute ionospheric vertical delays (the delay due to the ionosphere looking directly overhead) for a virtual set of ionospheric grid points that is spaced over much of North America. By using a bilinear interpolation algorithm, WAAS avionics use the ionospheric grid points to estimate the adjustments needed to correct for receiver-to-satellite ionospheric delays. Although WAAS was the first GPS-based augmentation aviation system in operation, similar systems are currently operational or in development in, for example, Europe, Japan, and India. Collectively, these systems are called satellite-based augmentation systems.

Doherty et al. (2004) documented that during the storm periods of October and November 2003, WAAS availability for precision approach service was degraded due to the rapid changes and strong spatial gradients in the ionosphere. Ionospheric gradients resulted in interruptions of WAAS precision approach services for 15 h on the 29 October and for 11.3 h on 30 October.

Skone et al. (2005) reported on the significant impacts on differential GPS systems (DGPS) used for a variety of marine applications during the October 2003 space weather events. DGPS involves calculating range errors at a reference station with known coordinates and relaying the error information to remote users within the region of coverage (generally within 500 km of the reference station). In this manner, orbital and atmospheric errors are reduced, and the satellite clock error may be eliminated, but receiver noise and multipath effects still remain. Various multipath mitigation techniques exist, including proper selection of antenna placement and receiver firmware and hardware.

Marine applications that use DGPS services include hydrographic surveying applications, exploration/exploitation of marine resources, assistance to vessel traffic management services, search and rescue operations, environmental assessment and clean-up, and underwater mine detection and disposal. Marine horizontal position accuracy requirements are 2–5 m (95%) and 8–20 m (95%) for safety of navigation in inland waterways and harbor entrances/approaches, respectively; horizontal position accuracies of 1–100 m (95%) are required for benefits of resource exploration in coastal regions (DOD/DOT, 2001). Horizontal positioning accuracies better than 2 m (95th percentile) are required for Special Order surveys (IHO, 1998). For the most part, marine DGPS services offer satisfactory performance in terms of accuracy, reliability, and availability. Under high levels of ionospheric activity, however, significant degradations in DGPS positioning accuracies can occur. During the October 2003 storms, marine DGPS positioning errors were observed to increase by factors of 10–30 under increased ionospheric activity. In particular, gradients of up to 50 ppm are associated with storm-enhanced density (SED; discussed in section 8.2.2). DGPS positioning errors of 20 m or more have persisted for hours during such events.

Additional examples of high-precision GNSS application include surface mining operations for natural resource extraction (Coster et al., 2017). These services were disrupted due to discontinuities in GPS precise positioning services during auroral activity. Similarly, precision farming and offshore oil drilling require high-precision GNSS. In places like Brazil and northern Canada, there have been numerous reports of disruptions.

8.5. SUMMARY

Society cannot become overly reliant on technology without an awareness and understanding of the effects of future space weather disruptions. As technology advances, societies of tomorrow are expected to increase their need for highly accurate communication and navigation systems. These systems are vulnerable to space weather, and yet many of these systems are critical to our ability to respond to other emergencies. Most space weather effects are global in nature. For society to properly address them, real-time global information will be required.

To achieve these goals, problems that must be overcome include

- Collecting space weather data over oceans and in remote, difficult to access locations
- Reliable and relatively inexpensive global communication links
- Real-time data transfer and access
- Establishment of open, free, easily accessible databases
- Data stored in standardized, common formats.

REFERENCES

Aarons, J., & Basu, S. (1994). Ionospheric amplitude and phase fluctuations at the GPS frequencies. *Proceedings of the 7th International Technical Meeting of the Satellite Division of The Institute of Navigation (ION GPS 1994)*, Salt Lake City, UT, September 1994, pp. 1569–1578.

Aarons, J, Klobuchar, J. A., Whitney, H. E., Austen, J., Johnson, A. L., & Rino, C. L. (1983). Gigahertz scintillations associated with equatorial patches. *Radio Sci, 18*(3), 421–434.

Alizadeh, M. M., Schuh, H., Todorova, S., & Schmidt, M. (2011). Global ionosphere maps of VTEC from GNSS, satellite altimetry, and formosat-3/COSMIC data. *J Geod, 85*, 975. https://doi.org/10.1007/s00190-011-0449-z

Anghel, A., Carrano, C., Komjathy, A., Astilean, A., & Letia, T. (2009). Kalman filter-based algorithms for monitoring the ionosphere and plasmasphere with GPS in near-real time. *J. Atmos. Sol. Terr. Phys., 71*, 158–174. doi:10.1016/j.jastp.2008.10.006

Basu, Su., Basu, S., Makela, J. J., MacKenzie, E., Doherty, P., Wright, J. W., et al. (2008). Large magnetic storm-induced nighttime ionospheric flows at midlatitudes and their impacts on GPS-based navigation systems. *J. Geophys. Res., 113*, A00A06. doi:10.1029/2008JA013076

Basu, S., Groves, K.M., Quinn, J. M., & Doherty, P. (1999). A comparison of TEC fluctuations and scintillations at Ascension Island. *Journal of Atmospheric and Solar-Terrestrial Physics, 61*(16), 1219–1226. https://doi.org/10.1016/S1364-6826(99)00052-8

Basu, S., Basu, S., Groves, K. M., Yeh, H.-C., Su, S.-Y., Rich, F. J., et al. (2001). Response of the equatorial ionosphere in the South Atlantic Region to the Great Magnetic Storm of July 15, 2000. *Geophys. Res. Lett., 28*, 3577–3580. *doi:10.1029/2001GL013259*

Betz, J. W., Lu, M., Morton, Y.T.J., & Yang, Y. (2019). Introduction to the special issue on the BeiDou navigation system. *NAVIGATION, 66*, 3–5. https://doi.org/10.1002/navi.293

Booker, H. G., & Wells, H. W. (1938). Scattering of radio waves by the F-region of the ionosphere. *J. Geophys. Res., 43*(3), 249– 256. doi: 10.1029/TE043i003p00249

Burke, W. J., Huang, C. Y., Gentile, L. C., & Bauer, L. (2004). Seasonal-longitudinal variability of equatorial plasma bubble occurrence. *Ann. Geophys., 22*, 3089.

Bust, G. S., & Mitchell, C. N. (2008). History, current state, and future directions of ionospheric imaging. *Rev. Geophys., 46*, RG1003. doi:10.1029/2006RG000212

Carrano, C. S., Groves, K. M., & Rino, C. L. (2019). On the relationship between the rate of change of total electron content index (ROTI), irregularity strength (CkL), and the scintillation index (S4). *J. Geophys. Res.,124*. doi: 10.1029/2018JA026353

Cerruti, A. P., Kintner, P. M., Gary, D. E., Lanzerotti, L. J., de Paula, E. R., & Vo, H. B. (2006). Observed solar radio burst effects on GPS/Wide Area Augmentation System carrier-to-noiseratio. *Space Weather, 4*, S10006. doi:10.1029/2006SW000254

Cerruti, A. P., Kintner, P. M., Gary, D. E., Mannucci, A. J., Meyer, R. F., Doherty, P., and Coster, A. J. (2008). Effect of intense December 2006 solar radio bursts on GPS receivers. *Space Weather, 6*, S10D07. doi:10.1029/2007SW000375

Conker, R. S., El-Arini, M. B., Hegarty, C. J., & Hsiao, T. (2003). Modeling the effects of ionospheric scintillation on GPS/Satellite-Based Augmentation System availability. *Radio Sci., 38*(1), *1001. doi:*10.1029/2000RS002604

Coster, A. J., & Foster, J. C. (2007). Space weather impacts of the sub-auroral polarization stream. *Radio Sci. Bulletin, 321*, 28–36.

Coster, A. J., Foster, J. C., & Erickson P. J. (2003). Monitoring the ionosphere with GPS. *GPS World*, May, pp. 40–45.

Coster, A. J., Foster, J. C., Erickson, P. J., & Rich, F. J. (2001). Regional GPS mapping of storm enhanced density during the 15–16 July 2000 geomagnetic storm. *Proceedings of ION GPS*, Salt Lake City, UT, 2001, pp. 2532–2539.

Coster, A., & Komjathy, A. (2008). Space weather and the Global Positioning System. *Space Weather, 6*.

Coster, A., Pratt, M., Burke, B., Misra, P. (1999). Addition of dual-frequency GLONASS Data to GPS Ionospheric Modeling. *Proceedings of Ionospheric Effects Symposium*, Alexandria, VA, 1999, pp. 476–483.

Coster, A. J., Skone, S., Hampton, D., Donovan, E. (2017). Monitoring space weather with GNSS networks: Expanding GNSS networks into Northern Alaska and Northwestern Canada. *Proceedings of the 30th International Technical Meeting of the Satellite Division of The Institute of Navigation (ION GNSS+ 2017)*. Portland, Oregon, September 2017, pp. 2055–2066. https://doi.org/10.33012/2017.15134

Datta-Barua, S., Doherty, P. H., & Delay, S. H. (2003). Ionospheric scintillation effects on single and dual frequency GPS Positioning. *Proceedings of the ION GPS 2003*, Portland, OR, 9–12 9–12 September, pp. 336–346.

Datta-Barua, S., Walter, T., Blanch, J., & Enge, P. (2005). Can WAAS availability be inferred from geomagnetic data? An analysis. *Proceedings of Ionospheric Effects Symposium 2005*, Springfield, VA, 3–5 May 2005.

Davies, K. (1990). *Ionospheric radio*. London: Peter Peregrinus.

Doherty, P., Coster, A. J., & Murtagh, W. (2004). Space weather effects of October–November 2003, *GPS Sol., 8*(3). doi:10.1007/s10291-004-0109-3

Doherty, P. H., Delay, S. H., Valladares, C. E., & Klobuchar, J. A. (2000). Ionospheric scintillation effects in the equatorial and auroral regions. *Proceedings of the ION GPS 2000*. Salt Lake City, UT, September 2000.

DOD/DOT (2001). Federal radio navigation plan. U.S. Department of Defense/Department of Transport (DODVNTSC-RSPA-01-3/DOD-4650.5).

El-Arini, M. B., Conker, R. S., Ericson, S. D., Bean, K. W., Niles, F., Matsunaga, K., & Hoshinoo, K. (2003). Analysis of the effects of ionospheric scintillation on GPS L2 in Japan. *Proceedings of the ION GPS 2003*. Portland, OR, September 2003.

Fejer, B. G., Scherliess, L., & de Paula, E. R. (1999). Effects of the vertical plasma drift velocity on the generation and evolution of equatorial spread *F. J. Geophys. Res., 104*(A9), 19859–19869. doi:10.1029/1999JA900271

Fernandes, M. J., Lazaro, C., Nunes, A. L., & Pires, N. (2010) GNSS-derived path delay: An approach to compute the wet tropospheric correction for coastal altimetry. *IEEE Geoscience and Remote Sensing Letters*, *7*(3), 596–600.

Foster, J. C. (1993). Storm time plasma transport at middle and high latitudes. *J. Geophys. Res.*, *98*(A2), 1695.

Foster, J. C., & Burke, W. J. (2002). SAPS: A new categorization for sub-auroral electric fields. *EOS Trans. AGU*, *83*(36), 393.

Foster, J. C., Coster, A. J., Erickson, P. J., Holt, J. M., Lind, F. D., Rideout, W., et al. (2005). Multiradar observations of the polar tongue of ionization. *J. Geophys. Res.*, *110*, A09S31. doi:10.1029/2004JA010928

Foster, J. C., Erickson, P. J., Coster, A. J., Goldstein, J., & Rich, F. J. (2002). Ionospheric signatures of plasmaspheric tails. *Geophys. Res. Lett.*, *29*(13), 1623. doi:10.1029/2002GL015067

Foster, J. C., Rideout, W., Sandel, B., Forrester, W. T., & Rich, F. J. (2007). On the relationship of SAPS to storm-enhanced density. *J. Atmos. Sol. Terr. Phys.*, *69*, 303–313.

Foster, J. C., & Vo, H. B. (2002)., Average characteristics and activity dependence of the subauroral polarization stream. *J. Geophys. Res.*, *107(*A12*)*, 1475. doi:10.1029/2002JA009409

Ganguly, S., Jovancevic, A., Brown, A., Kirchner, M., & Zigic, S. (2003). Ionospheric scintillation monitoring and mitigation using a software GPS receiver. *Radio Science*, *39*(1), 1–9.

Goncharenko, L. P., Coster, A. J., Chau, J. L., & Valladares, C. E. (2010). Impact of sudden stratospheric warmings on equatorial ionization anomaly. *J. Geophys. Res.*, *115*, A00G07. doi:10.1029/2010JA015400

Groves, K. M., Basu, S., Weber, E. J., Smitham, M., Kuenzler, H., Valladares, C. E., et al. (1997). Equatorial scintillation and system support. *Radio Sci.*, *32*, 2047.

Guier, W. H., & Weiffenbach, G. C. (1959). The Doppler determination of orbits. Johns Hopkins Univ Laurel MD Applied Physics Lab, Accession Number AD0409103. https://apps.dtic.mil/dtic/tr/fulltext/u2/409103.pdf

Guier, W. H., & Weiffenbach, G. C. (1997) Genesis of satellite navigation. *Johns Hopkins APL Technical Digest*, *18*(2).

Heelis, R. A., Sojka, J. J., David, M., & Schunk, R. W. (2009). Storm time density enhancements in the middle-latitude dayside ionosphere. *J. Geophys. Res.*, *114*, A03315. doi:10.1029/2008JA013690

Hei, M. A., Heelis, R. A., & McClure, J. P. (2005). Seasonal and longitudinal variation of large-scale topside equatorial plasma depletions. *J. Geophys. Res.*, *110*, A12315. doi:10.1029/2005JA011153

Heise, S., Jakowski, N., Wehrenpfennig, A., Reigber, C., & Luehr, H. (2002). Sounding of the topside ionosphere/plasmasphere based on GPS measurements from CHAMP: Initial results. *Geophys. Res. Lett.*, *29*(14), 1699. doi:10.1029/2002GL014738

Hlubek N., Berdermann, J., Wilken, V., Gewies, S., Jakowski, N., Wassaie, M., & Damtie, B. (2014). Scintillations of the GPS, GLONASS, and Galileo signals at equatorial latitude. *J. Space Weather Space Clim.*, *4*, A22. doi: 10.1051/swsc/2014020

Immel, T. J., Sagawa, E. England, S. L., Henderson, S. B., Hagan, M. E., Mende, S. B., et al. (2006). Control of equatorial ionospheric morphology by atmospheric tides. *Geophysical Research Letters*, *33*(15). http://doi.org/10.1029/2006GL026161

International Hydrographic Organization (IHO) (1998). IHO Standards for Hydrographic Surveys, International Hydrographic Organization, Special Publication No 44, 4th ed.

Jakowski, N., Wilken, V., & Mayer, C. (2007). Space weather monitoring by GPS measurements on board CHAMP. *Space Weather*, *5*, S08006. doi:10.1029/2006SW000271

Kelley, M. (1989). *The Earth's ionosphere, plasma physics and electrodynamics*. Academic Press, ISBN 0-12-404012-8.

Kelley, M. C., Vlasov, M. N., Foster, J. C., & Coster, A. J. (2004). A quantitative explanation for the phenomenon known as storm-enhanced density. *Geophys. Res. Lett.*, *31*, L19809. doi:10.1029/2004GL020875

Keskinen, M. J., Basu, S., & Basu, S. (2004). Midlatitude sub-auroral ionospheric small scalestructure during a magnetic storm. *Geophys. Res. Lett.*, *31*, L09811. doi:10.1029/2003GL019368

Kintner, P. M., Ledvina B. M., & de Paula E. R. (2007). GPS and ionospheric scintillations. *Space Weather*, *5*, S09003. doi:10.1029/2006SW000260.

Klobuchar, J.A., 1996. Ionospheric effects on GPS. In: Parkinson, B.W. and J.J. Spilker (Ed.), *Global Positioning System: Theory and application* Vol. 1. American Institute of Aeronautics and Astronautics INC.

Knight, M., Cervera, M., & Firm, A. (1999). A comparison of predicted and measured GPS performance in an ionospheric scintillation environment. *Proceedings of ION GPS-99*. Nashville, TN, September.

Langley, R. (2007). Solar burst impacts on GPS. *GPS World* (May). Available at http://sidt.gpsworld.com/gpssidt/article/articleDetail.jsp?id=421360.

Ledvina, B. M., Makela, J. J., & Kintner, P. M. (2002). First observations of intense GPS L1 amplitude scintillations at midlatitude. *Geophys. Res. Lett.*, *29*(14), 1659. doi:10.1029/2002GL014770

Ledvina, B. M., Makela J. J., and Kintner P. M. (2004). Temporal scales of the GPS L1 amplitude scintillations at midlatitude. *Radio Science*, *39*, 1. doi:10.1029/2002RS002832

Lee, I. T., Tsai, H. F., Liu, J. Y., Lin, C. H., Matsuo, T., & Chang, L. C. (2013). Modeling impact of FORMOSAT-7/COSMIC-2 mission on ionospheric space weather monitoring. *J. Geophys. Res. Space Physics*, *118*, 6518– 6523. doi:10.1002/jgra.50538

Mannucci, A. J., Tsurutani, B. T., Iijima, B. A., Komjathy, A., Saito, A., Gonzalez, W. D., et al. (2005). Dayside global ionosphericresponse to the major interplanetary events of October 29 –30, 2003 "Halloween Storms." *Geophys. Res. Lett.*, *3ˆ2*, L12S02. doi:10.1029/2004GL021467

Maruyama, N., Richmond, A. D., Fuller-Rowell, T. J., Codrescu, M. V., Sazykin, S., Toffoletto, F. R., et al. (2005). Interaction between direct penetration and disturbance dynamo electric fields in the storm-time equatorial ionosphere. *Geophys. Res. Lett.*, *32*, L17105. doi:10.1029/2005GL023763

Mayer, C., Becker, C., Jakowski, N., & Meurer, M. (2010). Ionosphere monitoring and inter-frequency bias determination using Galileo: First results and future prospects. *Adv. Space Res.*, *47*, 859–866.

Morrissey, T. N., Shallberg, K. W., Van Dierendonck, A. J., & Nicholson, M. J. (2004). GPS receiver performance characterization under realistic ionospheric phase scintillation environments. *Radio Science*, *39*(Rs1S20), 1–18.

Na, H. R., Hall, B., & Sutton, E. (1995). Ground station spacing effects in ionospheric tomography. *Annales Geophysicae*, *13*, 1288–1296.

Parkinson, B. W. (1980). *Overview, Global Positioning System*, Vol. I. Washington DC: Institute of Navigation.

Patra, A. K., Phanikumar, D. V., & Pant, T. K. (2009). Gadanki radar observations of F regionfield-aligned irregularities during June solstice of solar minimum: First results and preliminary analysis. *J. Geophys. Res.*, *114*, A12305. doi:10.1029/2009JA014437

Pi, X., Boulat, B. M., Mannucci, A. J., & Stowers, D. A. (2002). Latitudinal characteristics of L-band ionospheric scintillation. *Proceedings of the ION GPS 2002*. Portland, OR, 24–27September 2002.

Pi, X., Mannucci, A. J., Lindqwister, U. J., & Ho, C. M. (1997). Monitoring of global ionospheric irregularities using the worldwide GPS networ., *Geophys. Res. Lett.*, *24*(18), 2283–2286.

Pryse, S. E., Mitchell, C. N., Heaton, J.A.T., & Kersley, L. (1995). Travelling ionospheric disturbance imaged by tomographic techniques. *Annales Geophysicae*, *13*, 1325–1342.

Pullen, S., Opshaug, G., Hansen, A., Walter, T., Enge, P., & Parkinson, B. (1998). A preliminary study of the effect of ionospheric scintillation on WAAS user availability in equatorial regions. *Proceedings of the ION GPS 1998*. Alexandria, VA, September.

Raymund, T. D. (1995). Comparisons of several ionospheric tomography algorithms. *Annales Geophysicae*, *13*, 1254–1262.

Reinisch, B. W. and X. Huang (2001). Deducing topside profiles and total electron content from bottomside ionograms, Adv. *Space Res.*, Vol. *27*, No. 1, p. 22–30. 2001, 0273-1177/01.

Retterer, J. M. (2005). Physics-based forecasts of equatorial radio scintillation for C/NOFS. *Space Weather*, *3*, S12C03.

Retterer, J. M. (2010). Forecasting low-latitude radio scintillation with 3-D ionospheric plume models: 2. Scintillation calculation. *J. Geophys. Res.*, *115*, A03307. doi:10.1029/2008JA013840

Rideout, W., & Coster, A. (2006). Automated GPS processing for global total electron content data. *GPS Solutions*, *10*(3), 219–228.

Rocken, C., Kuo, Y. H., Schreiner, W., Hunt, D., & Sokolovskiy, S. (2000). COSMIC system description. *J Terr Atmos Ocean Sci TAO*, *11*(1), 21–52.

Secan, J. A. (1996). WBMOD: Ionospheric Radiowave Scintillation Model, Version 13.04. NorthWest Res. Assoc., Inc., Bellevue, WA.

Skone, S., Coster, A., Hoyle, V., & Laurin, C. (2003). WAAS availability and performance at high latitudes. Presented at ION GPS/GNSS. Inst. of Navig., Portland, OR, September 2003.

Skone, S., & Knudsen, K. (2000). Impact of ionospheric scintillations on SBAS performance. *Proceedings of the ION GPS 2000*. Salt Lake City, UT, September.

Skone, S., & Yousuf, R. (2007). Performance of satellite-based navigation for marine users during ionospheric disturbances. *Space Weather*, *5*, S01006. doi:10.1029/2006SW000246

Skone. S., Yousuf, R., & Coster, A. (2005). Mitigation of ionospheric errors for marine DGPS. *European Journal of Navigation*, *3*(3), 58–66.

Stankov, S. M., Jakowski, N., Heise, S., Muhtarov, P., Kutiev, I., & Warnant, R. (2003). A new method for reconstruction of the vertical electron density distribution in the upper ionosphere and plasmasphere. *J. Geophys. Res.*, *108*, 1164. doi:10.1029/2002JA009570, A5

Su, S.-Y., Liu, C. H., Ho, H. H., & Chao, C. K. (2006). Distribution characteristics of topside ionospheric density irregularities: Equatorial versus mid-latitude regions. *J. Geophys. Res.*, *111*, A06305. doi:10.1029/2005JA011330

Taur, R. R. (1973). Ionospheric scintillation at 4 and 6 GZ. *COMSAT Tech. Rev.*, *3*, 145–163.

Taur, R. R. (1976). Simultaneous 1.5- and 4-GHz ionospheric scintillation measurement. *Radio Sci.*, *11*, 1029–1036.

Thomas, E. G., Baker, J.B.H., Ruohoniemi, J. M., Clausen, L.B.N., Coster, A. J., Foster, J. C., & Erickson P. J. *(2013)*. Direct observations of the role of convection electric field in the formation of a polar tongue of ionization from storm enhanced density. *J. Geophys. Res. Space Physics*, *118*, 1180–1189. doi:10.1002/jgra.50116

Tsai, L. C., Liu, C. H., Tsai, W. H., & Liu, C. T. (2002). Tomographic imaging of the ionosphere using the GPS/MET and NNSS data. *Journal of Atmospheric and Solar-Terrestrial Physics 64*, 2003–2011.

Vierinen, J. A., Coster, J., Rideout, W. C., Erickson, P. J., & Norberg, J. (2015). Statistical framework for estimating GNSS bias. *AMT.* doi:10.5194/amtd-8-9373-2015

Vlasov, M., Kelley, M. C., & Kil, H. (2003). Analysis of ground-based and satellite observations of F-region behavior during the great magnetic storm of July, 2000. *J. Atmos. Sol.-Terr. Phys.*, *65*, 1223–1234.

Wickert, J., Beyerle, G., Hajj, G. A., Schwieger, V., & Reigber, C. (2002). GPS radio occultation with CHAMP: Atmospheric profiling utilizing the space-based single difference technique, *Geophys. Res. Lett.*, *29*(8), 1187. doi:10.1029/2001GL013982

Yizengaw, E. (2019). Ionospheric dynamics and its strong longitudinal dependences, In C-S. Huang & G. Lu (Eds.), *Advances in ionospheric research: Current understanding and challenges*. American Geophysical Union.

Yizengaw, E., & Carter, B. A. (2016). Imaging the global vertical density structure from the ground and space. *Ionospheric space weather: Longitude dependence and lower atmosphere forcing*, *220*, 105.

Yizengaw, E., Dewar, J., MacNeil, J., Moldwin, M. B., Galvan, D., Sanny, J., et al. (2008). The occurrence of ionospheric signatures of plasmaspheric plumes over different longitudinal sectors. *J. Geophys. Res.*, *113*, A08318. doi:10.1029/2007JA012925

Yizengaw, E., Dyson, P. L., & Essex, E. A. (2004). Tomographic reconstruction of the ionosphere using ground-based GPS data in the Australian region. *Workshop on Applications of Radio Science (WARS)*. Hobart, Australia, February 2004.

Yizengaw, E., & Groves, K. (2018). Longitudinal variability of equatorial electrodynamics, ionospheric irregularities, and ambient electron density. *Space Weather*, *14*. doi:10.1029/2018SW001980

Yizengaw, E., & Moldwin, M. B. (2005). The altitude extension of the mid-latitude trough and its correlation with plasmapause position. *Geophys. Res. Lett.*, *32*, L09105. doi:10.1029/2005GL022854.

Yizengaw, E., Moldwin, M. B., Dyson, P. L., Fraser, B. J., & Morley, S. K. (2006). First tomographic images of ionospheric outflows. *Geophys. Res. Lett.*, *33*, L20102. doi:10.1029/2006GL027698

Yizengaw, E., Moldwin, M. B., Zesta, E., Biouele, C. M., Damtie, B., Mebrahtu, A., et al. (2014). The longitudinal variability of equatorial electrojet and vertical drift velocity in the African and American sectors. *Ann. Geophys.*, *32*, 231–238.

Yizengaw, E., Retterer, J., Pacheco, E. E., Roddy, P., Groves, K., Caton, R., & Baki, P. (2013). Postmidnight bubbles and scintillations in the quiet-time June solstice. *Geophys. Res. Lett.*, *40*, 5592–5597. doi:10.1002/2013GL058307

Yizengaw, E., Zesta, E., Moldwin, M. B., Damtie, B., Mebrahtu, A., Valladares, C. E., & Pfaff, R. F. (2012). Longitudinal differences of ionospheric vertical density distribution and equatorial electrodynamics. *J. Geophys. Res.*, *117*, A07312. doi:10.1029/2011JA017454

9

Geomagnetic Field Impacts on Ground Systems

John Kappenman[1] and William Radasky[2]

ABSTRACT

This chapter covers the impacts of geomagnetic storms on ground systems with emphasis on the high voltage power grid. Some additional information is provided describing effects on undersea cable communications and mitigating corrosion on oil and pipelines. The chapter describes the variations of past geomagnetic storms in both locations and time variations. The chapter describes the process of converting incident magnetic fields to electric fields and finally to geomagnetically induced currents (GICs). The impacts of these currents on the operation of large transformers is discussed and how power grids may respond to these currents. The chapter also describes the impacts to the U.S. and Canadian power grids during the Great Geomagnetic Storm from March 13-14, 1989. There is also discussion about the possibility of future extreme storms and their potential impact on power systems. A bibliography is provided for those who wish to study these mechanisms and impacts on their own.

9.1. BACKGROUND FOR POWER SYSTEMS

The ability to comprehensively assess the vulnerability of a high-voltage power grid to the geomagnetic storm environment produced by solar activity stems from earlier investigations that have been underway to understand the complex interactions of variations in the magnetosphere and how these couple with deep Earth conductivity and finally to the equally complex infrastructures such as electric power grids, communication cable, and pipeline systems. The magnetic field produced by disturbances in the magnetosphere rises on the order of seconds and may last for hundreds of seconds or longer. Once the magnetic field is known, one must model the deep Earth conductivity profiles to compute the electric fields, and then couple these time and angular variations to models of the high-voltage power grid to compute the currents that will flow.

[1] *Storm Analysis Consultants, Duluth, Minnesota, USA*
[2] *Metatech Corporation, Goleta, California, USA*

Impacts to telecommunications were observed very soon after introduction of this technology, such as the telegraph systems of the 1850s. Electric power system impacts from geomagnetic storms were comprehensively reported as early as 1940 and have been growing in importance as the power system has grown over the intervening years. In order to understand the impacts on power systems, it is necessary to step through the generation process of the fields, the currents, and the responses of the power network.

Geomagnetic storms are a reaction to energy input into the strongly coupled Sun-Earth system from both kinetic pressure increases and interplanetary magnetic field changes (Gonzalez et al., 1999). Although there are different types of disturbances noted at the Earth surface (these types will be discussed later), the disturbances can be characterized as very slowly varying magnetic fields with rise times as fast as a few seconds, and pulse widths of up to an hour. The rate of change of the magnetic field is a major factor in creating significant electric fields in the Earth and thereby inducing quasi-dc current flow in the power transmission network. Since geomagnetic

Space Physics and Aeronomy Collection Volume 5: Space Weather Effects and Applications, Geophysical Monograph 262, First Edition.
Edited by Anthea J. Coster, Philip J. Erickson, and Louis J. Lanzerotti.

DOI: 10.1002/9781119815600.ch9

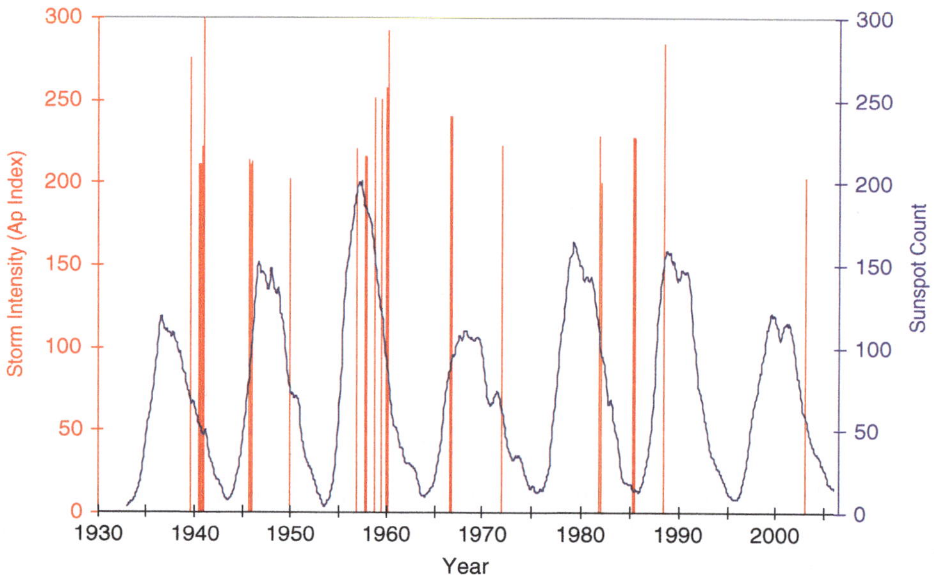

Figure 9.1 Sunspot cycles and the occurrence and intensity (using Ap index) of large geomagnetic storms.

storms (with different magnitudes) are a frequent occurrence, this allows for extensive opportunities to fully benchmark each component of complex simulation models and therefore provide greater confidence in the impacts of plausible severe future threats such as the threat posed by an extreme geomagnetic storm scenario.

The context of the evolution of power system vulnerability discoveries can be further understood by an overview of the geomagnetic storm phenomenon, which is closely associated with the more familiar variability of the sunspot cycle. Figure 9.1 provides a plot of the sunspot count as well as large geomagnetic storms over a ~70-year period in which evolution of the modern-day electric power grid happened to occur. Because each sunspot cycle is typically ~11 years in duration, this plot provides the sunspot count of Cycle 23 back to Cycle 17, which began in ~1932. As can be seen, not all sunspot cycles are of equal intensity, and Cycle 19 (late 1950s–early 1960s) is in fact the largest sunspot cycle of human record. The most recent cycle, Cycle 23, exhibited a profile similar to that of Cycle 17. At the time of the largest cycle, Cycle 19, much of the present US power grid high-voltage transmission system did not exist. To further explore the importance of the evolution of the power grid and growing vulnerability, it is necessary to look at specific large geomagnetic storms, as the sunspot count does not provide sufficient correlations to impacts observed at the Earth. In classification of the intensity of geomagnetic storms, an index called the Ap index is used, which provides a planetary measure of storm activity. Many of these storms caused notable impacts to various terrestrial technology systems of their respective eras. The storms of March 1989 and several in 1991 produced large and unprecedented operational impacts to power grids in the US and at other world locations. Also noteworthy is that large geomagnetic storms generally have not occurred around the peaks of sunspot activity. For example, a storm in February 1986 caused power system problems all across the eastern US and actually occurred at the absolute minimum between Solar Cycle 21 and 22. Because sunspots only provide a gross measure of overall solar activity, it does not accurately reflect the discrete eruptive events from violent solar active regions that, when Earth-directed, can trigger large geomagnetic storms. Rather, it is clear from this comparison that large and threatening geomagnetic storms can occur at any time during the sunspot cycle and pose a near-continuous threat probability.

When reviewing the occurrence of large storms, it is important to recognize that the evaluation of power system impacts is compounded by growing vulnerability of this infrastructure to geomagnetic disturbances. The extent of the growth in vulnerability over time is due to factors stemming from the growth of the high-voltage transmission grid in the US as well as changes within the grid that introduce new or enhance existing impact problems to the power grid. Figure 9.2 shows the growth of the US high-voltage transmission grid over the last 50 years (it should be noted that electric grid development in other locations around the world would have generally occurred at similar rates). This geographically widespread infrastructure readily couples through multiple ground points to the geoelectric field produced by disturbances in the geomagnetic field. As shown, from Cycle 19 through Cycle 22, the high-voltage grid has grown nearly tenfold. In essence, the antenna that is sensitive to disturbances has grown dramatically over time. As this network has grown in size, it has also grown in complexity. As will be discussed in later sections, one of the more important changes in technology base for the US power grid that can increase impacts to geomagnetic storms is the evolution to

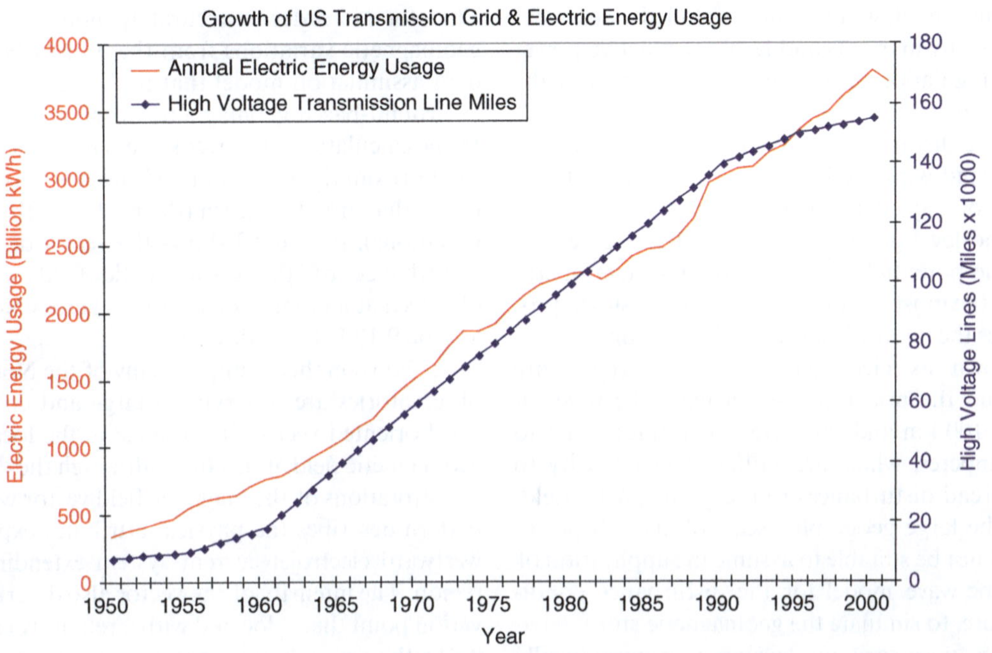

Figure 9.2 Growth of the US high-voltage transmission network and the annual electric energy usage over the past 50 years. In addition to increasing total network size, the network has grown in complexity with introduction of higher kV-rated lines that subsequently also tend to carry larger GIC[?] flows. (Grid size derived from data in EHV Transmission Line Reference Book and NERC Electricity Supply and Demand Database; energy usage statistics from US Dept of Energy, Energy Information Agency).

higher operating voltages of the network. The operating levels of the high-voltage network have increased from the 115–230 kV levels of the 1950s to networks that operate from 345 kV, 500 kV, and 765 kV across the continent.

In order to quantify the impacts of severe geomagnetic storm threats to the US power grid, is it necessary to develop a series of models that translate the disturbed space environment or geomagnetic field environment into specific impacts to the operation of the electric power grid. This requires the following steps:

• Modeling in detail the geographically widespread disturbances to the geomagnetic field from natural geomagnetic storm processes.

• Modeling the electromagnetic coupling between the disturbed space environment and the deep-Earth ground conductivity that produces a geoelectric field across the surface of the Earth.

• Modeling the interaction between the geoelectric field and the complex power grid topology to calculate the flow of geomagnetically induced current (GIC) throughout the exposed power grid infrastructure.

• Modeling of the operational impacts in the US power grid due to GIC flows caused by severe geomagnetic storm conditions.

While each of these models and associated environments are complex, these modeling efforts have been highly successful in accurately replicating geomagnetic storm events and performing detailed forensic analysis of geomagnetic storm impacts to electric power systems (Kappenman & Radasky, 2017). This capability has also been successfully applied towards providing predictive geomagnetic storm forecasting services to the electric power industry. To further describe the methodology used in this analysis, a brief overview is provided for each of the key modeling steps that were undertaken.

9.2. GEOMAGNETIC STORM ENVIRONMENT MODEL

An important facet of this investigation is the simulation of geomagnetic storm events and the impacts that these storms caused to electric power grid operation; it also investigates the potential impacts of very large storm events that have not recently been experienced by today's power systems. Electric power system operators realize that large and severe geomagnetic storms (such as the March 1989 storm) have the potential to cause important power system impacts. Therefore, comprehensive models allow the development of improved understandings of the extremes of the geomagnetic environment and consequential impacts that future severe storms may pose to the integrity of this important infrastructure. The forensic analysis of prior storm events provides two benefits:

1. it allows for validation of the overall electric power grid model accuracy by reproducing past observations of storm impacts on the power infrastructure, and

2. the simulation allows for a more detailed assessment of past critical storm events and levels of reactive power and voltage regulation stress that occurred across the electric power grid.

Geomagnetic disturbances are caused by interactions of the solar wind with the Earth's magnetic field. There are a number of ways that a geomagnetic storm can produce a ground-level geomagnetic field disturbance that could have the potential to impact power system operations. One of the most important geomagnetic storm processes involves the intensification and flow of ionospheric currents known as electrojets. These electrojets are formed around the north and south magnetic poles at altitudes of ~100 km and can have magnitudes of up to ~1 million amperes, which are sufficient in intensity to cause widespread disturbances to the geomagnetic field. Because of the large geographic scale of the US power grid, it would not be suitable to assume the application of a simple plane-wave model for the disturbance conditions. Therefore, to simulate the geomagnetic storm environment, it is necessary to develop a geographically gridded specification of the complex spatial and temporal dynamics of the disturbances as they propagate across North American locations that are to be modeled. Electrons injected into the ionosphere along geomagnetic field lines produce the ionospheric current systems and the resulting auroral/midlatitude electrojet current. Geomagnetic storms can propagate geographically rapidly and in complex and widespread disturbance patterns. Therefore, in order to accurately model the ground-level geomagnetic variations from these disturbance events, a data assimilation model that utilizes a widespread array of ground-based geomagnetic observatories is necessary. In the calculation of power system impacts, it is primarily the horizontal components of the magnetic field disturbance that are of concern (the north-south and east-west variations). Figure 9.3 shows the vector description of a disturbance of the geomagnetic field simultaneously observed at a number of locations across North America at time 9:10 UT, 10 May 1992.

As shown in this example, many of the North American observatories are observing a large and coherent southward-oriented vector disturbance in the locally observed geomagnetic field at this time, although the data will often show rotations of the magnetic field vector with time. This pattern describes the physical attributes expected from a westward electrojet current system extending across the region. The intensity of the vector also describes an observation point that is located with a relatively closer proximity to the ionospheric current source as well. Using this data, an assimilative model has been developed to provide a more detailed definition to the disturbance environment that would be suitable for application on the US power grid model. Figure 9.4 provides a depiction of these model results for the same 9:10 UT time step. This graphic more clearly defines the complex spatial disturbance pattern of the westward electrojet intensification that is occurring at this time. By assimilating other simultaneous observation

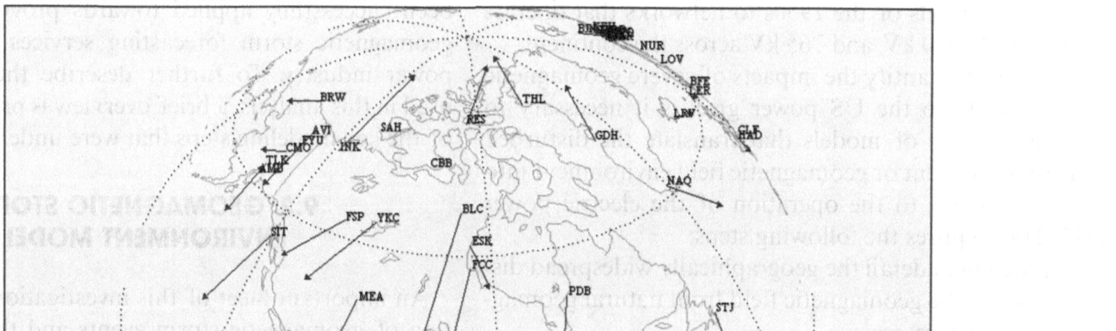

Figure 9.3 Vector intensity of geomagnetic field disturbances at numerous magnetic observatories at a single time point. Reproduced with permission of Metatech Applied Power Solutions.

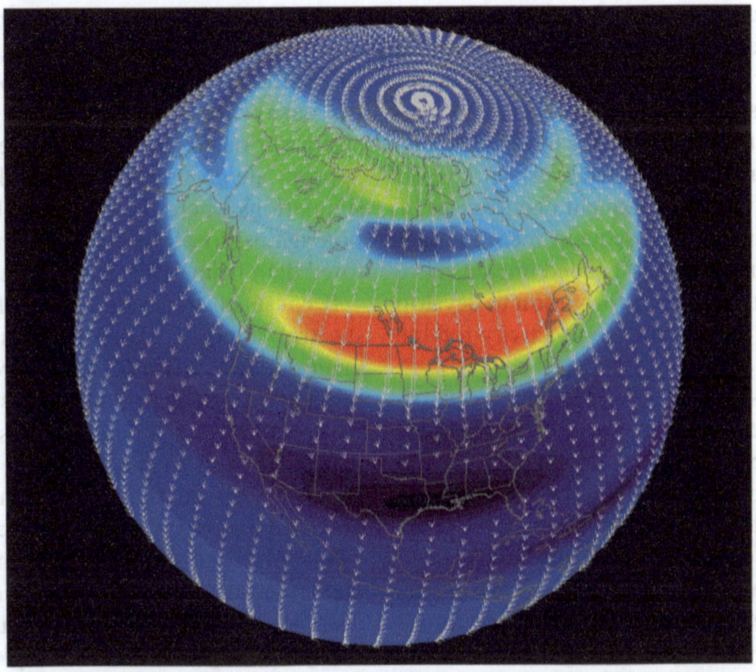

Figure 9.4 Magnetic field environment model using the vector data from Figure 9.3. Reproduced with permission of Metatech Applied Power Solutions.

data, it is possible to also accurately characterize the temporal dynamics of the storm event as well. This modeling approach provides an averaging between interobservatory intensities. It is possible that there could be small regions of higher intensity between observation locations, but this was not assumed for modeling purposes. In many of the storms that have been selected for evaluation, this uncertainty has been minimized by using as many observatories as possible and avoiding sparse data conditions.

For all the geomagnetic storm simulations, discrete magnetic observatory data are used and depicted at the highest available data cadence available from observatories. In addition to important benchmark storms, this same storm environment modeling capability is also used to develop the storm environments, and a detailed understanding of the climatology concerns for large and historically important storm threat conditions such as the Great Geomagnetic Storm of 13–14 March 1989 and possible larger storm threats to the US power grid.

9.3. GROUND MODELS AND ELECTRIC FIELD CALCULATION

In order to compute the induced currents flowing in a power network, it is necessary to be able to establish the electric fields induced in the Earth by variations in the geomagnetic field. Numerical models developed by Metatech require knowledge of the layered Earth conductivity to depths greater than 300 km. Past experience has indicated that 1D Earth conductivity models are sufficient to compute the local electric fields (Gilbert et al., 2012). Because there is considerable heterogeneity in conductivity over North America, multiple 1D models can be used where the conductivity variations are extremely large.

Four different techniques are used to develop ground conductivity models.

- Measurements of the conductivity vs. depth for the top 10s of km.
- Descriptions of the local geology, which can be used to infer an average conductivity vs. depth.
- Measurements of the deep conductivity for the region of interest (although continental scale data exist for most of the world).
- Actual measurements of geomagnetic and electric fields along with resultant GICs at locations and frequencies of interest.

In general across the US, information is available in each category, although the last technique is the most useful, when it is available (Radasky et al., 1993).

Ground conductivity models need to accurately reproduce geoelectric field variations that are caused by the very low frequency ranges of geomagnetic storms. These electromagnetic disturbances require models accurate over a frequency range from 0.3 Hz to as low as 0.00001 Hz. Because of the low-frequency content of the disturbance environments, it is necessary to take into consideration ground conductivities to appropriate depths. Numerous studies confirm that depths required are more than several hundred kilometers, although the exact depth is a function of the layers of conductivities present in a specific region. In most locations, ground conductivity varies substantially at the surface; these conductivity variations with depth can range 3 to 5 orders of magnitude. While surface conductivity can exhibit considerable lateral heterogeneity across the US, conductivity at depth is more uniform. Because of this, models of ground conductivity can be successfully applied over meso-scale distances and can be accurately represented by use of layered conductivity profiles or models. Frequent occurrences of geomagnetic storm events and subsequent measurement of these storm environments and associated impacts have provided opportunities to use this information to develop and validate models of the ground conductivity for the US power grid model.

A severe electrojet disturbance can produce a rate-of-change intensity profile (or dB/dt) of 2400 nT/min or greater. This disturbance is a very severe one that could be possible at high- to midlatitude locations throughout the US. A waveform of this intensity was used to evaluate differing ground model responses in terms of the peak geoelectric field that would be produced by such a disturbance. The comparative results of this evaluation for 18 differing ground models is shown in Figure 9.5. In the case of peak magnitude alone, the difference between the most responsive and the least responsive ground model can be more than a factor of 7. For the most responsive ground model, the peak geoelectric field exceeded 15 volts/km in intensity.

Figure 9.6 provides an overview map of the ground models selected for each color-coded region in the US. As shown, these models are applied in meso-scale dimensions throughout the US. The specific models were assigned based upon selecting a group of ground conductivity models that were appropriate to the known geological profiles of the region. The geological profiles provided can narrow the search to several candidate ground models for various regions. Further refining to select the most appropriate ground model was done through a series of simulations where the best model was selected by validation against local observations through the use of monitored data from prior geomagnetic storms.

Because of the wide range of magnitude as well as spectral response characteristics of the impulsive geomagnetic field disturbance environment, it is critical to select appropriate ground models for each region in order to produce valid simulation results. Validation of the ground models against forensic data provides the best means to assure that appropriate responses are being observed and to provide assurance that correct responses will result when simulations of more severe disturbances are undertaken. Several examples are provided in this overview of model response versus actual observations.

Figure 9.7 provides an example of the validation of a ground model. This figure shows a comparison of the observed north-south and east-west geoelectric field during a minor geomagnetic storm on 4 Nov 1993 in northern Minnesota (Kappenman et al., 1997). Also provided are the results of the simulation, which indicate good agreement in magnitudes as well as wave shape. These observations and simulations were conducted using a 1-second data cadence. The ability to replicate over a broad frequency range provides for good overall fit between

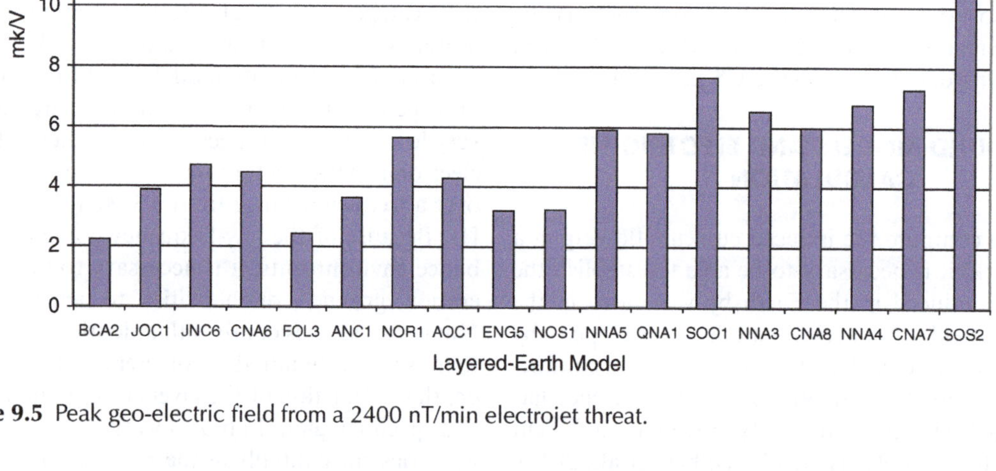

Figure 9.5 Peak geo-electric field from a 2400 nT/min electrojet threat.

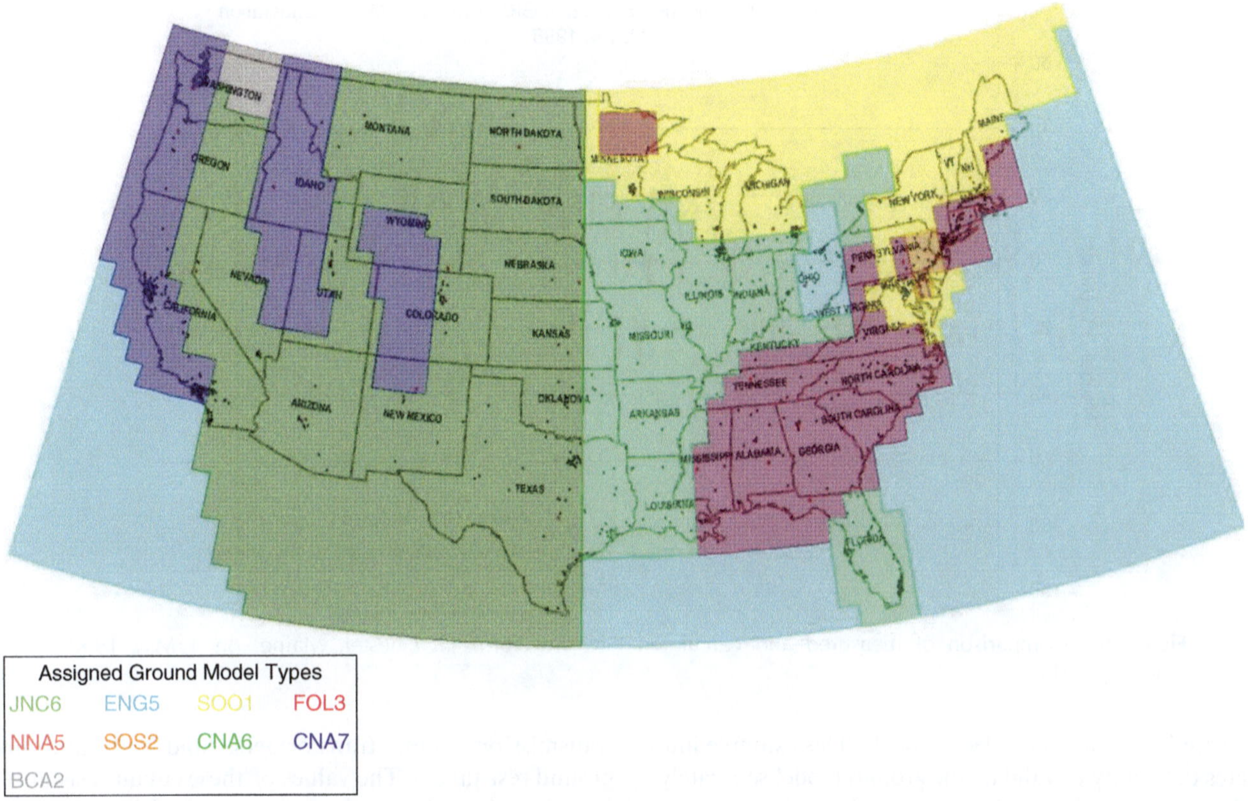

Figure 9.6 Multiple 1-D ground models for the US grid.

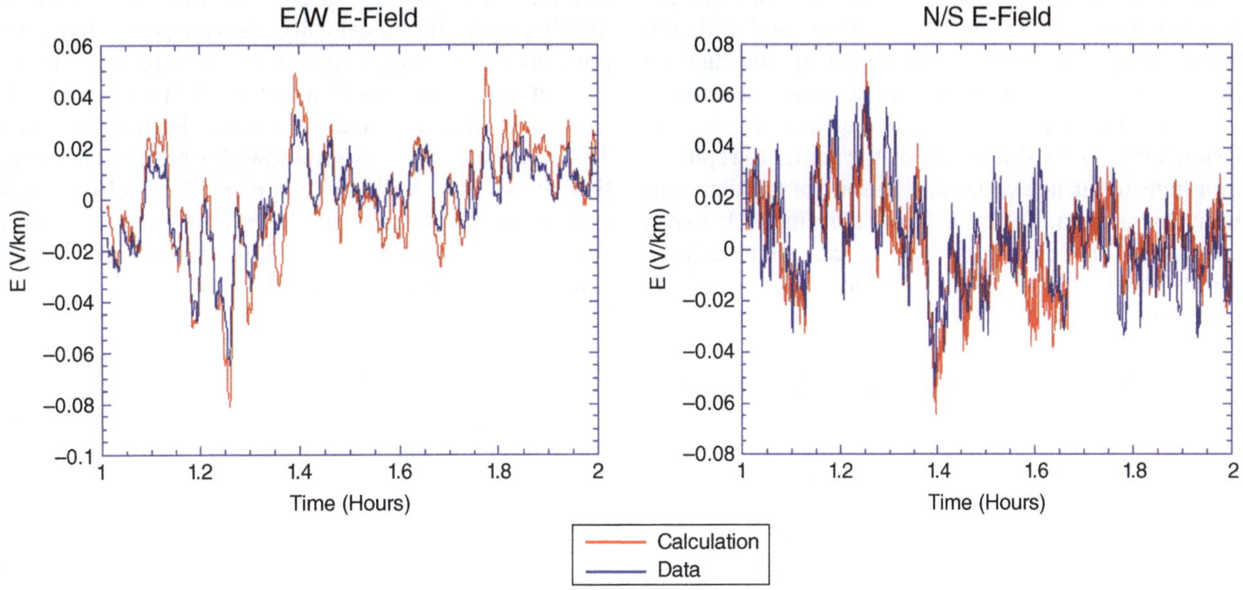

Figure 9.7 Comparison of calculated and measured electric fields for 4 Nov 1993 (Kappenman et al., 1997).

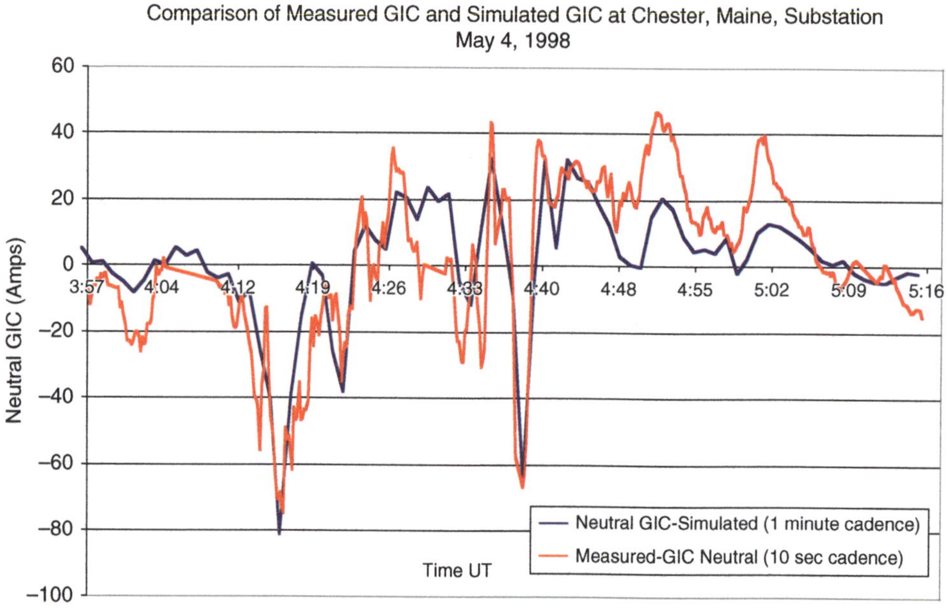

Figure 9.8 Comparison of measured and calculated GIC waveforms at Chester, Maine, on 4 May 1998 (Radasky, 2011).

observed and simulated electric fields. This example illustrates the ability to validate the ground model separately.

It is also possible to demonstrate the validation using measurements of GIC with the combined ground model and power grid model. An example of this is provided in Figure 9.8 for GIC measurements made in northern Maine during a storm on 4 May 1998. In this case, the measured data was observed at a 10-second cadence, while the calculation was limited to a 1-minute cadence due to availability of regional geomagnetic field disturbance data. This will also limit the frequency response of the simulation and reduce some of the ability to replicate minor high-frequency variations. In spite of the data limitation for simulations, the close agreement between observed GIC and calculated GIC indicates that both the ground model and the power grid model are accurately replicating the storm's behavior.

9.4. US ELECTRIC POWER GRID CIRCUIT MODEL

As one of the four component models that are necessary to calculate GIC flows and the impacts on power system reliability due to GIC flows, the power grid model is important for two reasons in determining the simulation results. This portion of the model in particular defines the complex topology of the network circuit, as this topology couples with the equally complex geoelectric field environment that occurs from natural geomagnetic storm processes. The other important aspect of the power grid circuit model is contained in the resistive impedance of the network circuit elements (primarily the transmission lines, transformers, and substation-to-ground resistance). The values of these circuit resistances combined with the resultant geoelectric fields across the network define the pattern and magnitude of GIC flows.

Because of the mature nature of these data and the many years of power industry experience in modeling AC flows in the network, it is expected that this portion of the US power grid model will be less uncertain than other portions of the overall model. In order to calculate the flow of GIC, the model needs to define not only the impedance of all circuit elements but also their locations. The definition of locations allows for proper determination of coupling with the complex 2D geoelectric field patterns that will occur due to various threat conditions that are to be examined. As will be shown in subsequent analysis, the majority of all GIC flows will occur in the highest voltage portions of the network; hence, this limitation provides most of the GIC flows and resulting impacts from those flows.

Figure 9.9 provides a map of the overall transmission network included in the CONUS region model of the US grid. The major transmission voltages are color highlighted by operating voltage with the three operating voltages of 345 kV, 500 kV, and 765 kV (there are also several lines in the Washington state region that are operated at 300 kV and that are included in the model). Figure 9.10 provides the mileage statistics for each of these three voltage classes that are represented in the US model for the CONUS region. As shown, the most common transmission voltage is the 345 kV, which makes up about 64% of total transmission line miles. The highest operating voltage is the

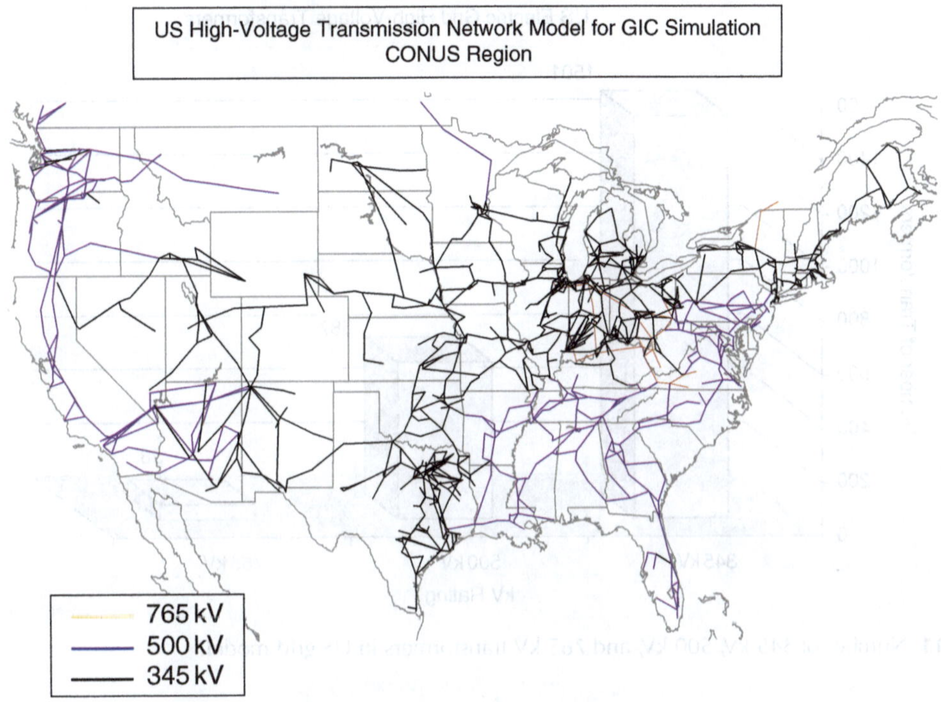

Figure 9.9 Map of 345 kV, 500 kV, and 765 kV substations and transmission network in US grid model.

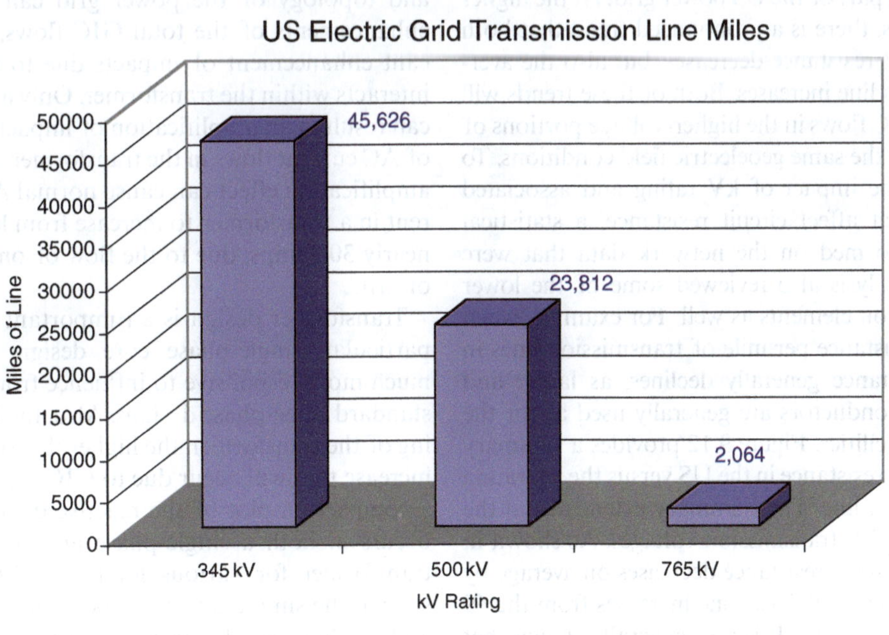

Figure 9.10 Miles of 345 kV, 500 kV, and 765 kV transmission lines in US grid model.

765 kV and is primarily located in the Illinois, Ohio, Indiana, West Virginia, and upstate New York regions of the US. Both the 345 kV and 500 kV portions of the network are more widely distributed across the US.

A total of 2146 high-voltage transformers are modeled in the continental US (CONUS) region for the US power grid model. The population of transformers by kV rating is shown in Figure 9.11. The population of transformers by kV rating generally follows the population of transmission line miles, as previously shown. The simulation model is able to calculate and estimate the flow of GIC in each of these transformers and provides estimates of the degree of half-cycle saturation and system impacts this saturation could cause to the operation of the US Power Grid.

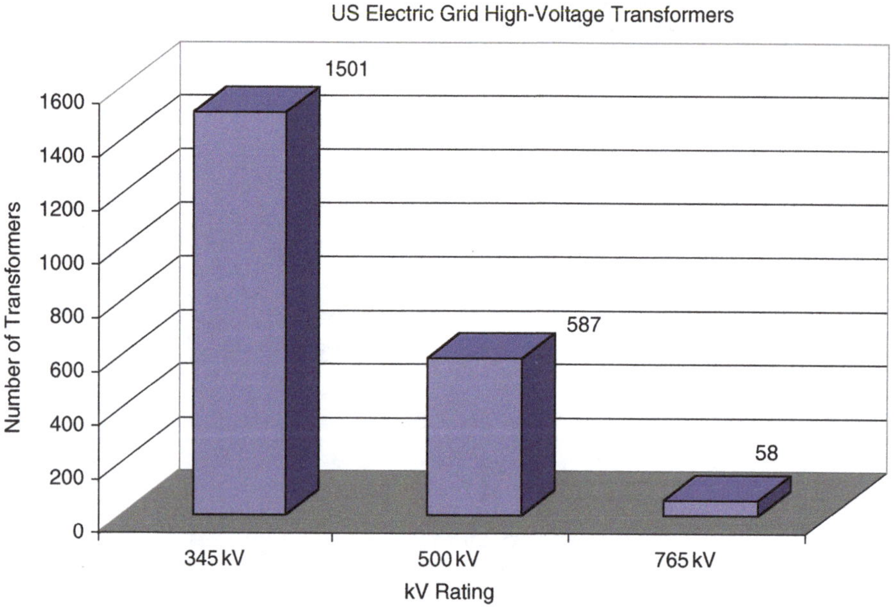

Figure 9.11 Number of 345 kV, 500 kV, and 765 kV transformers in US grid model.

The operating voltage of the transmission network is an important factor in determining the level of GIC flow that will occur on each part of the US power grid. At the higher operating voltages, there is a pronounced trend that both the average circuit resistance decreases, but also the average length of each line increases. Both of these trends will result in larger GIC flows in the higher-voltage portions of the network given the same geoelectric field conditions. To better illustrate the impact of kV rating and associated design factors that affect circuit resistance, a statistical analysis was performed on the network data that were available. This analysis also reviewed some of the lower voltage transmission elements as well. For example, when looking at the resistance per mile of transmission lines in the US, the resistance generally declines, as larger and lower resistance conductors are generally used as for the higher kV rates facilities. Figure 9.12 provides a summary of the average line resistance in the US versus the kV rating of the transmission line. This summary extends from the 115 kV to the 765 kV transmission voltages. As shown in this figure, the per-mile resistance decreases on average by approximately a factor of 10 as one increases from the 69 kV line to the 765 kV line. It is also generally shown that for operating voltages of 345 kV and above the resistances are generally very small (less than 0.1 ohm per mile).

9.5. TRANSFORMER AND AC POWER GRID PERFORMANCE MODEL

The flow of the GIC in transformers is the root cause of all power system problems, as the GIC causes half-cycle saturation to occur in the exposed transformers. While the extremes of the threat environment, the conductivity of the deep-Earth ground, and the kV rating and topology of the power grid can cause significant enhancements of the total GIC flows, the most significant enhancement of impacts due to GIC is how GIC interacts within the transformer. Only a few amps of GIC can result in an amplification of impacts in the operation of AC current flows in the transformer. In some cases, the amplification effect can cause normal AC excitation current in a transformer to increase from less than 1 amp to nearly 300 amps, due to the flow of only 25 amps/phase of GIC.

Transformer design is an important consideration; in particular, single-phase core design transformers are much more responsive to influence from GIC than most standard three-phase designs. Also, the higher the kV rating of the transformer, the higher the total reactive power increase that will occur due to GIC. Figure 9.13 provides a comparison plot of the reactive demand increase that occurs in both a single-phase and three-phase 500 kV transformer for various levels of GIC per phase. As shown, the simple difference between three-phase design and single-phase design causes a factor of 4 increase in total reactive power demand over the range of GIC flow levels.

A similar increase in the impacts of GIC flows on transformer reactive power demands occurs for increases in transformer kV rating. Figure 9.14 provides a comparison of single-phase 345 kV, 500 kV, and 765 kV transformers. As shown in this comparison, for the same level of GIC flow, the 765 kV transformer will have nearly two times higher reactive power demands than the

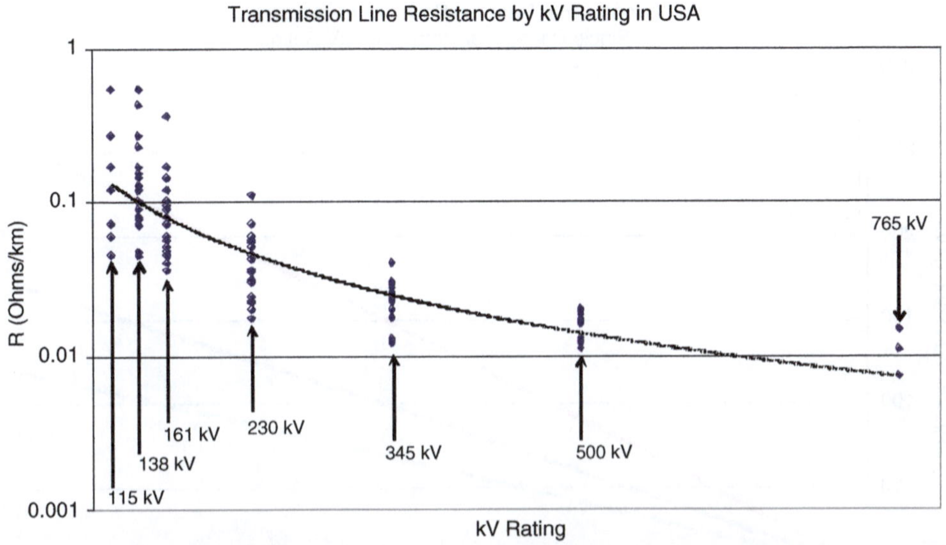

Figure 9.12 Range of transmission line resistance for the major kV-rating classes for transmission lines in the US electric power grid infrastructure population. Also shown is a trend line of resistance weighted to population averages. The lower R for the higher voltage lines will also cause proportionately larger GIC flows in this portion of the power grid. (Derived from data in EHV Transmission Line Reference Book and from US Dept of Energy–Energy Information Agency and FERC Form 1 Database.)

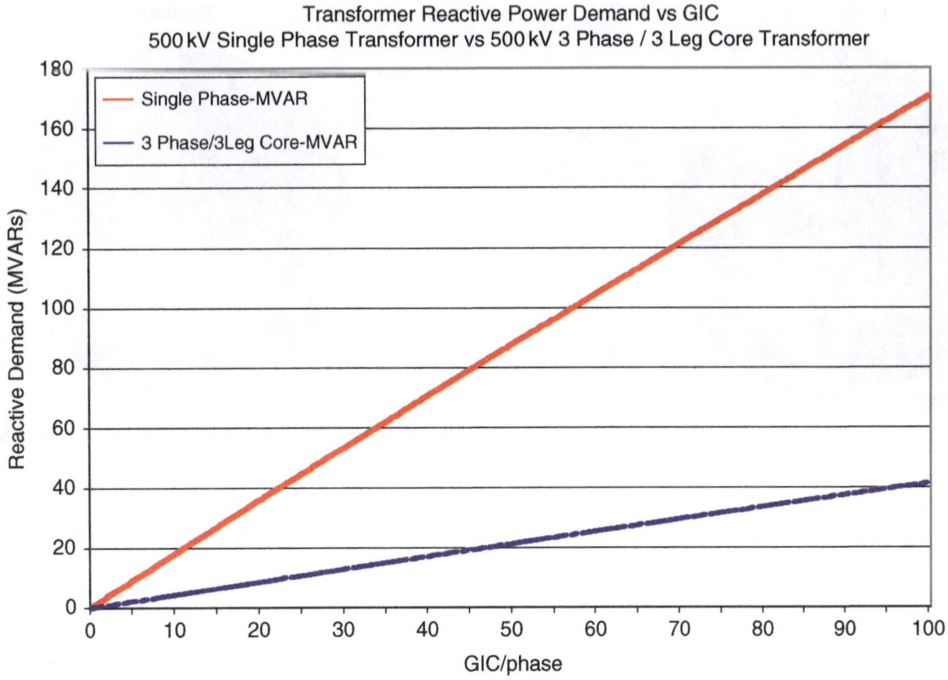

Figure 9.13 Transformer MVAr increase versus GIC for 500 kV single phase and three-phase, three-legged core form.

345 kV transformer. A similar ratio of reactive power demand versus kV rating would also occur for three-phase transformers, though total reactive power levels will be smaller than those in comparable single-phase design transformers.

As evident from these behavior characteristics, transformer design factors can be an important determinant in system behavior. While transformer kV rating is fully defined for all transformers included in the US power grid model, there is some uncertainty about the context

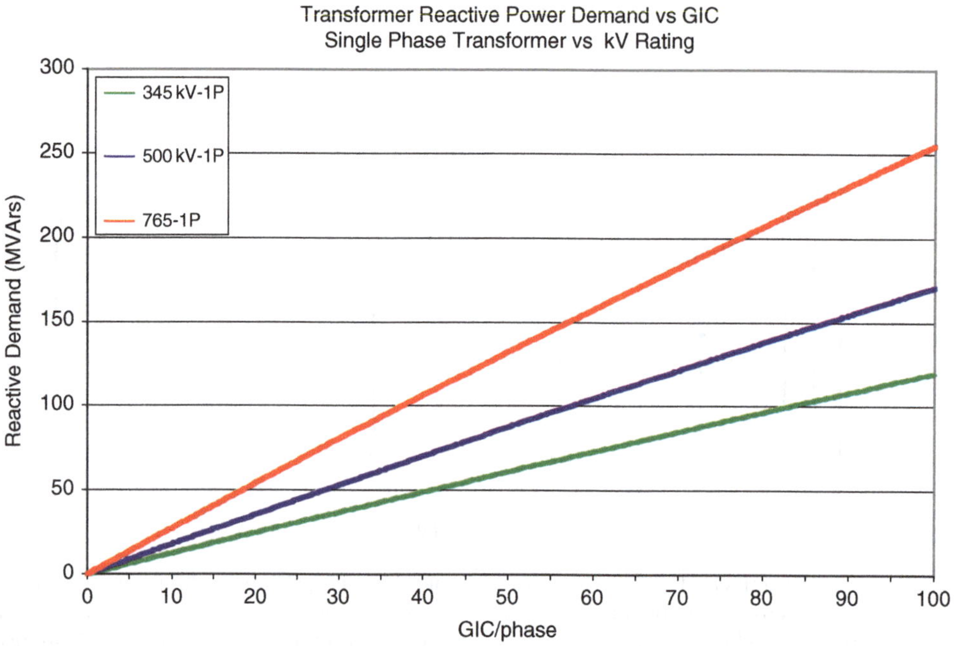

Figure 9.14 Transformer MVAr increase versus GIC for 345 kV, 500 kV, and 765 kV transformers.

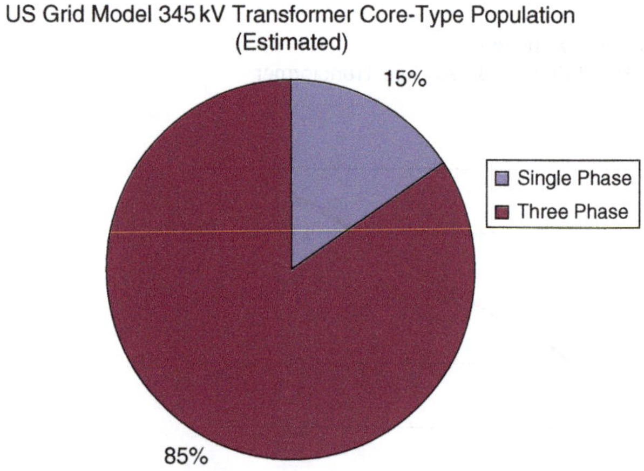

Figure 9.15 Demographic estimates of 345 kV transformers: single phase vs. three phase.

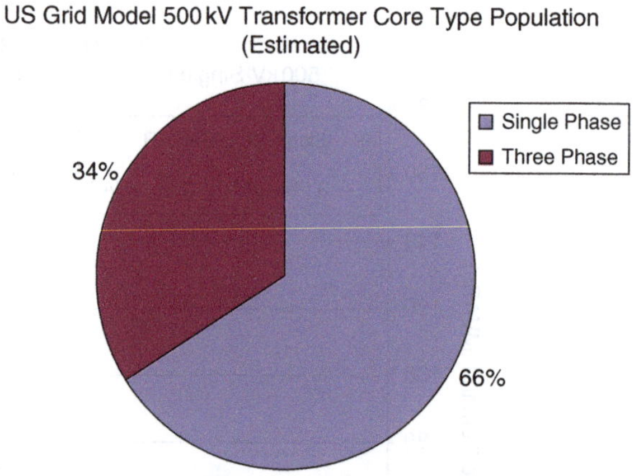

Figure 9.16 Demographic estimates of 500 kV transformers: single phase vs. three phase.

of the ratio of single-phase and three-phase transformers in the overall transformer population. Because the design of transformers was not readily available in all cases, estimates of single-phase/three-phase populations were therefore based upon MVA rating, in that transformers of 600 MVA or more were assumed to be single-phase design. Based upon these estimates, it is estimated that 85% of the 345 kV transformers will be three phase (Figure 9.15). For the 500 kV population, only 34% of these transformers are three phase (Figure 9.16), while the bulk of the population is the more susceptible single-phase design. At the 765 kV level, nearly all of these transformers are expected to be of single-phase design (Figure 9.17).

Using this method of allocating transformer design is a possible source of uncertainty, and the desire for this analysis is to not produce overly pessimistic results, which might occur if population estimates of single-phase transformer design in the US grid are overestimated. In order to test the assumptions applied to the US grid as a whole regarding the transformer design population, a comparison was made with available transformer data from the BPA power grid in the Pacific Northwest region. The BPA system is predominantly a 500 kV and 230 kV transmission system with also a small number of 345 kV facilities, and BPA has ~6% of the US ownership of all

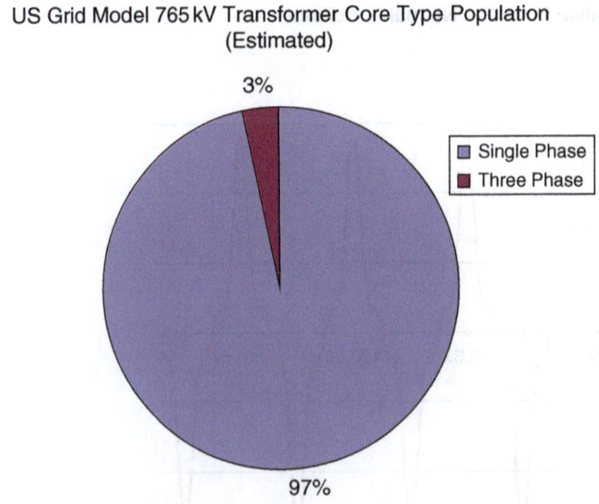

Figure 9.17 Demographic estimates of 765 kV transformers: single phase vs. three phase.

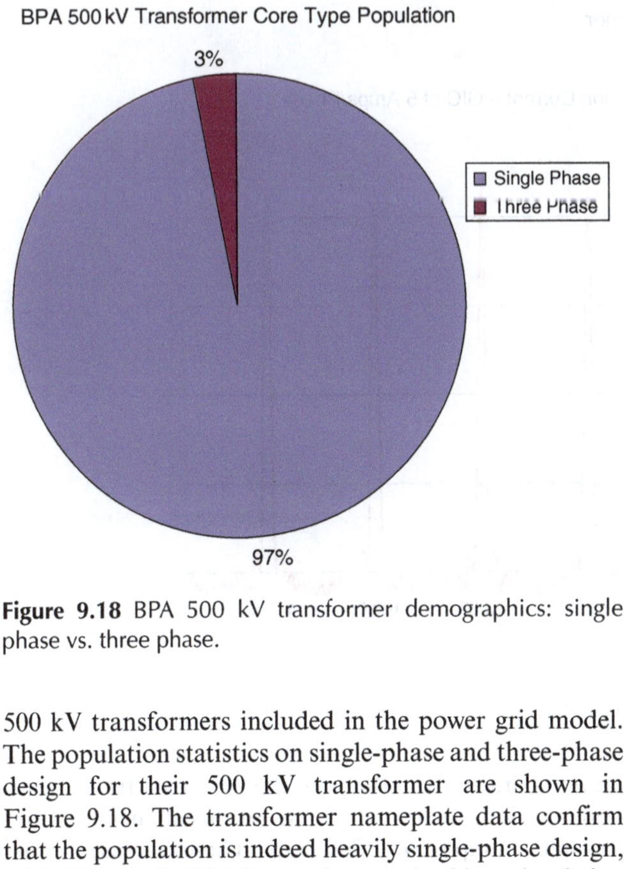

Figure 9.18 BPA 500 kV transformer demographics: single phase vs. three phase.

500 kV transformers included in the power grid model. The population statistics on single-phase and three-phase design for their 500 kV transformer are shown in Figure 9.18. The transformer nameplate data confirm that the population is indeed heavily single-phase design, with 97% of all 500 kV transformers in this region being single-phase units. This population ratio is actually much higher than the US 500 kV population as a whole estimate, which is modeled as being only 66% single-phase design. BPA also has a large population of 230 kV transformers, and therefore population ratios of these transformers can be a somewhat relevant comparison for the

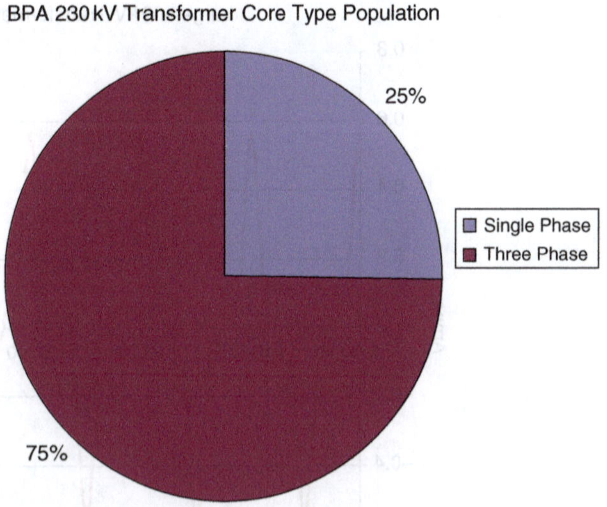

Figure 9.19 BPA 230 kV transformer demographics: single phase vs. three phase.

estimated populations for the 345 kV transformers estimated for the US model. As shown in Figure 9.19, the ratio of single-phase design in the BPA 230 kV transformers is 25%; this is also a higher ratio of single-phase units than assumed for the entire US 345 kV population, which is estimated to be only 15%. This comparison suggests that the estimated 345 kV population of single-phase design transformers may actually be larger than currently represented in the US model.

Publicly available data from the ECAR pool (East Central Area Reliability Coordination Agreement) were also used to confirm the very high population of 765 kV single-phase design transformers. The ECAR pool contained nearly all the 765 kV transformers in the US, with the exception of a handful of 765 kV transformers operating in upstate New York. (This pool is now part of the Mid-Continent Independent System Operator). These verifications provide a level of confidence that simulation results, which are in part based upon population estimates of transformer design, will generally not be overly pessimistic and may in actuality be somewhat optimistic for this expedited review.

Transformers under half-cycle saturation pose a dual threat to system reliability: system-wide voltage collapse by increase in reactive power in each exposed transformer and the disruptive effects of harmonics and AC waveform distortion on relay and protective systems. Accurate estimates of both reactive power demand and AC waveform distortions are therefore necessary in models of transformer behavior. Figures 9.13 and 9.14 (previously discussed) provided an overview of reactive power demand behavior of transformers. The AC waveform distortions are more difficult to estimate and need to be considered on a case-by-case basis.

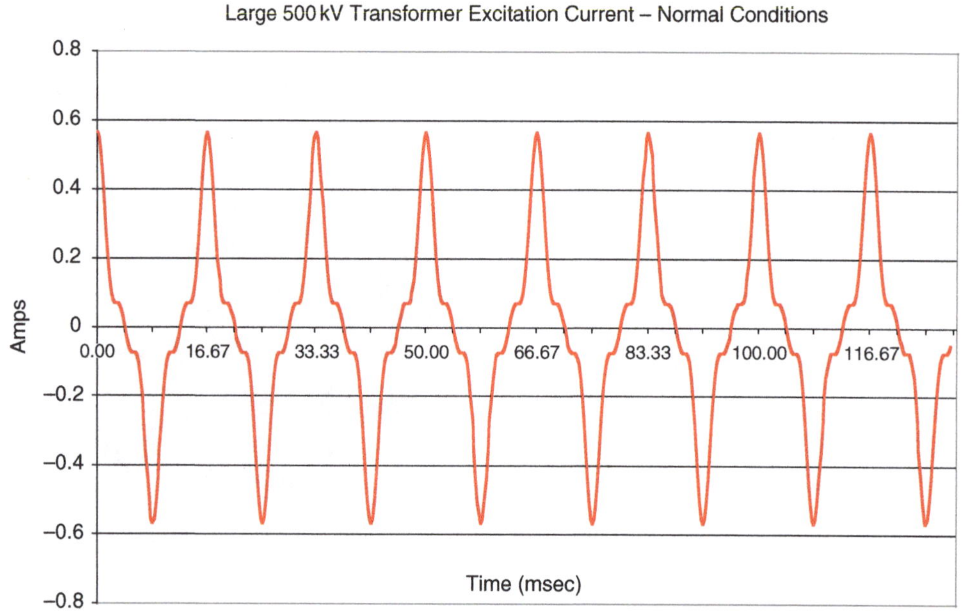

Figure 9.20 Normal excitation current in 500 kV transformer.

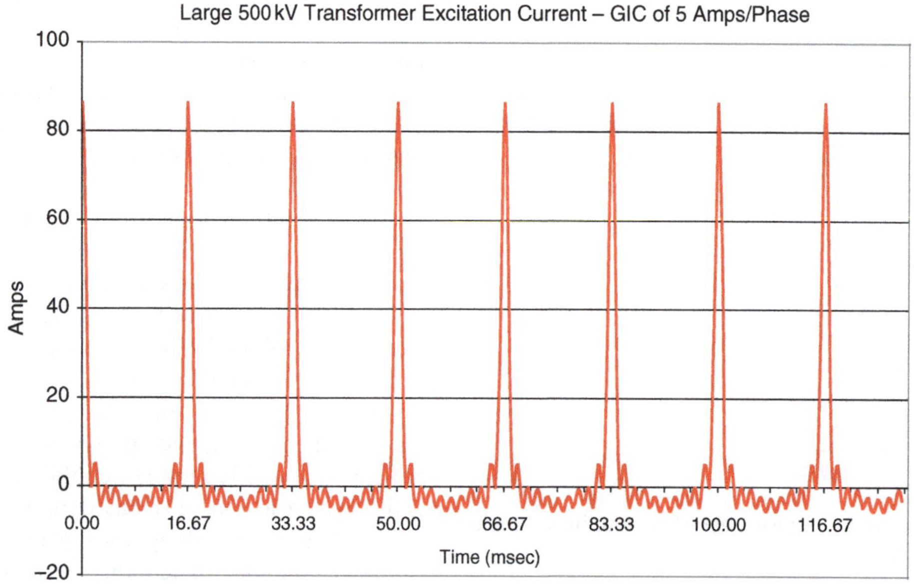

Figure 9.21 Distorted excitation current with 5 Amps/phase of GIC.

Some general examples can be provided to illustrate the nature of impacts that can occur on a wide scale during disturbance threats. Figure 9.20 provides a plot of the normal AC excitation current that would be observed in the high side of a 500 kV transformer, while Figures 9.21, 9.22, and 9.23 show the significant increases and distortion of transformer excitation current for 5 Amps/phase of GIC (Figure 9.21) and also for 25 Amp (Figure 9.22) and 100 Amp GIC events (Figure 9.23). These examples illustrate that AC peak excitation currents drawn from the power system by a transformer under various levels of GIC excitation can increase from less than 1 amp to over 800 amps. These increased transformer excitation currents are highly distorted and rich in both even and odd harmonics, and they combine with the normal load current on the transformer and propagate distortions throughout the power grid. To illustrate the potential impact of these distortions, the primary current (including load of 300 amps under normal conditions) can be synthesized during several severe GIC

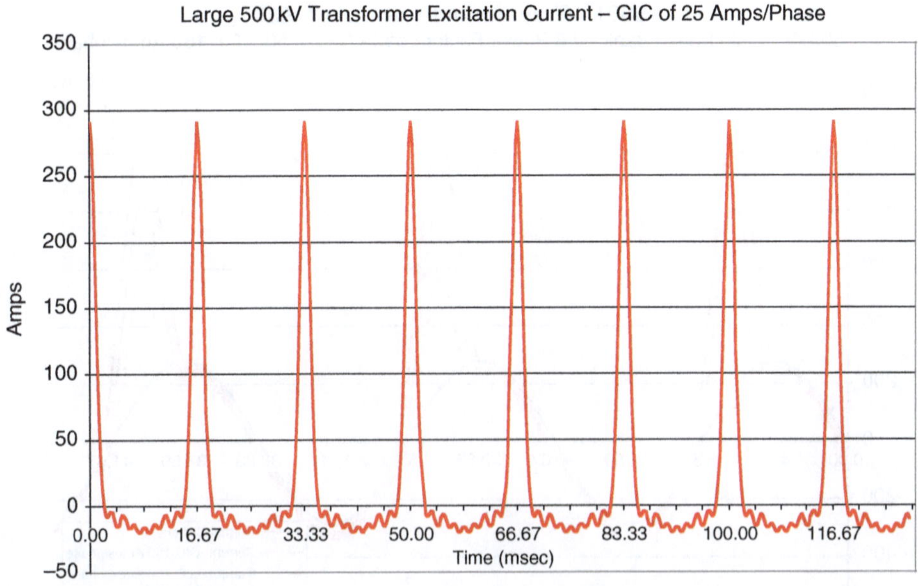

Figure 9.22 Distorted excitation current with 25 Amps/phase of GIC.

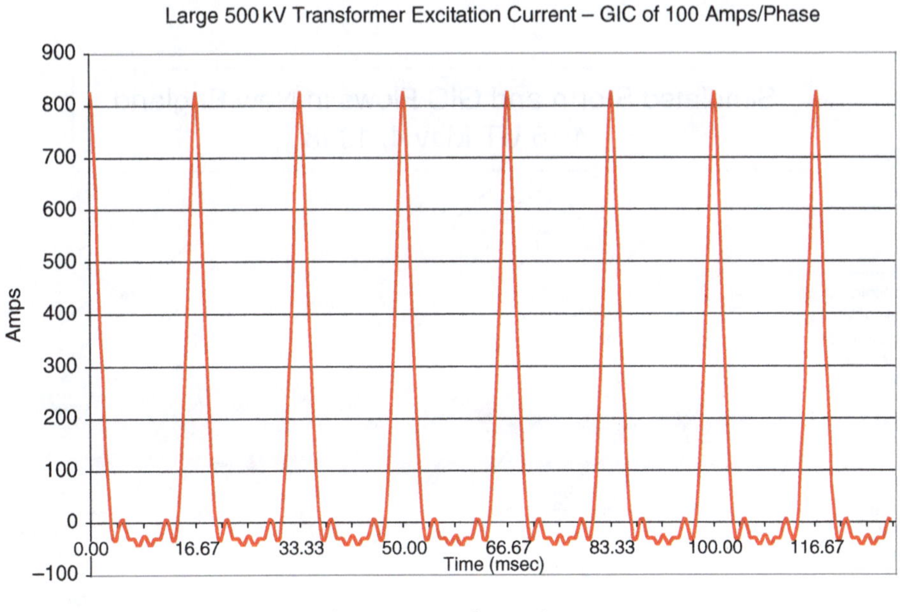

Figure 9.23 Distorted excitation current with 100 Amps/phase of GIC.

exposures. Figure 9.24 shows the normal AC load current for a 500 kV transformer (blue waveform) and the resulting AC currents under three different GIC conditions. It is evident that not only are large distortions occurring, but also a significant peak of overcurrent will result. Total waveform distortions of over 200% are expected to occur; moreover, these simulations do not take into account additional waveform distortions that could occur due to simultaneous saturation of adjacent transformers and local resonances, which could increase distortions further.

In addition to the details of behavior by individual transformers on the power grid, the disturbance conditions can be summarized for the network as a whole, which provides a descriptive overview of the disturbance impact regions. Figure 9.25 provides a summary of geomagnetic storm conditions for the mid-Atlantic through New England regions of the US for a storm event on 4 May 1998 at time 4:18 UT. This graphic shows the geographic location of all transmission lines and stations across the region. Also shown are the storm conditions and GIC flows at this moment in time. The storm

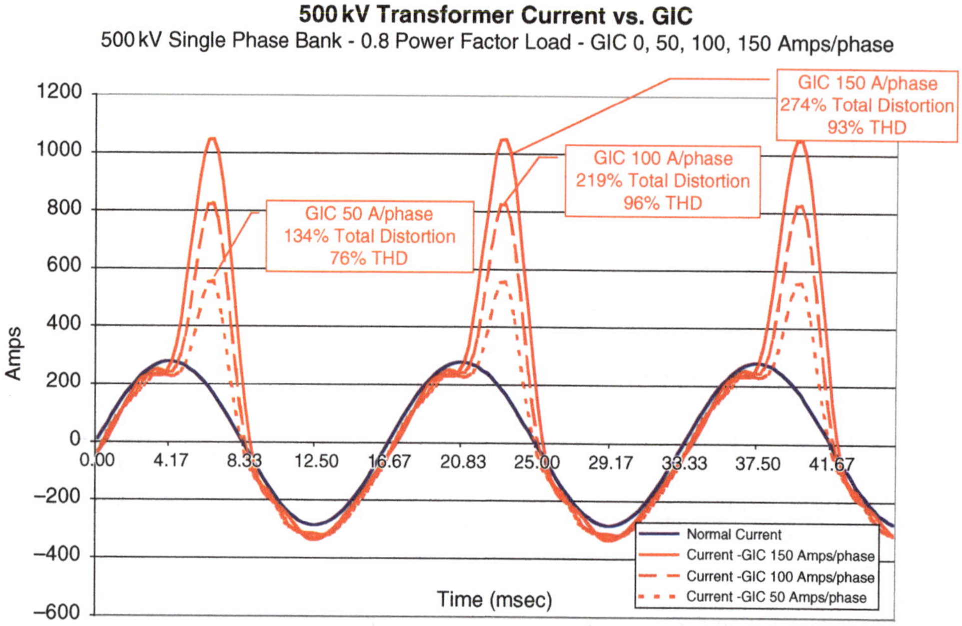

Figure 9.24 Transformer total load current: normal conditions (blue) and with 50, 100, and 150 Amps/phase of GIC (red).

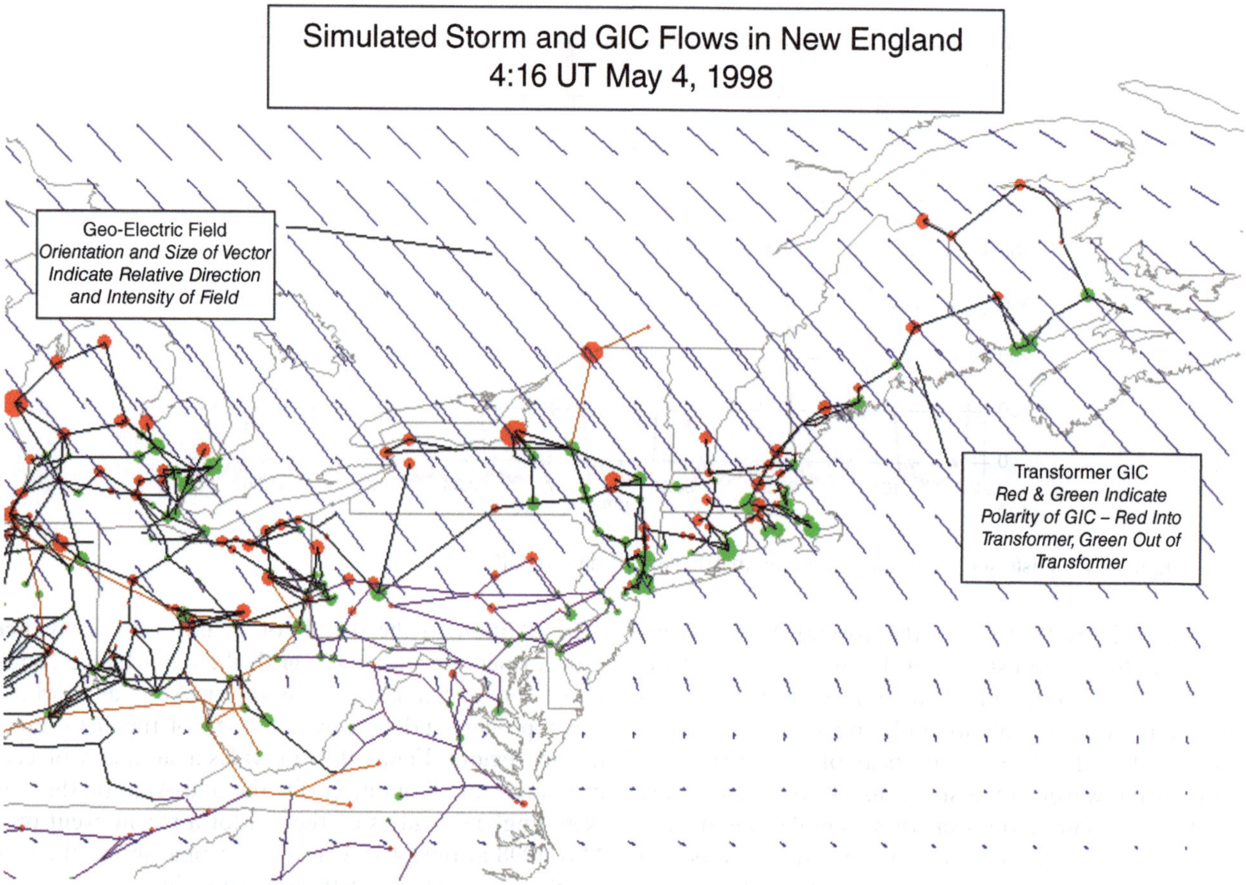

Figure 9.25 Pattern of geoelectric field and GIC flows in New England region of power grid model for 4 May 1998 storm at 4:16 UT.

conditions in this example are illustrated by the blue vectors that depict the spatial complexity of the geoelectric field across the region. These vectors depict both the orientation of the field and the relative intensity. As shown in this figure, there is a strong southeasterly oriented geoelectric field over the New England region, and this field rapidly diminishes in intensity and changes in orientation over mid-Atlantic locations. The relative levels of GIC flows are shown at each substation by red- or green-colored dots. The colors indicate polarity of GIC flow in transformers located at each of these stations; red means a flow of current into the transformer from ground, while green indicates a flow of GIC from the transformer to ground. The polarity of the GIC flow does not have any particular significance on the impact of that GIC to individual transformers, as either polarity of GIC will produce half-cycle saturation effects in a transformer. The sizes of the colored dots also change in proportion to the magnitude of the GIC flow at each location. Therefore, the size of dots indicates the relative level of regional impacts.

9.6. THE EVOLVING VULNERABILITY OF ELECTRIC POWER GRIDS AND IMPLICATIONS FOR HISTORICALLY LARGE STORMS

A simple way to summarize much of the discussion in this overview of the power grid model is that today's sprawling, high-voltage power grids are more susceptible to space weather impacts than ever before. While almost all research into space weather impacts on technology systems has focused on the dynamics of the space environment, the role of the design and operation of the technology system in introducing or enhancing vulnerabilities to space weather is often overlooked. In the case of electric power grids, both the manner in which systems are operated and the accumulated design decisions engineered into present-day networks around the world have tended to significantly enhance geomagnetic storm impacts. The result is to increase the vulnerability of this critical infrastructure to space weather disturbances.

The space weather community and the power industry have not fully understood these operation and design implications. The application of detailed simulation models has provided tools for forensic analysis of recent storm activity and when adequately validated can be readily applied to examine impacts due to historically large storms, as will be described in subsequent sections of this report. However, even simple empirical extrapolations provide some perspective on the level of threat and the possible consequences of storms on present-day infrastructures.

Historic records of geomagnetic disturbance conditions and, more important, geoelectric field measurements provide a perspective on the ultimate driving force that can produce large GIC flows in power grids. Because geoelectric fields and resulting GIC are caused by the rate of change of the geomagnetic field, one of the most meaningful methods to measure the severity of impulsive geomagnetic field disturbances is by the magnitude of the geomagnetic field change per minute, measured in nanoteslas per minute (nT/min). For example, the regional disturbance intensity that triggered the Hydro Quebec collapse during the 13 March 1989 storm only reached an intensity of ~480 nT/min. Large numbers of power system impacts in the United States were also observed for intensities that ranged from 300 to 600 nT/min during this storm. However, the most severe rate of change in the geomagnetic field observed during this storm reached a level of ~2000 nT/min over the lower Baltic. The last such disturbance with an intensity of ~2000 nT/min over North America was observed during a storm on 4 August 1972, when the power grid infrastructure was less than half its current size.

An analysis of the AT&T telecom cable (L4) failure in northern Illinois in a region of ~800 nT/min disturbance intensity caused a geoelectric field of at least 7 V/km (Anderson et al., 1974). A dB/dt disturbance of >2000 nT/min was observed over central and southern Sweden on 13–14 July 1982. A coincident peak geoelectric field of 9.1 V/km was observed in central Sweden during this storm on railroad communication circuits. Similar observations from that region of 20 V/km occurred for a storm in May 1921 (Kappenman, 2006), suggesting that a peak rate of change of ~5000 nT/min is possible. This disturbance level is nearly 10 times larger than the levels that precipitated the North American power system impacts of 13 March 1989.

While the magnetospheric drivers and deep-Earth conductivities that shape the geoelectric field response have not changed significantly over this period of time, power system infrastructures have experienced dramatic changes in size and complexity. A recurrence of historically large storms, even a storm as recently as 1982, could have entirely different outcomes in terms of impacts on today's power systems.

For any natural hazard, prudence requires that design and operational adjustments be made to mitigate such risks to infrastructures. Power system designers and operators expect these systems to be challenged by the elements, and where those challenges were fully understood in the past, the system design has worked extraordinarily well. Geomagnetic storm environments can develop almost instantaneously over large geographic footprints, with the ability to essentially blanket the continent with an intense threat environment and produce significant collateral damage to critical infrastructures. Power networks are operated using what is termed an "N-1" operation criterion. That is, the system must always be operated to withstand the next credible disturbance contingency without causing a cascading collapse of the system as a whole. This criterion normally works very well for the well-understood terrestrial environment challenges, which usually propagate more slowly and are more geographically confined. When a routine weather-related single-point failure occurs, the power system needs to be rapidly adjusted (10 minutes being the allowed time limit) and positioned to survive the next possible contingency. Space weather disturbances, however, can have a sudden onset and cover large geographic regions. They therefore cause near-simultaneous, correlated, multipoint failures in power system infrastructures, allowing little or no time for meaningful human interventions. In contrast to well-conceived design standards that have been successfully applied for more conventional threats, no comprehensive design criteria have ever been considered to check the impact of the geomagnetic storm environments. Further, as this analysis demonstrates, the design actions that have occurred over many decades have greatly escalated the dangers posed by these storm threats.

9.7. SIMULATIONS AND REVIEW OF STORM IMPACTS ON THE US POWER GRID

The effects of the 13–14 March 1989 Great Geomagnetic Storm were also widely felt on power grids throughout the US. The consequences of the storm in the US were not as severe, compared to the complete blackout that occurred on the Hydro Quebec grid. However, the storm did pose serious and widespread threats to the operational integrity of the US power pool at times. The storm also caused several cases of severe transformer damage. As previously discussed, four different substorm intervals were selected for detailed simulation and assessment using the US power grid model. The selected intervals also happened to coincide with the bulk of the power system problems reported. This forensic analysis allows for needed perspective on the levels of disturbance threat and resultant GIC flows that could trigger widespread power system problems. This simulation provides a means of further calibrating the assessments that are necessary to evaluate the potential consequences of future and potentially larger geomagnetic storm events on the US power grid infrastructure.

9.7.1. Substorm Interval 7:40–8:00 UT

As previously described, the disturbance intensification primarily occurred in the vicinity of the US-Canada border region. In the eastern part of North America, the disturbance triggered the collapse of the Hydro Quebec grid (Czeck et al., 1992) but also propagated disturbances effects into the neighboring US regions as well (NERC, 1990). At the time of the Quebec collapse, the US was importing 1949 MW from Quebec. All of this capacity was lost due to the collapse. The most important loss involved the 765 kV AC tie line that runs from upstate New York to the Montreal region, where a DC interconnection is made with the Quebec system.

As shown in Figure 9.26, the highest intensity observed from this storm in North America occurred in the southern

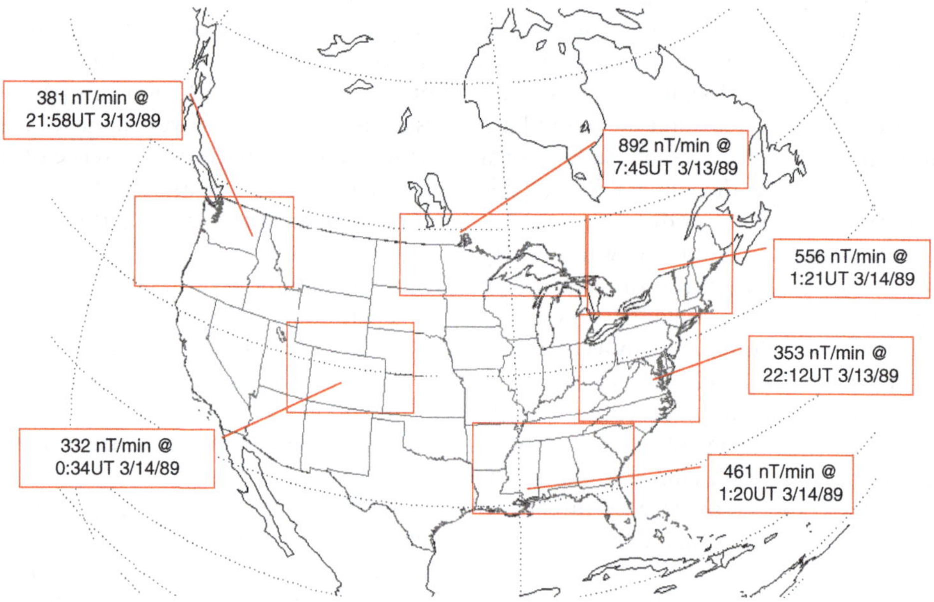

Figure 9.26 Peak dB/dt observed by region on 13–14 March 1989.

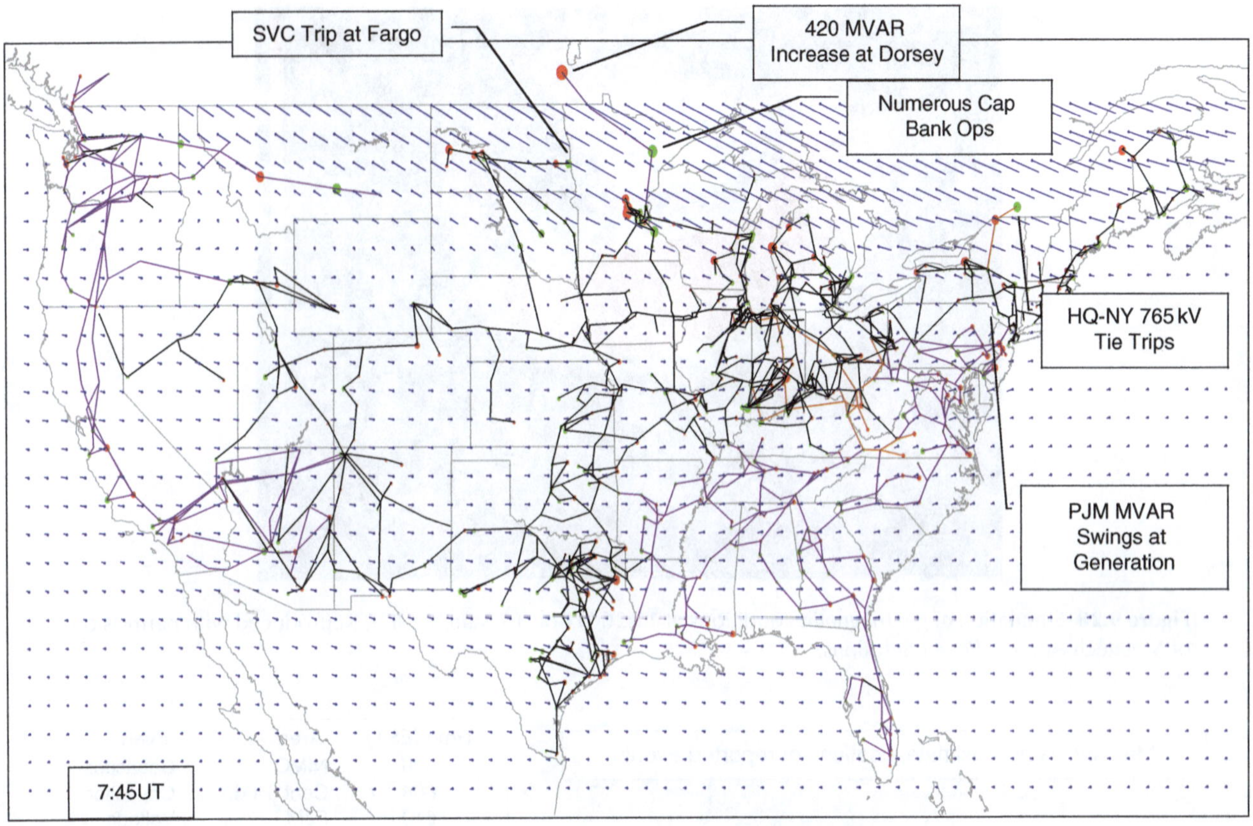

Figure 9.27 GIC and geoelectric field conditions: 7:45 UT, 13 March 1989.

Manitoba and northern Minnesota regions, reaching a peak of 892 nT/min at time 7:45 UT. This intense disturbance drove an increase in reactive demands that was also observed; in one case the reactive power output of synchronous condensers increased by 420 MVARs at one Manitoba 500 kV substation. Figure 9.27 provides a map of the simulated geoelectric field conditions and resulting GIC flow conditions in the US grid at the peak disturbance time of 7:45 UT. Also noted in this map are the locations of important reports of power system anomalies at this time, which clearly indicates they were caused by the storm environment. It is likely that less severe system impacts resulted due to the time of day of this substorm, in that this was 2:45 EST to 23:45 PST. These are all times of minimum system load across the US, and operating postures that would, in most cases, provide for highest reactive power reserves on the system.

9.7.2. Substorm Interval 10:50–12:00 UT

The substorm event of this time interval had location characteristics similar to those of the previously discussed interval. The source of the ground-level geomagnetic field disturbance was an intensification of the westward electrojet, and the footprint of the disturbance primarily centered on the US-Canada border. Figure 9.28 provides a depiction of the area at time 11:26 UT. While the region is large, the intensity of this substorm was generally 20%–50% less severe than the previous substorm. The most intense portion of interest was over portions of the eastern US. Figure 9.29 provides a summary map and table of a large number of reported power system events that were concentrated in this region over the time period from 6:60 to 6:30 EST (11:06–11:30 UT). These events largely involve capacitor switching to provide MVARs in response to the storm. Various alarms and associated event recorders were initiated, due to the highly unusual system events.

9.7.3. Substorm Interval 21:20–22:30 UT

The substorm activity during this time interval produced some of the largest and most widespread impacts observed across the US power grid. Figure 9.30 provides a map and table summary of reported events over the time period 16:00–17:23 EST (21:00–22:23 UT). As shown, a large number of events were reported across the entire continent, even as far south as Los Angeles (NERC, 1990). The power system problems were caused by a more complex pattern of storm intensification during this time

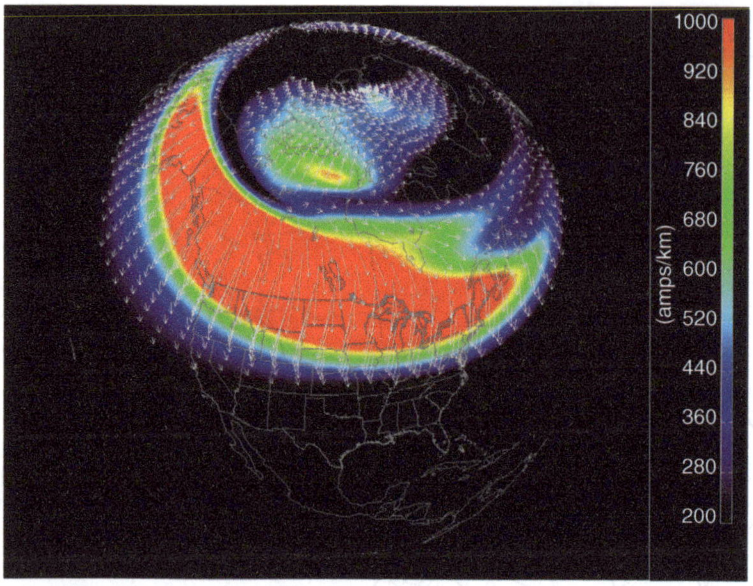

Figure 9.28 Simulation of geomagnetic conditions at 11:26 UT on 13 March 1989. Reproduced with permission of Metatech Applied Power Solutions.

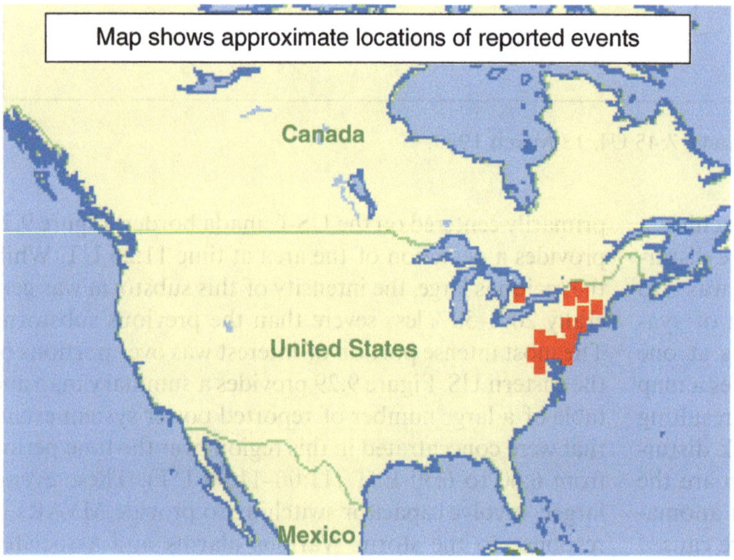

Time (EST)	Area	Event
606	NIMO	Capacitor
608	Cent. Hud.	Capacitor
610	PJM	Voltage
613	NIMO	Capacitor
615	APS	Capacitor
615	Va. Pwr.	Capacitor
617	PJM	Oscillograph
618	NIMO	Capacitor
618	PJM	Recorder
618	PJM	MW
618	Va. Pwr.	Capacitor
618	Va. Pwr.	Capacitor
619	Va. Pwr.	Capacitor
619	PJM	Recorder
619	Cent. Hud.	Capacitor
620	Va. Pwr.	Capacitor
624	Va. Pwr.	Capacitor
624	Va. Pwr.	Capacitor
624	OH	Oscillograph
625	Va. Pwr.	Capacitor
630	Atl. Elec.	MVAR

Figure 9.29 Reported North American power system impacts, 13 March 1989 for times 6:06–6:30 EST (11:06–11:30 UT).

interval than prior intervals. Also, the disturbance regions expanded to midlatitude locations over the US, which exposed large portions of the US grid to moderately intense disturbance conditions. The disturbance conditions at times 21:50 UT and 22:00 UT are shown in Figures 9.31 and 9.32. As shown, there was both a large westward electrojet extending from Europe to the center of the North American continent and also a large eastward electrojet from the eastern US through the Pacific Northwest. It is also evident from comparison that these disturbance regions were very dynamic, as they significantly expanded in area over the time interval shown. The environment was particularly harsh because of the lower latitude position of the eastward electrojet. The disturbance region exposed a large fraction of the US grid, causing the large number of events reported. Since the reporting of these events was voluntary, it is likely that further events occurred that were unreported. This would

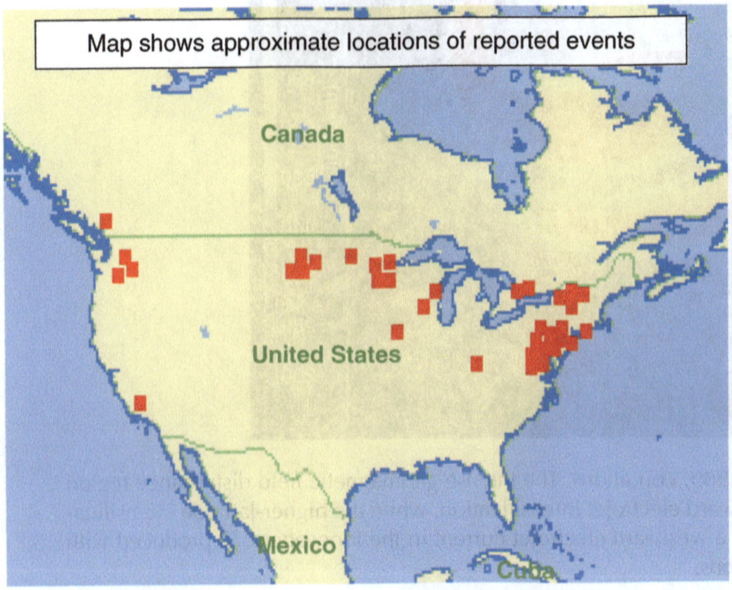

Time (EST)	Area	Event
1600	Atl. Elec.	MVAR
1602	Va. Pwr.	Capacitor
1610	PJM	Noise
1615	PJM	Generator
1625	PJM	Oscillograph
1626	PJM	Oscillograph
1630	SC Edison	Current
1630	SC Edison	Current
1630	SC Edison	Noise
1640	PJM	Voltage
1644	PJM	Alarm
1644	PJM	Capacitor
1645	WPL	Voltage
1649	PJM	Recorder
1651	NIMO	Capacitor
1653	NIMO	Capacitor
1654	PJM	Alarm
1655	Minn. Power	Voltage
1655	Atl. Elec.	Voltage
1665	Atl. Elec.	MVAR
1658	BC Hydro	Voltage
1658	OH	Demand
1658	WAPA	Converter
1658	BPA	Noise
1658	WAPA	Line
1658	WKPL	Alarm
1658	BPA	Capacitor
1658	BPA	Transformer
1700	UPA	Voltage
1700	LILCO	Voltage
1700	IIGE	Voltage
1700	WEP	Noise
1701	PJM	Capacitor
1701	NIMO	Capacitor
1701	Va. Pwr.	Capacitor
1701	Va. Pwr.	Capacitor
1701	OH	Voltage
1701	OH	Oscilligraph
1703	Va. Pwr.	Capacitor
1708	UPA	Capacitor
1709	WAPA	Converter
1709	WAPA	Transformer
1709	WAPA-Fargo	Voltage
1709	WAPA	Line
1709	WAPA	Relay
1711	NIMO	Capacitor
1720	UPA	Voltage
1723	Va. Pwr.	Capacitor

Figure 9.30 Reported North American power system impacts, 13 March 1989 for times 16:00–17:23 EST (21:00–22:23 UT).

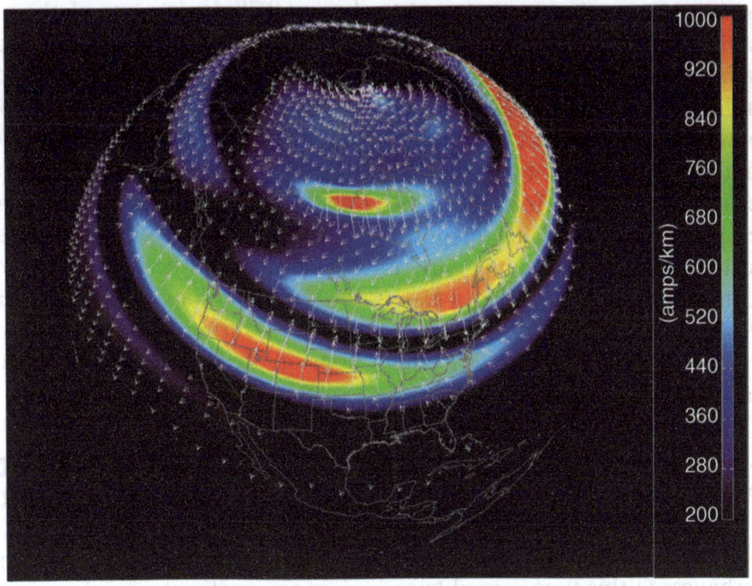

Figure 9.31 Simulated 21:50 UT, 13 March 1989, conditions. Reproduced with permission of Metatech Applied Power Solutions.

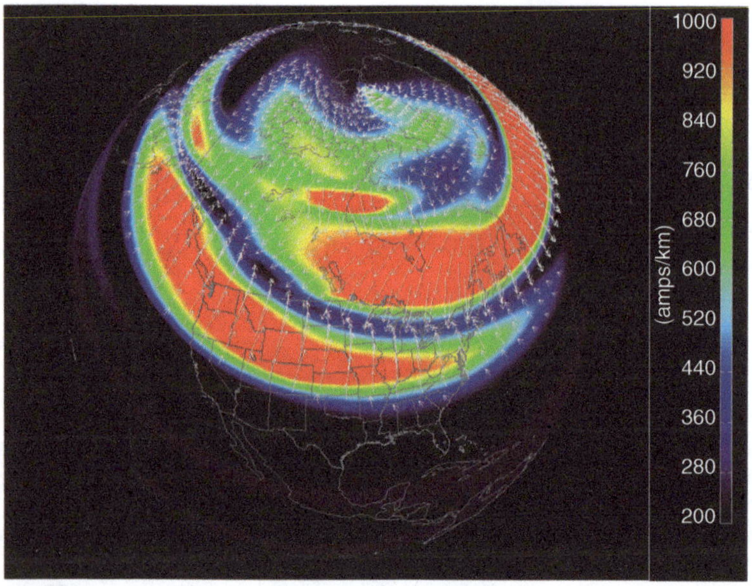

Figure 9.32 Simulated 22:00 UT, 13 March 1989, conditions. The intense geomagnetic field disturbance region at midlatitudes of the US is caused by an eastward electrojet intensification, while the higher-latitude intensification extending back towards Europe is due to a westward electrojet current in the ionosphere. Reproduced with permission of Metatech Applied Power Solutions.

be especially true for low-latitude regions of the US, where most power system operators are much less familiar with GIC problems and may have dismissed anomalies to "cause unknown." The other likely aspect explaining the higher relative impacts was the time of the disturbance intensification. The 16:00–18:00 EST time interval is during the peak electrical demand periods of the day for the entire US, whereas the other two time intervals previously discussed were during early AM hours and minimum system load conditions.

The storm impacts can be assessed further by reviewing the geoelectric field conditions and the estimated GIC flows across the US power grid at several key times. Figure 9.33 shows the conditions at time 21:44 UT; at this time an intense geoelectric field extends from the mid-Atlantic region into the upper Midwest. Voltage regulation problems were reported in Wisconsin, along with a number of problem reports in the Pennsylvania and New Jersey regions. Figure 9.34 shows the conditions at time 21:51 UT; here, the storm is particularly intense over the Northeastern US. This also caused four capacitor banks to trip in New York. By time 21:57 UT (Figure 9.35), the intense geoelectric field extends from the mid-Atlantic/New England regions to the Pacific Northwest. This triggered a large number of problems all across the US, as noted in the figure. Figure 9.36 provides the conditions at time 22:09 UT; the disturbances at this time caused trips of transmission facilities in the western North Dakota/Montana region and also trips of capacitor banks in upstate New York.

9.7.4. Substorm Interval 0:30–2:00 UT 14 March 1989

The last interval of intense substorm activity occurred between 0:30 and 2:00 UT on 14 March. As shown in Figure 9.37 for time period from 20:00–20:32 EST, a large number of power system events were reported (NERC, 1990), with the impacts being particularly significant from the New England region through the upper Midwest. Figure 9.38 depicts one of the more intense disturbance periods during this interval. The large disturbance over the New England/mid-Atlantic to Midwest regions of the US is due to a westward electrojet intensification. There are also less severe intensifications developing along the Florida panhandle and along the US-Mexico border region at this time. It is at this time that a dB/dt intensity of 461 nT/min was observed at Bay St. Louis, Mississippi, a level that was very close to the intensity that initiated the Quebec collapse. The EST time of this substorm also occurred during early evening peak load demand conditions on the US grid; this may have acted to trigger additional power system problems.

9.8. DESCRIPTIONS OF DIFFERENT TYPES OF GEOMAGNETIC STORMS

While this chapter has emphasized the electrojet type of geomagnetic storm, there are two other major types that have characteristics that are different than the well-known electrojet storm. They are the sudden impulse (SI) and the coronal hole high speed stream (CHHSS) storms.

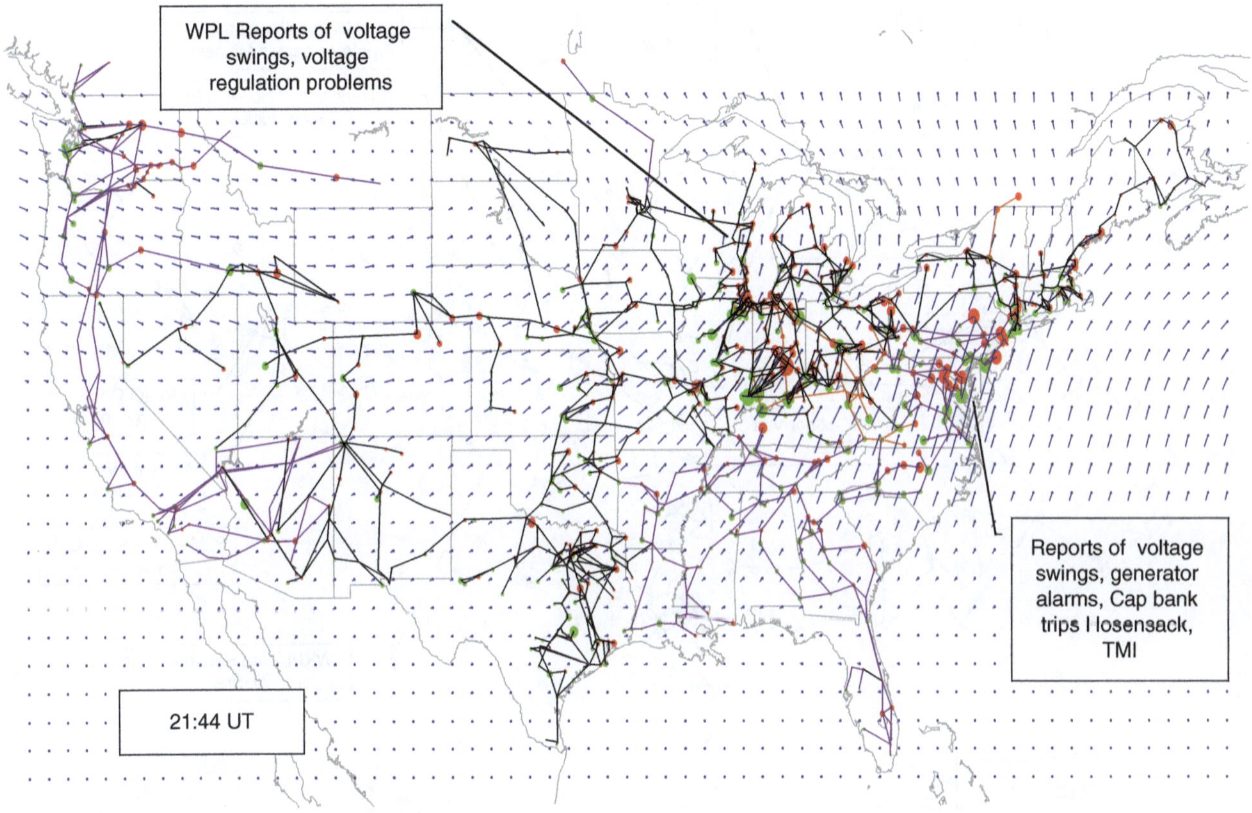

Figure 9.33 Simulation of US power grid conditions at 21:44 UT on 13 March 1989.

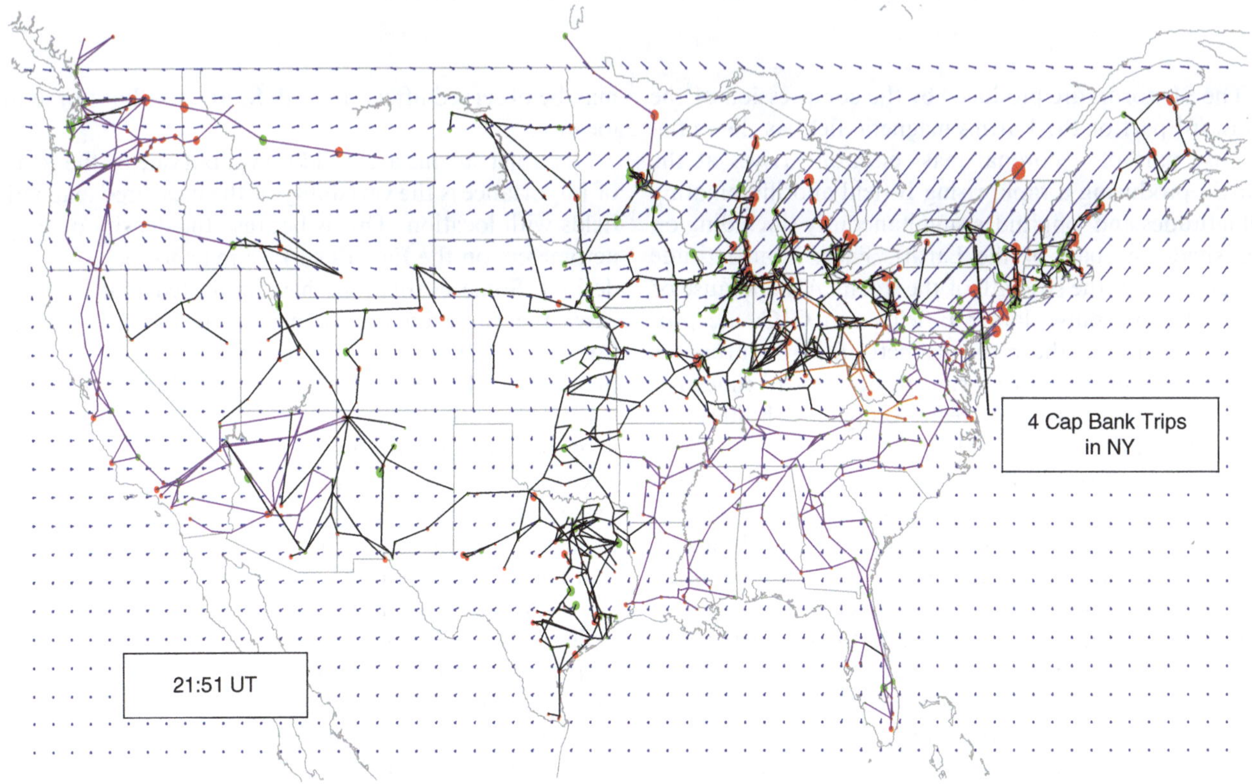

Figure 9.34 Simulation of US power grid conditions at 21:51 UT on 13 March 1989.

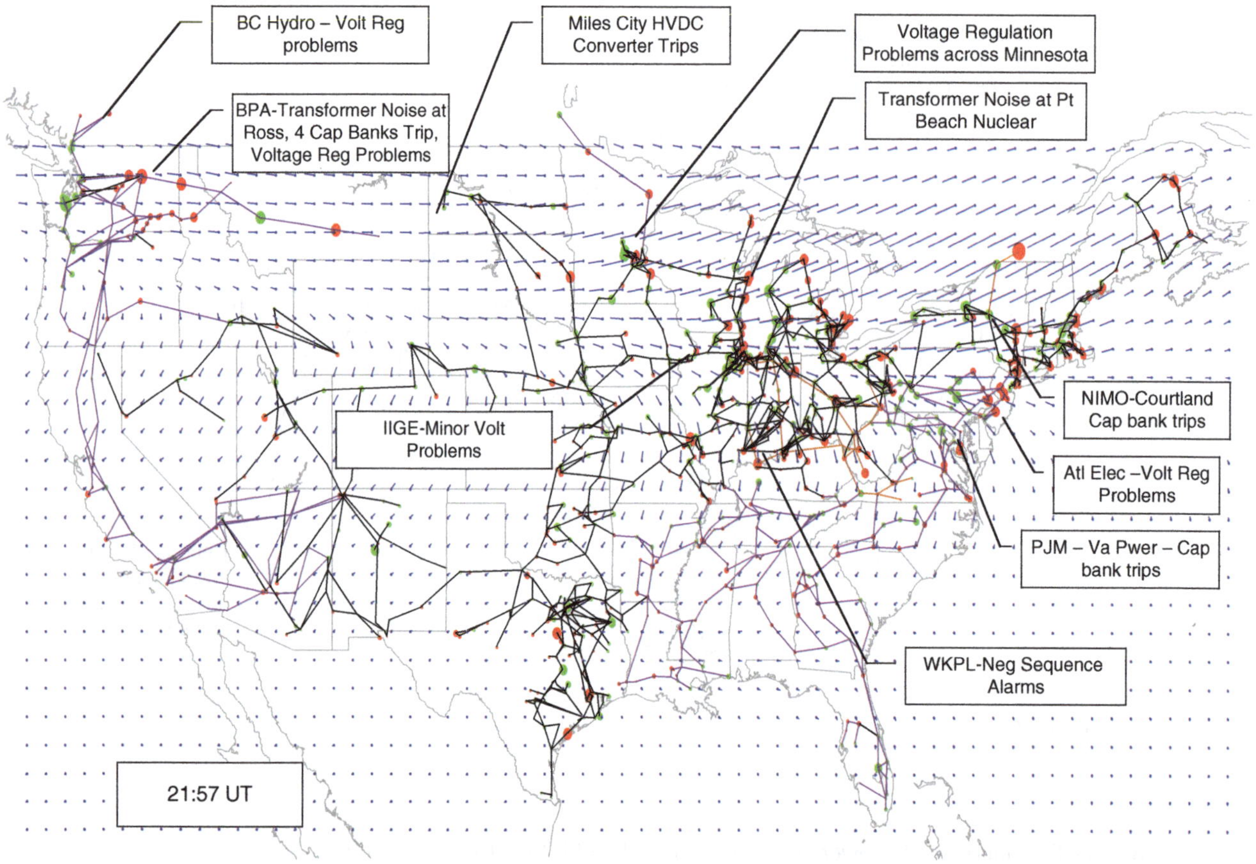

Figure 9.35 Simulation of US power grid conditions at 21:57 UT on 13 March 1989.

The SI storms are produced by the compression of the sunward part of the Earth's magnetic field as the initial intense part of a coronal mass ejection arrives at the Earth, producing a rapid rising B-field and illuminating all latitudes and longitudes simultaneously (including on the shadowed part of the Earth). There is no latitude dependence of the strength of the storm. Also because of the fast-rising pulse, larger E-fields and GICs are produced relative to the peak magnetic field disturbances measured. This type of storm can occur 5–10 times per year and is most likely to create voltage instabilities in a power network, even at low latitudes.

Recent studies of SI events have considered many worldwide magnetometer locations (Savage & Radasky, 2019). In particular, a set of locations as shown in Figure 9.39 was considered. One example of the measured magnetic fields at the Earth's surface is shown in Figure 9.40. Note the correlation of the fast-rising portions of the measurements, which can induce significant electric fields in spite of the low value of the peak magnetic fields. There are occasionally differences in the waveforms after the peak, depending on the location of the measurement, as the beginning of an electrojet storm may be observed after the peak SI at some but not at all locations.

Electrojet storms are well known, but one of the issues of importance is the variability of the measured magnetic fields with location. This is because the density of magnetometers on the Earth's surface may not be as high as desired. Some evaluations have indicated that in a few cases the variation of the peak B-fields with latitude (locations ~1000 km apart were very similar) was less than expected during a large storm (see Figures 9.41 and 9.42). It has also been observed in the past that the peak magnetic field longitudinal consistency for a significant electrojet storm is also on the order of 1000 km. While a few observed cases are interesting, more data are needed to ensure that this general trend is correct. This information has an impact on the proper modeling of electrojet storms for power grids in the future.

The CHHSS storms are less understood, but an examination of measured data in 1994 indicates that it is possible for multiple long-lasting coronal holes to appear simultaneously on the Sun, and to continuously produce charged particles that will induce magnetic fields at the Earth nearly every night. As the B-field peak levels are

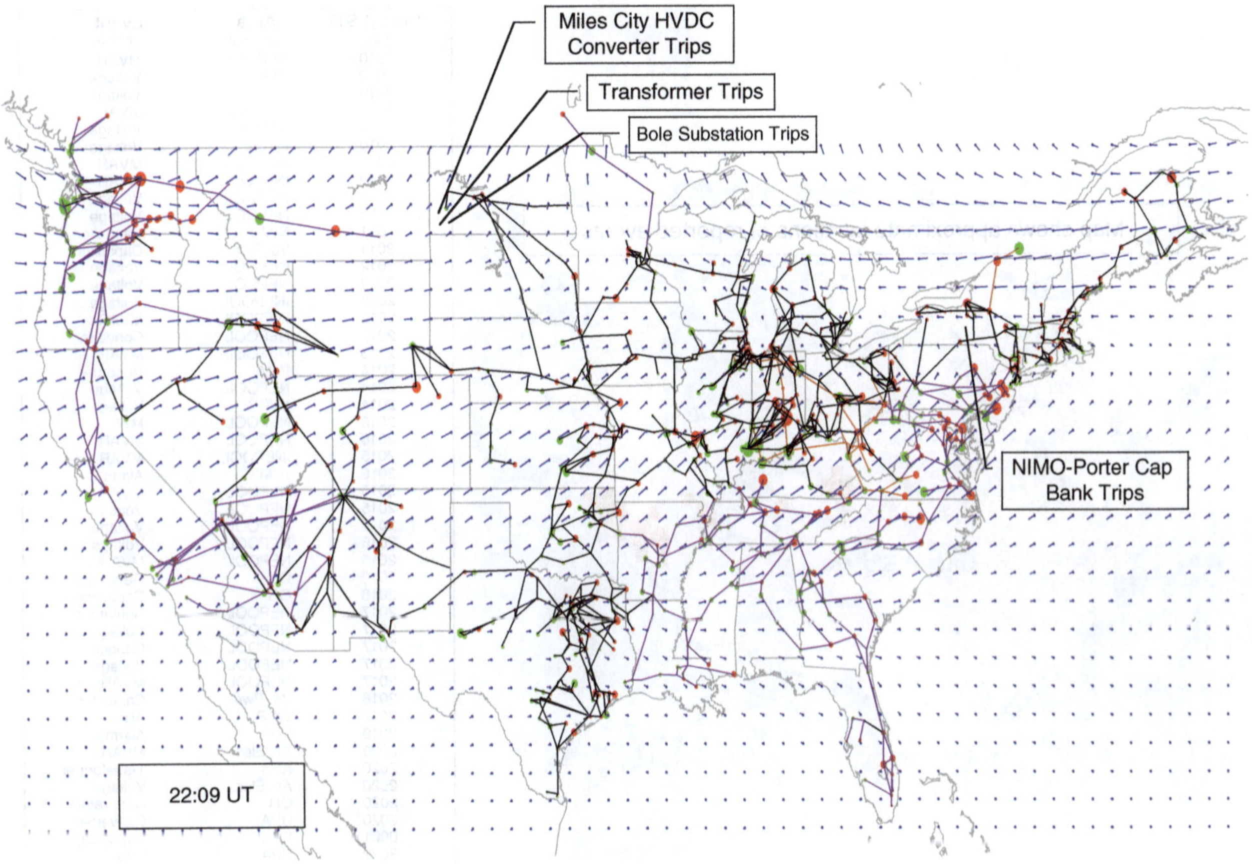

Figure 9.36 Simulation of US power grid conditions at 22:09 UT on 13 March 1989.

not as high as electrojet storms, the CHHSS storms tend to produce lower-level GICs that are high enough to create transformer saturation but are unlikely to cause voltage collapse. In fact, the data shown below in Figure 9.43 indicate that it is possible for low-level GICs to occur nearly every night for 3–4 months, creating a cumulative heating issue for exposed transformers (Savage & Radasky, 2019).

9.9. IMPACTS ON OTHER TYPES OF GROUND SYSTEMS

The main emphasis in this chapter has been with respect to the impacts of geomagnetic storms on the high-voltage electric power system. It turns out that other ground systems can be affected by the geomagnetic storm magnetic fields and the electric fields induced in the Earth. This section will briefly review some of those infrastructures and will indicate in the bibliography some references that can be viewed for more details.

One major infrastructure of interest includes undersea communications cables, which are powered from each end in order that undersea repeaters can operate properly for the communications signals. The concern is that the DC voltages fed from each end can be perturbed by the induced electric fields in the Earth at the bottom of the oceans. This can create overvoltages or undervoltages in the power system. While seawater is highly conducting as compared to the rock and soil of the Earth, many geomagnetic storms have very low frequency content and can penetrate the water with little attenuation (although the depth of the water is an important factor). Several studies have been performed by Lanzerotti and others to examine the performance of the TAT-8 cable over many years. This cable extends from New Jersey to the UK under the Atlantic Ocean.

Lanzerotti observed on 13 March 1989 and later on 15 July 2000 that there were noticeable voltage spikes measured in the power system of TAT-8 during the intense parts of the two electrojet storms (Lanzerotti et al., 2001). In the case of March 1989, the voltage level reached ~60 V and in July 2000 the voltage spike peaked at ~150 V. For long undersea cables, one important factor is that an electrojet storm magnetic and electric field typically vary significantly over the length of an undersea cable, so a 6300 km long cable will not observe a constant electric field over its entire length at the same time. An exception to this case is when there is an SI event; then the entire

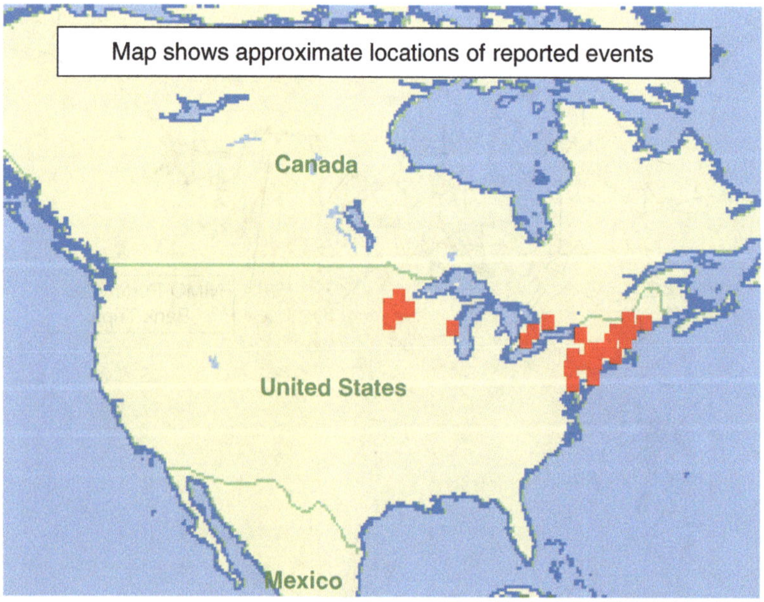

Time (EST)	Area	Event
2000	NEPOOL	MVAR
2010	NEPOOL	MVAR
2010	NEPOOL	Voltage
2010	NEPOOL	Voltage
2010	NEPOOL	MVAR
2010	NEPOOL	Voltage
2010	LILCO	Voltage
2010	NEPOOL	MVAR
2010	INEPOOL	MVAR
2010	NEPOOL	Voltage
2010	NEPOOL	Voltage
2011	NIMO	Capacitor
2011	Va. Pwr.	Capacitor
2012	NEPOOL	Voltage
2012	NEPOOL	Voltage
2012	INEPOOL	Voltage
2012	INEPOOL	MVAR
2012	NEPOOL	Converter
2013	NEPOOL	MVAR
2014	PJM	Recorder
2014	NEPOOL	Voltage
2014	NYPP	Voltage
2015	NEPOOL	MW
2016	NEPOOL	MVAR
2015	INEPOOL	MVAR
2016	PJM	Alarm
2015	NEPOOL	MVAR
2015	NEPOOL	Voltage
2015	NEPOOL	Voltage
2016	NEPOOL	Voltage
2015	NEPOOL	MVAR
2016	OH	Generator
2016	OH	Capacitor
2017	NEPOOL	Converter
2017	NEPOOL	Voltage
2017	NEPOOL	Capacitor
2017	NEPOOL	Voltage
2017	NEPOOL	MVAR
2018	Va. Pwr.	Capacitor
2019	WEP	Alarm
2019	UPA	Alarms
2020	Atl. Elec.	MVAR
2020	APS	Transformer
2020	Atl. Elec.	Voltage
2020	OH	Generator
2020	UPA	Converter
2021	PJM	Capacitor
2022	UPA	Line
2022	PJM	Alarm
2024	CPA	Voltage
2024	CPA	Capacitor
2032	PJM	Capacitor

Figure 9.37 Reported North American power system impacts, 13–14 March 1989 for times 20:00–20:32 EST (1:00–1:32 UT).

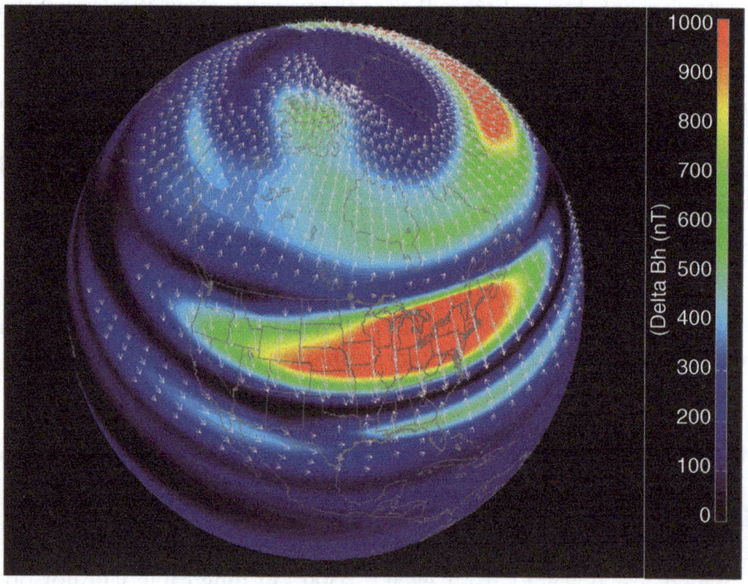

Figure 9.38 Simulated 01:20 UT, 14 March 1989, conditions. Reproduced with permission of Metatech Applied Power Solutions.

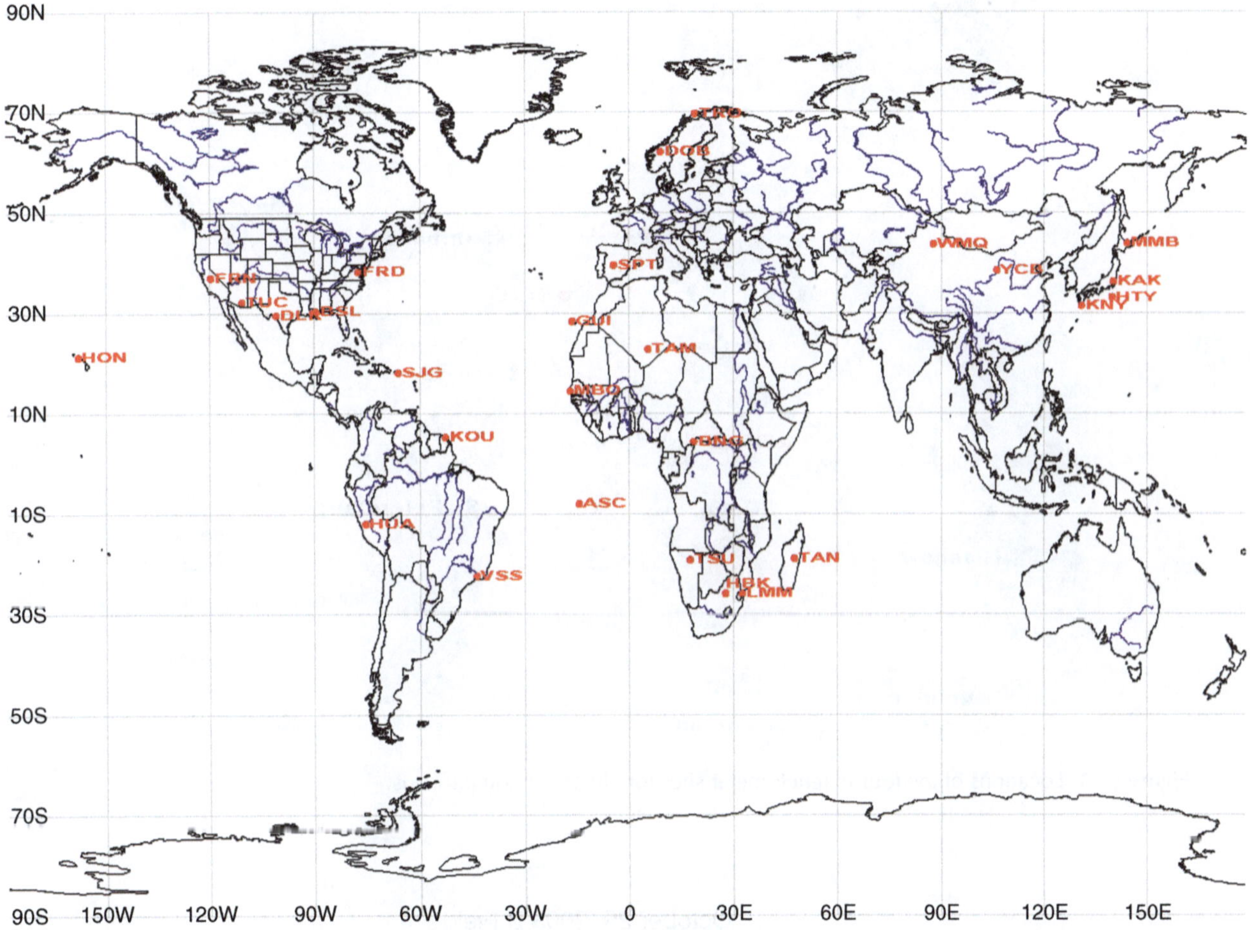

Figure 9.39 World map of measurement sites considered for the evaluation of SI events.

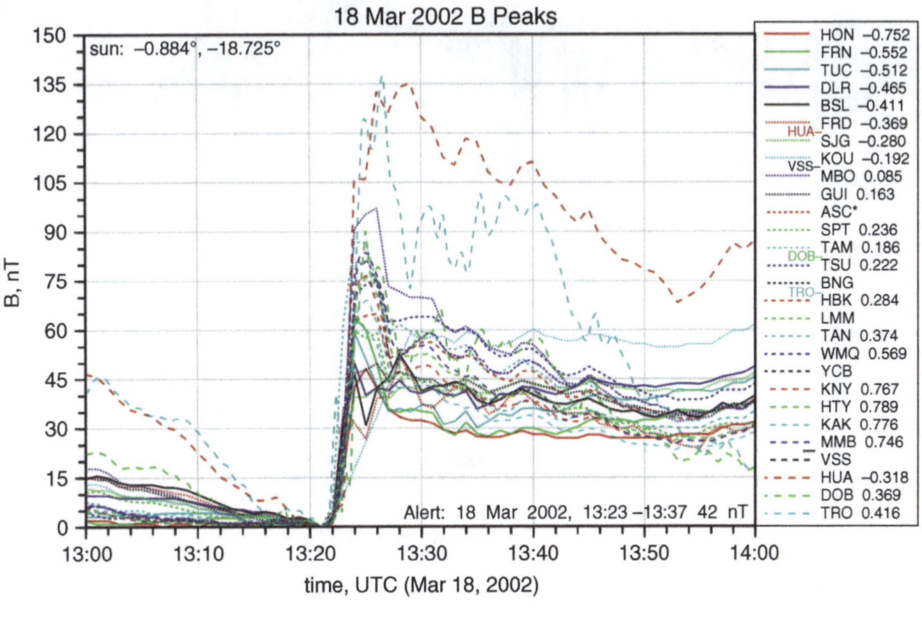

Figure 9.40 Comparison of the measured SI event on 18 March 2002 at 13:23 UT.

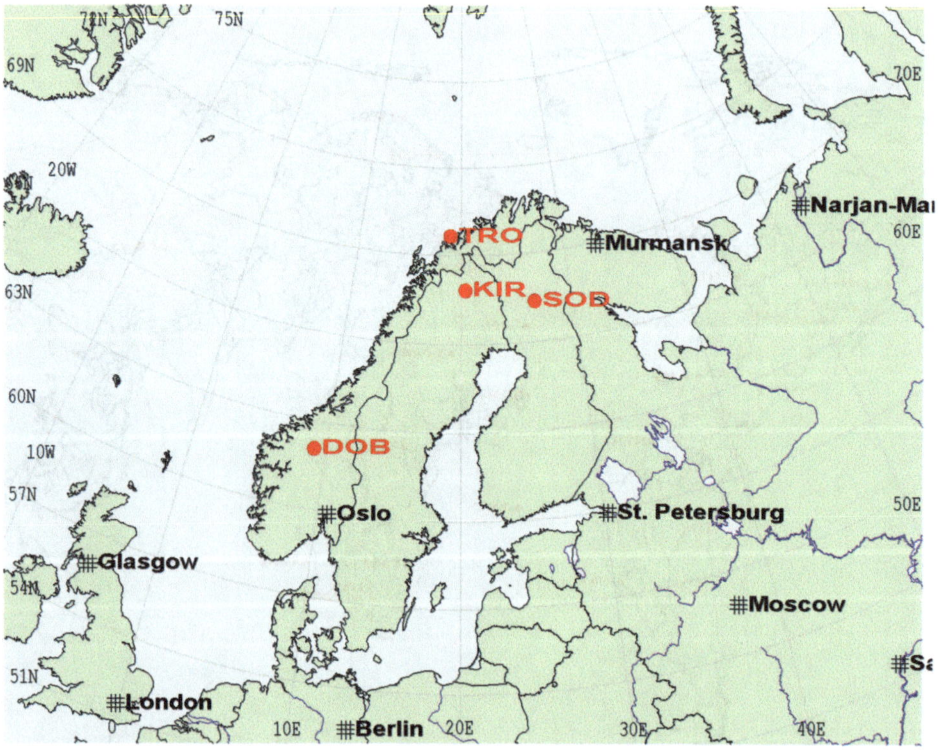

Figure 9.41 Locations of the four magnetometer sites for electrojet comparisons.

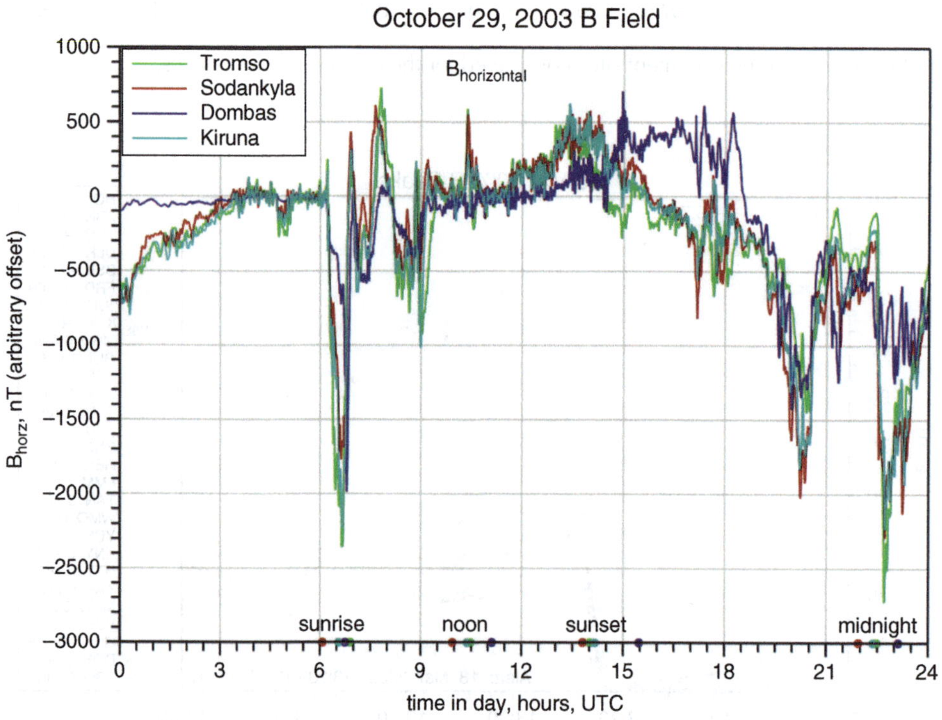

Figure 9.42 Horizontal B fields at four sites for 29 October 2003.

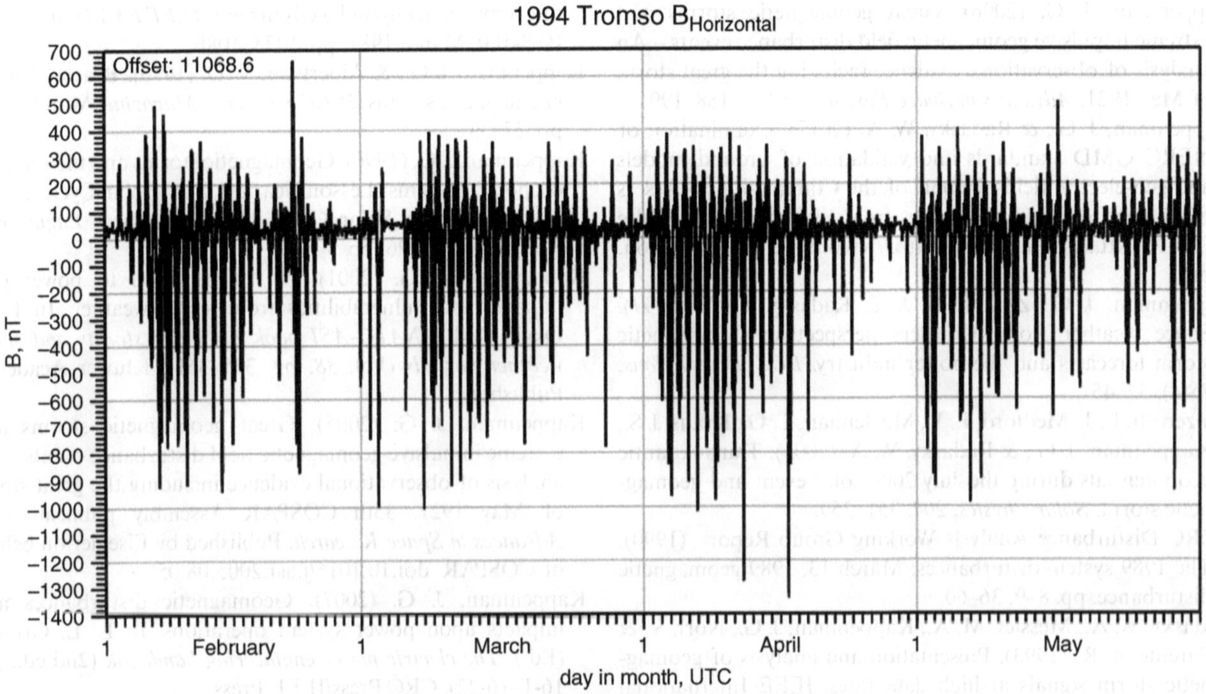

Figure 9.43 Tromso horizontal B-fields for 4-month period in 1994.

length of an undersea cable will be illuminated simultaneously but at a lower peak field level. Also there is another mitigating factor, as the SI magnetic field has a higher frequency content than the magnetic field from an electrojet storm, so the attenuation through seawater to the depth of the cable will be higher. This in fact was observed in July 2000 when the TAT-8 induced cable produced nearly the same peak voltage from the SI part of the event and the electrojet portion.

In addition to undersea cables, another technology affected by geomagnetic storms includes oil and gas pipelines (Campbell, 1980; Gummow et al., 2002). The problem with these systems is that in order to reduce the corrosion of these pipelines, cathodic protection systems are used to impress a given voltage between the pipelines and ground to prevent corrosion. During a geomagnetic storm, long pipelines will observe induced voltages from the storm, which have been found to be higher than the level of voltage protection. Since the intensity and frequency of geomagnetic storms increase if a pipeline is located at high magnetic latitudes, this problem is worse in North American locations such as Canada and Alaska.

There are many factors that influence the rate and amount of corrosion produced by geomagnetic storms, and these include the magnetic latitude of the pipeline, the length of the pipeline, the local Earth resistivity, the type of coating on the pipelines, and the voltage level of the cathodic system. While one solution might be to increase the voltage level of the cathodic system, this solution introduces increased costs. Studies to measure the rate of corrosion due to geomagnetic storms are not easy to achieve, so it may be difficult to obtain a cost-benefit analysis. The bibliography provides a selection of references if one wishes to examine this issue more closely.

REFERENCES

Anderson III, C. W., Lanzerotti, L. J., & Maclennan, C. G. (1974). Outage of the L4 system and the geomagnetic disturbance of 4 August 1972. *Bell System Tech. J.*, *53*, 1817.

Campbell, W. H. (1980). Observation of electric currents in the Alaska oil pipeline resulting from auroral electrojet current sources. *Geophys. J. R. Astron. Soc.*, *61*, 437–449.

Czeck, P., Huynh, H., & Dutil, A. (1992). The HydroQuebec system blackout of 13 March 1989: System response to geomagnetic disturbance. *Proc. Geomagnetically Induced Currents Conf.*, EPRI, Palo Alto, CA, June 1992.

Gilbert, J. L., Radasky, W. A., & Savage, E. B. (2012). A technique for calculating the currents induced by geomagnetic storms on large high voltage power grids. IEEE EMC Symposium, Pittsburgh, August 2012, pp. 323–328.

Gonzalez, W. D., Tsurutani, B. T., & Clúa de Gonzalez, A. L. (1999). Interplanetary origin of geomagnetic storms. *Space Science Reviews*, *88*(3–40, 529–562.

Gummow, R. A., Boteler, D. H., & Trichtchenko, L. (2002). Telluric and ocean current effects on buried pipelines and their cathodic protection systems. PRCI Contract PR-262-0030, Final Report, December 2002, pp. 33–64.

Kappenman, J. G. (2006). Great geomagnetic storms and extreme impulsive geomagnetic field disturbance events – An analysis of observational evidence including the great storm of May 1921. *Advances in Space Research, 38*(2), 188–199.

Kappenman, J. G., & Radasky, W. A. (2017). Examination of NERC GMD standards and validation of ground models and geo-electric fields. Report of the Commission to Assess the Threat to the United States from Electromagnetic Pulse (EMP) Attack, July 2017. http://www.firstempcommission.org.

Kappenman, J. G., Zanetti, L. J., & Radasky W. A. (1997). Space weather from the users perspective: Geomagnetic storm forecasts and the power industry. *EOS Transactions, 78*(4), 37–45.

Lanzerotti, L. J., Medford, L. V., Maclennan, C. G., Kraus, J. S., Kappenman, J. G., & Radasky, W. A. (2001). Trans-Atlantic geopotentials during the July 2000 solar event and geomagnetic storm. *Solar Physics, 204*, 351–359.

NERC Disturbance Analysis Working Group Report. (1990). The 1989 system disturbances: March 13, 1989 geomagnetic disturbance. pp. 8–9, 36–60.

Radasky, W. A., Messier, M. A., Kappenman, J. G., Norr, S., & Parenteau, R. (1993). Presentation and analysis of geomagnetic storm signals at high data rates. IEEE International Symposium on EMC, August 1993, pp. 156–157.

Savage, E. B., & Radasky, W. A. (2019). Overview of geomagnetic storm coupling to a power grid. EMC Sapporo and APEMC 2019, June 2019, pp. 116–119.

APPENDIX: BIBLIOGRAPHY

These papers are recommended for further reading as they provide some of the early information in this field, and in a few cases they provide new developments that may be important to some readers.

Albertson, V. D., & Thorson, J. M. (1973). Power system disturbances during a K-8 geomagnetic storm: August 4, 1972. IEEE PES Transactions Paper T 73 369-6, IEEE PES Summer Meeting, Vancouver, BC, July 15–20, 1973.

Albertson, V. D., Kappenman, J. G., Mohan, N., & Skarbakka, G. A. (1981). Load-flow studies in the presence of geomagnetically-induced currents. *IEEE PAS Transactions*, PAS-100, February 1981, pp. 594–607.

Campbell, W. H. (1978). Induction of auroral zone electric currents within the Alaska pipeline. *Pure and Applied Geophysics, 116*, 1143–1173.

Campbell, W. H. (1986). An interpretation of induced electrical currents in long pipelines caused by natural geomagnetic sources of the upper atmosphere. *Surveys in Geophys., 8*, 239–259.

Erinmez, I. A., Majithia, S., Rogers, C., Yasuhiro, T., Ogawa, S., Swahn, H., & Kappenman, J. G. (2002). Application of modeling techniques to assess geomagnetically induced current risks on the NGC transmission system. CIGRE, Session 2002, paper 39–304.

Kappenman, J. G., Albertson, V. D., & Mohan, N. (1981). Current transformer and relay performance in the presence of geomagnetically-induced currents. *IEEE PAS Transactions*, PAS-100, March 1981, pp. 1078–1088.

Kappenman, J. G., & Albertson, V. D. (1990). Bracing for the geomagnetic storms. *IEEE Spectrum Magazine*, March 1990, pp. 27–33.

Kappenman, J. G. (1996). Geomagnetic storms and their impact on power systems: Lessons learned from Solar Cycle 22 and the outlook for Solar Cycle 23. *IEEE Power Engineering Review*, May 1996, pp. 5–8.

Kappenman, J. G. (2001). An introduction to power grid impacts and vulnerabilities from space weather. In I. A. Daglis (Ed.), *NATO-ASI book on space storms and space weather hazards* (Vol. *38*, pp. 335–361). Kluwer Academic Publishers.

Kappenman, J. G. (2005). Great geomagnetic storms and extreme impulsive geomagnetic field disturbance events – An analysis of observational evidence including the great storm of May 1921. 35th COSPAR Assembly publication in *Advances in Space Research*. Published by Elsevier on behalf of COSPAR. doi:10.1016/j.asr.2005.08.055

Kappenman, J. G. (2007). Geomagnetic disturbances and impacts upon power system operations. In L. L. Grigsby (Ed.), *The electric power engineering handbook* (2nd ed., pp. 16-1–16-22). CRC Press/IEEE Press.

Kappenman, J. G., & Radasky, W. A. (2008). A white paper providing an overview of the geomagnetic storm environment and electromagnetic coupling to undersea cables. Technical Report, Metatech Corporation.

Lanzerotti, L. J. (1983). Geomagnetic induction effects in ground-based systems. *Space Science Reviews, 34*, 347–356.

Lanzerotti, L. J., & Medford, L. V. (1992). Geomagnetic disturbances and long-haul telecommunications cables. EPRI Technical Report (EPRI-TR-10050).

Lanzerotti, L. J., Chave, A. D., Sayres, C. H., Medford, L. V., & Maclennan, C. G. (1993). Large-scale electric field measurements on the Earth's surface: A review. *J. Geophys. Res., 98*, 23525–23534.

Lanzerotti, L. J., Sayres, C. H., Medford, L. V., Kraus, J. S., Maclennan, C. G. & Thomson, D. J. (1993). Statistical examination of induced voltages across long oceanic telecommunications cables. *Proc. of Solar-Terrestrial Predictions - IV Workshop*, Ottawa, Canada, September 1993, pp. 224–232.

Lanzerotti, L. J., Medford, L. V., Maclennan, C. G., &Thomson, J. (1995). Studies of large-scale earth potentials across oceanic distances. *AT&T Technical Journal, 74*, 73–84.

Lanzerotti, L. J., Medford, L. V., Sayres, D. S., Maclennan, C. G., Lepping, R. P., & Szabo, A. (1998). Space weather: Response of large-scale geopotentials to an interplanetary magnetic cloud. *J. Geophys. Res., 103*, 9351–9356.

Lanzerotti, L. J., Thomson, D. J., & Maclennan, C. G. (1999). Engineering issues in space weather. In M. A. Stuchly (Ed.), *Modern radio science* (pp. 23–50). Oxford University Press.

Medford, L. V., Lanzerotti, L. J., Kraus, J. S., & Maclennan, C. G. (1989). Trans-Atlantic Earth potential variations during the March 1989 magnetic storms. *Geophysical Research Letters, 16*, 1145–1148.

NERC. (2010). High-impact, low-frequency event risk to the North American bulk power system. A Jointly-Commissioned

Summary Report of the North American Electric Reliability Corporation and the U.S. Department of Energy's November 2009 Workshop, NERC, June 2010.

Pirjola, R. (2001). Geoelectromagnetic disturbances and their effects on technological systems. *Proc. URSI-E/Zurich EMC*, Zurich, Switzerland, February 2001, pp. 163–169.

Pirjola, R., Viljanen, A., Pulkkinen, A., & Amm, O., (2000). Space weather risk in power systems and pipelines. *Phys. Chem. Earth*, *25*, 333–337.

Pulkkinen, A., Viljanen, A., & Pirjola, R. (2000). Large geomagnetically induced currents in the Finnish high-voltage power system. Technical report, Finnish Meteorological Institute.

Radasky, W. A. (2011). Overview of the impact of intense geomagnetic storms on the U.S. high voltage power grid. IEEE Electromagnetic Compatibility Symposium, Long Beach, California, 15–19 August 2011, pp. 300–305.

Sayres, C. H., Lanzerotti, L. J., Medford, L. V., & Maclennan, C. G. (1998). Measurements and simulations of daily ionosphere-driven geopotentials across Pacific cables. *Adv. Space Res.*, *22*, 135–138.

Epilogue: The Road to Future Progress in Space Weather Understanding

Philip J. Erickson[1], Louis J. Lanzerotti[2], and Anthea J. Coster[1]

At the time of formation of the AGU in 1919, with the combining of the Committee on Geophysics of the National Research Council of the National Academy of Sciences and the American National Committee of the International Union of Geodesy and Geophysics, the predecessor to the *Journal of Geophysical Research, Terrestrial Magnetism and Atmospheric Electricity* had already been published for nearly 25 years (originally established as *Terrestrial Magnetism* in 1886). The founder of the journal, Louis A. Bauer, was a well-recognized American student and researcher of terrestrial magnetism and the first director of the Department of Terrestrial Magnetism of the Carnegie Institution of Washington.

While *Terrestrial Magnetism and Atmospheric Electricity* was largely devoted to fundamental issues and studies in geophysics, as is the *Journal of Geophysical Research* today, there appeared on occasion articles that could be used by applications-oriented researchers to understanding the more practical aspects that geophysics posed to the technologies of the day. In the late 19th century, as outlined in the introduction to this volume, the main practical problem related to the successful operations of telegraph systems that were beginning to span much of the world. Understanding possible telegraph problems meant that better understanding of the Earth and its profiles and conductivities was needed. Such data were occasionally available from the more basic research results published under Bauer's editorship.

At the time of the founding of the AGU a century ago, wireless communications were becoming big business for commerce and for government. AGU's founding was nearly two decades after Marconi's successful wireless transmission of the Morse Code "S" from Poldhu Bay, Cornwall, to Saint John's, Newfoundland, and a decade after his recognition with the Nobel Prize. The ionosphere, which both facilitated and adversely affected wireless communications, had not as yet been physically measured (this only occurred by the studies of Sir Edward Appleton in 1924). However, even before Appleton's radiophysics research, the existence of a reflecting layer in the atmosphere had been postulated by the theoretical work of British researchers Balfour Stewart, Oliver Heaviside, and Arthur Kennelly. And on occasion, some research appeared in *Terrestrial Magnetism and Atmospheric Electricity* under the rubric of atmospheric electricity that might be applicable to those pursuing the practical objectives of successful wireless communications.

As the chapters in this volume testify, much progress has occurred since these formative days of space weather research. Indeed, the technologies invented and developed by human ingenuity over the last century have contributed greatly to many human applications, ranging from communications by various means, to navigation, detection and surveillance, and human space flight. Many of these new technologies have been found to be affected by heliospheric and geophysical processes, as the various chapters relate here. In turn, the technological desire to mitigate space weather has reflected back on fundamental research directions. In particular, the needs for successful design and operations of these technologies have quite often spurred advances in fundamental understandings of the Sun, the interplanetary medium, and Earth's

[1] *Haystack Observatory, Massachusetts Institute of Technology, Westford, Massachusetts, USA*

[2] *New Jersey Institute of Technology and Alcatel Lucent Bell Laboratories, New Jersey, USA (ret.)*

Space Physics and Aeronomy Collection Volume 5: Space Weather Effects and Applications, Geophysical Monograph 262, First Edition.
Edited by Anthea J. Coster, Philip J. Erickson, and Louis J. Lanzerotti.

DOI: 10.1002/9781119815600.epilogue

space environment. The duality of this dialogue has in other contexts been described as a simultaneous requirement for both "research to operations" and "operations to research" efforts. The volume's descriptions of space weather impacts vividly demonstrate this diversity across commercial power grids, communications disruptions in space and near the planet's surface, orbiting platform atmospheric drag variations, navigation systems, and space radiation impacts on systems and even on humans above the surface of the Earth.

Ultimately, human interest in space weather effects, on this planet and eventually others, will, by the laws of physics and chemistry, be assured in the future. Humans are inexorably progressing towards a fully spacefaring society. Current trends make this setpoint clear, for example, through the large increases in affordable commercial space launch services targeting low Earth orbit, the rise of lightweight yet flexible space platforms such as cubesats and nano satellites, and the desire to establish reliable electromagnetic communications for outposts on other worlds. All these applications, and many others of similar nature, share a common and absolutely essential need to understand their system's surrounding medium, and in particular the natural plasma and atmospheric environment, not only in its climatology but also in a quantification of its variations (weather). Physics dictates that this goal will remain a significant challenge, due to the vastness of space and the fundamental long-distance, multiscale action provided by electromagnetic coupling forces. These factors severely tax both space and time observational requirements and the needed fidelity of both theoretical and computational modeling.

In the end, a thorough, comprehensive, and quantitative knowledge of space weather and climate in both their fundamental and applied forms is required to move beyond today's islands of understanding towards a full map of the landscape. Human navigation of that landscape will provide technical and scientific capabilities only dimly perceived today. We hope that the comprehensive chapters in this volume provide a beneficial snapshot of that full map, and a strong and inexorable motivation to progress forward the frontiers of science and engineering knowledge.

INDEX

Space Physics and Aeronomy Collection Volume 5: Space Weather Effects and Applications, Geophysical Monograph 262, First Edition.
Edited by Anthea J. Coster, Philip J. Erickson, and Louis J. Lanzerotti.

DOI: 10.1002/9781119815600.index